*John Ingham, Irving J. Dunn,
Elmar Heinzle, Jiří E. Přenosil,
Jonathan B. Snape*
Chemical Engineering Dynamics

1807–2007 Knowledge for Generations

Each generation has its unique needs and aspirations. When Charles Wiley first opened his small printing shop in lower Manhattan in 1807, it was a generation of boundless potential searching for an identity. And we were there, helping to define a new American literary tradition. Over half a century later, in the midst of the Second Industrial Revolution, it was a generation focused on building the future. Once again, we were there, supplying the critical scientific, technical, and engineering knowledge that helped frame the world. Throughout the 20th Century, and into the new millennium, nations began to reach out beyond their own borders and a new international community was born. Wiley was there, expanding its operations around the world to enable a global exchange of ideas, opinions, and know-how.

For 200 years, Wiley has been an integral part of each generation's journey, enabling the flow of information and understanding necessary to meet their needs and fulfill their aspirations. Today, bold new technologies are changing the way we live and learn. Wiley will be there, providing you the must-have knowledge you need to imagine new worlds, new possibilities, and new opportunities.

Generations come and go, but you can always count on Wiley to provide you the knowledge you need, when and where you need it!

William J. Pesce
President and Chief Executive Officer

Peter Booth Wiley
Chairman of the Board

John Ingham, Irving J. Dunn, Elmar Heinzle,
Jiří E. Přenosil, Jonathan B. Snape

Chemical Engineering Dynamics

An Introduction to Modelling and Computer Simulation

Third, Completely Revised Edition

WILEY-VCH Verlag GmbH & Co. KGaA

The Authors

Dr. John Ingham
74, Manor Drive
Bingley BD16 1PN
United Kingdom

Dr. Irving J. Dunn
Chalet Maedi
Quellweg 3
8841 Gross
Switzerland

Prof. Dr. Elmar Heinzle
Universität des Saarlandes
Technische Biochemie
Am Stadtwald Geb. II
66123 Saarbrücken
Germany

Dr. Jiří E. Přenosil
ETH Zürich
Department of Chem. Engineering
8092 Zürich
Switzerland

Dr. Jonathan B. Snape
Mylnefield Research Services Ltd.
Errol Road
Invergowrie
Dundee DD2 5DA
Scotland

Cover illustration
The background picture was directly taken from MADONNA, the program provided with this book, in the foreground, a betch reactor is shown with kindly permission of Pete Csiszar.
www.postmixing.com

All books published by Wiley-VCH are carefully produced. Nevertheless, authors, editors, and publisher do not warrant the information contained in these books, including this book, to be free of errors. Readers are advised to keep in mind that statements, data, illustrations, procedural details or other items may inadvertently be inaccurate.

Library of Congress Card No.: applied for

British Library Cataloguing-in-Publication Data
A catalogue record for this book is available from the British Library.

Bibliographic information published by the Deutsche Nationalbibliothek
The Deutsche Nationalbibliothek lists this publication in the Deutsche Nationalbibliografie; detailed bibliographic data are available in the Internet at http://dnb.d-nb.de.

© 2007 WILEY-VCH Verlag GmbH & Co. KGaA, Weinheim

All rights reserved (including those of translation into other languages). No part of this book may be reproduced in any form – by photoprinting, microfilm, or any other means – nor transmitted or translated into a machine language without written permission from the publishers. Registered names, trademarks, etc. used in this book, even when not specifically marked as such, are not to be considered unprotected by law.

Typesetting K+V Fotosatz GmbH, Beerfelden
Printing Strauss GmbH, Mörlenbach
Bookbinding Litges & Dopf GmbH, Heppenheim
Wiley Bicentennial Logo Richard J. Pacifico

Printed in the Federal Republic of Germany
Printed on acid-free paper

ISBN 978-3-527-31678-6

Contents

Preface XIII

Nomenclature XVII

1   **Basic Concepts**
1.1   Modelling Fundamentals   *1*
1.1.1   Chemical Engineering Modelling   *1*
1.1.2   General Aspects of the Modelling Approach   *3*
1.1.3   General Modelling Procedure   *3*
1.2   Formulation of Dynamic Models   *4*
1.2.1   Material Balance Equations   *4*
1.2.2   Balancing Procedures   *6*
1.2.2.1   Case A: Continuous Stirred-Tank Reactor   *7*
1.2.2.2   Case B: Tubular Reactor   *7*
1.2.2.3   Case C: Coffee Percolator   *8*
1.2.3   Total Material Balances   *16*
1.2.3.1   Case A: Tank Drainage   *17*
1.2.4   Component Balances   *18*
1.2.4.1   Case A: Waste Holding Tank   *19*
1.2.4.2   Case B: Extraction from a Solid by a Solvent   *20*
1.2.5   Energy Balancing   *22*
1.2.5.1   Case A: Continuous Heating in an Agitated Tank   *27*
1.2.5.2   Case B: Heating in a Filling Tank   *28*
1.2.5.3   Case C: Parallel Reaction in a Semi-Continuous Reactor with Large Temperature Changes   *29*
1.2.6   Momentum Balances   *31*
1.2.7   Dimensionless Model Equations   *31*
1.2.7.1   Case A: Continuous Stirred-Tank Reactor (CSTR)   *32*
1.2.7.2   Case B: Gas-Liquid Mass Transfer to a Continuous Tank Reactor with Chemical Reaction   *34*
1.3   Chemical Kinetics   *35*
1.3.1   Rate of Chemical Reaction   *35*
1.3.2   Reaction Rate Constant   *38*

Chemical Engineering Dynamics: An Introduction to Modelling and Computer Simulation, Third Edition
J. Ingham, I. J. Dunn, E. Heinzle, J. E. Prenosil, J. B. Snape
Copyright © 2007 WILEY-VCH Verlag GmbH & Co. KGaA, Weinheim
ISBN: 978-3-527-31678-6

1.3.3	Heat of ReactionHeat of Reaction 39
1.3.4	Chemical Equilibrium and Temperature 39
1.3.5	Yield, Conversion and Selectivity 39
1.3.6	Microbial Growth Kinetics 41
1.4	Mass Transfer Theory 43
1.4.1	Stagewise and Differential Mass Transfer Contacting 43
1.4.2	Phase Equilibria 45
1.4.3	Interphase Mass Transfer 46

2	**Process Dynamics Fundamentals**
2.1	Signal and Process Dynamics 51
2.1.1	Measurement and Process Response 51
2.1.1.1	First-Order Response to an Input Step-Change Disturbance 51
2.1.1.2	Case A: Concentration Response of a Continuous Flow, Stirred Tank 52
2.1.1.3	Case B: Concentration Response in a Continuous Stirred Tank with Chemical Reaction 54
2.1.1.4	Case C: Response of a Temperature Measuring Element 55
2.1.1.5	Case D: Measurement Lag for Concentration in a Batch Reactor 57
2.1.2	Higher Order Responses 58
2.1.2.1	Case A: Multiple Tanks in Series 58
2.1.2.2	Case B: Response of a Second-Order Temperature Measuring Element 60
2.1.3	Pure Time Delay 61
2.1.4	Transfer Function Representation 62
2.2	Time Constants 63
2.2.1	Common Time Constants 64
2.2.1.1	Flow Phenomena 64
2.2.1.2	Diffusion–Dispersion 65
2.2.1.3	Chemical Reaction 65
2.2.1.4	Mass Transfer 65
2.2.1.5	Heat Transfer 67
2.2.2	Application of Time Constants 67
2.3	Fundamentals of Automatic Control 68
2.3.1	Basic Feedback Control 68
2.3.2	Types of Controller Action 69
2.3.2.1	On/Off Control 70
2.3.2.2	Proportional Integral Derivative (PID) Control 70
2.3.2.3	Case A: Operation of a Proportional Temperature Controller 72
2.3.3	Controller Tuning 73
2.3.3.1	Trial and Error Method 74
2.3.3.2	Ziegler–Nichols Open-Loop Method 74
2.3.3.3	Cohen–Coon Controller Settings 75
2.3.3.4	Ultimate Gain Method> 75
2.3.3.5	Time Integral Criteria 76

2.3.4	Advanced Control Strategies	76
2.3.4.1	Cascade Control	76
2.3.4.2	Feedforward Control	77
2.3.4.3	Adaptive Control	78
2.3.4.4	Sampled Data or Discrete Control Systems	78
2.4	Numerical Aspects of Dynamic Behaviour	79
2.4.1	Optimisation	79
2.4.1.1	Case A: Optimal Cooling for a Reactor with an Exothermic Reversible Reaction	79
2.4.2	Parameter Estimation	81
2.4.2.1	Non-Linear Systems Parameter Estimation	82
2.4.2.2	Case B: Estimation of Rate and Equilibrium Constants in a Reversible Esterification Reaction Using MADONNA	83
2.4.3	Sensitivity Analysis	85
2.4.4	Numerical Integration	88
2.4.5	System Stability	91
3	**Modelling of Stagewise Processes**	
3.1	Introduction	93
3.2	Stirred-Tank Reactors	93
3.2.1	Reactor Configurations	93
3.2.2	Generalised Model Description	95
3.2.2.1	Total Material Balance Equation	95
3.2.2.2	Component Balance Equation	95
3.2.2.3	Energy Balance Equation	95
3.2.2.4	Heat Transfer to and from Reactors	96
3.2.2.5	Steam Heating in Jackets	99
3.2.2.6	Dynamics of the Metal Jacket Wall	100
3.2.3	The Batch Reactor	102
3.2.3.1	Case A: Constant-Volume Batch Reactor	103
3.2.4	The Semi-Batch Reactor	104
3.2.4.1	Case B: Semi-Batch Reactor	106
3.2.5	The Continuous Stirred-Tank Reactor	106
3.2.5.1	Case C: Constant-Volume Continuous Stirred-Tank Reactor	109
3.2.6	Stirred-Tank Reactor Cascade	109
3.2.7	Reactor Stability	110
3.2.8	Reactor Control	115
3.2.9	Chemical Reactor Safety	117
3.2.9.1	The Runaway Scenario	118
3.2.9.2	Reaction Calorimetry	118
3.2.10	Process Development in the Fine Chemical Industry	119
3.2.11	Chemical Reactor Waste Minimisation	120
3.2.12	Non-Ideal Flow	123
3.2.13	Tank-Type Biological Reactors	124
3.2.13.1	The Batch Fermenter	126

3.2.13.2	The Chemostat *126*
3.2.13.3	The Feed Batch Fermenter *128*
3.3	Stagewise Mass Transfer *129*
3.3.1	Liquid–Liquid Extraction *129*
3.3.1.1	Single Batch Extraction *130*
3.3.1.2	Multisolute Batch Extraction *132*
3.3.1.3	Continuous Equilibrium Stage Extraction *133*
3.3.1.4	Multistage Countercurrent Extraction Cascade *136*
3.3.1.5	Countercurrent Extraction Cascade with Backmixing *137*
3.3.1.6	Countercurrent Extraction Cascade with Slow Chemical Reaction *139*
3.3.1.7	Multicomponent Systems *140*
3.3.1.8	Control of Extraction Cascades *141*
3.3.1.9	Mixer-Settler Extraction Cascades *142*
3.3.1.10	Staged Extraction Columns *149*
3.3.1.11	Column Hydrodynamics *152*
3.3.2	Stagewise Absorption *153*
3.3.3	Stagewise Distillation *156*
3.3.3.1	Simple Overhead Distillation *156*
3.3.3.2	Binary Batch Distillation *158*
3.3.3.3	Continuous Binary Distillation *162*
3.3.3.4	Multicomponent Separations *165*
3.3.3.5	Plate Efficiency *166*
3.3.3.6	Complex Column Simulations *167*
3.3.4	Multicomponent Steam Distillation *168*

4 **Differential Flow and Reaction Applications**
4.1	Introduction *173*
4.1.1	Dynamic Simulation *173*
4.1.2	Steady-State Simulation *174*
4.2	Diffusion and Heat Conduction *175*
4.2.1	Unsteady-State Diffusion *175*
4.2.2	Unsteady-State Heat Conduction and Diffusion in Spherical and Cylindrical Coordinates *178*
4.2.3	Steady-State Diffusion *179*
4.3	Tubular Chemical Reactors *180*
4.3.1	The Plug-Flow Tubular Reactor *181*
4.3.2	Liquid-Phase Tubular Reactors *185*
4.3.3	Gas-Phase Tubular Reactors *186*
4.3.4	Batch Reactor Analogy *189*
4.3.5	Dynamic Simulation of the Plug-Flow Tubular Reactor *190*
4.3.6	Dynamics of an Isothermal Tubular Reactor with Axial Dispersion *193*
4.3.6.1	Dynamic Difference Equation for the Component Balance Dispersion Model *193*

4.3.7	Steady-State Tubular Reactor Dispersion Model 196
4.4	Differential Mass Transfer 199
4.4.1	Steady-State Gas Absorption with Heat Effects 199
4.4.1.1	Steady-State Design 200
4.4.1.2	Steady-State Simulation 201
4.4.2	Dynamic Modelling of Plug-Flow Contactors: Liquid–Liquid Extraction Column Dynamics 202
4.4.3	Dynamic Modelling of a Liquid–Liquid Extractor with Axial Mixing in Both Phases 205
4.4.4	Dynamic Modelling of Chromatographic Processes 207
4.4.4.1	Axial Dispersion Model for a Chromatography Column 208
4.4.4.2	Dynamic Difference Equation Model for Chromatography 209
4.5	Heat Transfer Applications 213
4.5.1	Steady-State Tubular Flow with Heat Loss 213
4.5.2	Single-Pass, Shell-and-Tube, Countercurrent-Flow Heat Exchanger 214
4.5.2.1	Steady-State Applications 214
4.5.2.2	Heat Exchanger Dynamics 215
4.6	Difference Formulae for Partial Differential Equations 219
4.7	References Cited in Chapters 1 to 4 220
4.8	Additional Books Recommended 222
5	**Simulation Tools and Examples of Chemical Engineering Processes**
5.1	Simulation Tools 226
5.1.1	Simulation Software 226
5.1.2	Teaching Applications 227
5.1.3	Introductory MADONNA Example: BATSEQ-Complex Reaction Sequence 227
5.2	Batch Reactor Examples 232
5.2.1	BATSEQ – Complex Batch Reaction Sequence 232
5.2.2	BATCHD – Dimensionless Kinetics in a Batch Reactor 235
5.2.3	COMPREAC – Complex Reaction 237
5.2.4	BATCOM – Batch Reactor with Complex Reaction Sequence 240
5.2.5	CASTOR – Batch Decomposition of Acetylated Castor Oil 243
5.2.6	HYDROL – Batch Reactor Hydrolysis of Acetic Anhydride 247
5.2.7	OXIBAT – Oxidation Reaction in an Aerated Tank 250
5.2.8	RELUY – Batch Reactor of Luyben 253
5.2.9	DSC – Differential Scanning Calorimetry 258
5.2.10	ESTERFIT – Esterification of Acetic Acid with Ethanol. Data Fitting 261
5.3	Continuous Tank Reactor Examples 265
5.3.1	CSTRCOM – Isothermal Reactor with Complex Reaction 265
5.3.2	DEACT – Deactivating Catalyst in a CSTR 268
5.3.3	TANK and TANKDIM – Single Tank with Nth-Order Reaction 270

5.3.4	CSTRPULSE – Continuous Stirred-Tank Cascade Tracer Experiment	273
5.3.5	CASCSEQ – Cascade of Three Reactors with Sequential Reactions	276
5.3.6	REXT – Reaction with Integrated Extraction of Inhibitory Product	280
5.3.7	THERM and THERMPLOT – Thermal Stability of a CSTR	283
5.3.8	COOL – Three-Stage Reactor Cascade with Countercurrent Cooling	287
5.3.9	OSCIL – Oscillating Tank Reactor Behaviour	290
5.3.10	REFRIG1 and REFRIG2 – Auto-Refrigerated Reactor	295
5.3.11	REVTEMP – Reversible Reaction with Variable Heat Capacities	299
5.3.12	REVREACT – Reversible Reaction with Temperature Effects	305
5.3.13	HOMPOLY Homogeneous Free-Radical Polymerisation	310
5.4	Tubular Reactor Examples	315
5.4.1	TUBE and TUBEDIM – Tubular Reactor Model for the Steady State	315
5.4.2	TUBETANK – Design comparison for Tubular and Tank Reactors	317
5.4.3	BENZHYD – Dehydrogenation of Benzene	320
5.4.4	ANHYD – Oxidation of O-Xylene to Phthalic Anhydride	324
5.4.5	NITRO – Conversion of Nitrobenzene to Aniline	329
5.4.6	TUBDYN – Dynamic Tubular Reactor	332
5.4.7	DISRE – Isothermal Reactor with Axial Dispersion	335
5.4.8	DISRET – Non-Isothermal Tubular Reactor with Axial Dispersion	340
5.4.9	VARMOL – Gas-Phase Reaction with Molar Change	344
5.5	Semi-Continuous Reactor Examples	347
5.5.1	SEMIPAR – Parallel Reactions in a Semi-Continuous Reactor	347
5.5.2	SEMISEQ – Sequential-Parallel Reactions in a Semi-Continuous Reactor	350
5.5.3	HMT – Semi-Batch Manufacture of Hexamethylenetetramine	353
5.5.4	RUN Relief of a Runaway Polymerisation Reaction	355
5.5.5	SELCONT Optimized Selectivity in a semi-continuous reactor	362
5.5.6	SULFONATION Space-Time-Yield and Safety in a Semi-Continuous Reactor	365
5.6	Mixing-Model Examples	374
5.6.1	NOCSTR – Non-Ideal Stirred-Tank Reactor	374
5.6.2	TUBEMIX – Non-Ideal Tube-Tank Mixing Model	378
5.6.3	MIXFLO1 and MIXFLO2 – Mixed–Flow Residence Time Distribution Studies	381
5.6.4	GASLIQ1 and GASLIQ2 – Gas–Liquid Mixing and Mass Transfer in a Stirred Tank	385
5.6.5	SPBEDRTD – Spouted Bed Reactor Mixing Model	390

5.6.6	BATSEG, SEMISEG and COMPSEG – Mixing and Segregation in Chemical Reactors 394	
5.7	Tank Flow Examples 406	
5.7.1	CONFLO 1, CONFLO 2 and CONFLO 3 – Continuous Flow Tank 406	
5.7.2	TANKBLD – Liquid Stream Blending 409	
5.7.3	TANKDIS – Ladle Discharge Problem 412	
5.7.4	TANKHYD – Interacting Tank Reservoirs 416	
5.8	Process Control Examples 420	
5.8.1	TEMPCONT – Control of Temperature 420	
5.8.2	TWOTANK – Two Tank Level Control 424	
5.8.3	CONTUN – Controller Tuning Problem 427	
5.8.4	SEMIEX – Temperature Control for Semi-Batch Reactor 430	
5.8.5	TRANSIM – Transfer Function Simulation 435	
5.8.6	THERMFF – Feedforward Control of an Exothermic CSTR 437	
5.9	Mass Transfer Process Examples 442	
5.9.1	BATEX – Single Solute Batch Extraction 442	
5.9.2	TWOEX – Two-Solute Batch Extraction with Interacting Equilibria 444	
5.9.3	EQEX – Simple Equilibrium Stage Extractor 447	
5.9.4	EQMULTI – Continuous Equilibrium Multistage Extraction 449	
5.9.5	EQBACK – Multistage Extractor with Backmixing 453	
5.9.6	EXTRACTCON – Extraction Cascade with Backmixing and Control 456	
5.9.7	HOLDUP – Transient Holdup Profiles in an Agitated Extractor 459	
5.9.8	KLADYN, KLAFIT and ELECTFIT – Dynamic Oxygen Electrode Method for $K_L a$ 462	
5.9.9	AXDISP – Differential Extraction Column with Axial Dispersion 468	
5.9.10	AMMONAB – Steady-State Design of a Gas Absorption Column with Heat Effects 471	
5.9.11	MEMSEP – Gas Separation by Membrane Permeation 475	
5.9.12	FILTWASH – Filter Washing 479	
5.9.13	CHROMDIFF – Dispersion Rate Model for Chromatography Columns 483	
5.9.14	CHROMPLATE – Stagewise Linear Model for Chromatography Columns 486	
5.10	Distillation Process Examples 490	
5.10.1	BSTILL – Binary Batch Distillation Column 490	
5.10.2	DIFDIST – Multicomponent Differential Distillation 494	
5.10.3	CONSTILL – Continuous Binary Distillation Column 496	
5.10.4	MCSTILL – Continuous Multicomponent Distillation Column 501	
5.10.5	BUBBLE – Bubble Point Calculation for a Batch Distillation Column 504	
5.10.6	STEAM – Multicomponent, Semi-Batch Steam Distillation 508	

5.11	Heat Transfer Examples 511	
5.11.1	HEATEX – Dynamics of a Shell-and-Tube Heat Exchanger 511	
5.11.2	SSHEATEX – Steady-State, Two-Pass Heat Exchanger 515	
5.11.3	ROD – Radiation from Metal Rod 518	
5.12	Diffusion Process Examples 521	
5.12.1	DRY – Drying of a Solid 521	
5.12.2	ENZSPLIT – Diffusion and Reaction: Split Boundary Solution 525	
5.12.3	ENZDYN – Dynamic Diffusion with Enzymatic Reaction 529	
5.12.4	BEAD – Diffusion and Reaction in a Spherical Catalyst Bead 533	
5.13	Biological Reaction Examples 538	
5.13.1	BIOREACT – Process Modes for a Bioreactor 538	
5.13.2	INHIBCONT – Continuous Bioreactor with Inhibitory Substrate 543	
5.13.3	NITBED – Nitrification in a Fluidised Bed Reactor 547	
5.13.4	BIOFILM – Biofilm Tank Reactor 551	
5.13.5	BIOFILT – Biofiltration Column for Removing Ketone from Air 555	
5.14	Environmental Examples 560	
5.14.1	BASIN – Dynamics of an Equalisation Basin 560	
5.14.2	METAL – Transport of Heavy Metals in Water Column and Sediments 565	
5.14.3	OXSAG – Classic Streeter-Phelps Oxygen Sag Curves 569	
5.14.4	DISCHARGE – Dissolved Oxygen and BOD Steady-State Profiles Along a River 572	
5.14.5	ASCSTR – Continuous Stirred Tank Reactor Model of Activated Sludge 577	
5.14.6	DEADFISH – Distribution of an Insecticidein an Aquatic Ecosystem 581	
5.14.7	LEACH – One-Dimensional Transport of Solute Through Soil 584	
5.14.8	SOIL – Bioremediation of Soil Particles 591	

Appendix 597

1. A Short Guide to MADONNA 597
2. Screenshot Guide to BERKELEY-MADONNA 602
3. List of Simulation Examples 606

Subject Index 609

Preface

The aim of this book is to teach the use of modelling and simulation as a discipline for the understanding of chemical engineering processes and their dynamics. This is done via a combination of basic modelling theory and computer simulation examples, which are used to emphasise basic principles and to demonstrate the cause-and-effect phenomena in complex models. The examples are based on the use of a powerful and easy-to-use simulation language, called BERKELEY-MADONNA, that was already successfully used in the second edition of this book. Developed at the University of California for Windows and Macintosh, MADONNA represents almost all we have ever wanted in simulation software for teaching. The many programmed examples demonstrate simple modelling procedures that can be used to represent a wide range of chemical and chemical engineering process phenomena. The study of the examples, by direct computer experimentation, has been shown to lead to a positive improvement in the understanding of physical systems and confidence in the ability to deal with chemical rate processes. Quite simple models can often give realistic representations of process phenomena. The methods described in the text are applicable to a range of differing applications, including process identification, the analysis and design of experiments, process design and optimisation, process control and plant safety, all of which are essential aspects of modern chemical technology.

The book is based on the hands-on use of the computer as an integral part of the learning process. Although computer-based modelling procedures are now commonplace in chemical engineering, our experience is that there still remains a considerable lack of ability in basic modelling, especially when applied to dynamic systems. This has resulted from the traditional steady state approach to chemical engineering and the past emphasis on flow-sheeting for large-scale continuous processes. Another important contributing factor is the perceived difficulty in solving the large sets of simultaneous differential equations that result from any realistic dynamic modelling description. With modern trends towards more intensive high-value batch processing methods, the need for a better knowledge of the plant dynamics is readily apparent. This is also reinforced by the increased attention that must now be paid to proper process control, process optimisation and plant safety. Fortunately the PC and Macintosh computers with suitable simulation software now provide a fast and convenient means of solution.

Chemical Engineering Dynamics: An Introduction to Modelling and Computer Simulation, Third Edition
J. Ingham, I. J. Dunn, E. Heinzle, J. E. Prenosil, J. B. Snape
Copyright © 2007 WILEY-VCH Verlag GmbH & Co. KGaA, Weinheim
ISBN: 978-3-527-31678-6

The excellent software BERKELEY-MADONNA enables a more modern, Windows-based (also Macintosh compatible) and menu driven solution.

In this third edition we have revised the theoretical part and introduced a number of new simulation examples. Some examples deal with safety problems in chemical reactors and others are related to modelling of environmental systems and are located in a new Environmental Process section.

Organisation of the Book

The book consists of an introduction to basic modelling presented in Chapters 1 to 4. An introduction to simulation principles and methods and the simulation examples are found in Chapter 5. The first four chapters cover the basic theory for the computer simulation examples and present the basic concepts of dynamic modelling. The aim is not to be exhaustive, but simply to provide sufficient introduction, for a proper understanding of the modelling methodology and computer-based examples. Here the main emphasis is placed on understanding the physical meaning and significance of each term in the resulting model equations. Chapter 5, constituting the main part of the book, provides the MADONNA-based computer simulation exercises. Each of the examples is self-contained and includes a model description, the model equations, exercises, nomenclature, sample graphical output and references. The combined book thus represents a synthesis of basic theory and computer-based simulation examples. The accompanying CD includes the MADONNA simulation language for Windows and Macintosh and the ready-to-run simulation example programs. Each program is clearly structured with comments and complete nomenclature. Although not included within the main body of the text, the MADONNA solution programs provided on the CD are very simple both to write and to understand, as evidenced by the demonstration program BATSEQ in Section 5.1.3. All the programs are clearly structured and are accompanied by clear descriptions, nomenclature and details of any special items of programming that might be included. All programs are therefore very easy to understand, to apply and, if needed, to modify. Further, a clear connection between the model relationships described in the text and the resulting program is very apparent.

Chapter 1 deals with the basic concepts of modelling, and the formulation of mass and energy balance relationships. In combination with other forms of relationship, these are shown to lead to a systematic development for dynamic models. Though the concepts are simple, they can be applied equally well to very complex problems.

Chapter 2 is employed to provide a general introduction to signal and process dynamics, including the concept of process time constants, process control, process optimisation and parameter identification. Other important aspects of dynamic simulation involve the numerical methods of solution and the resulting stability of solution; both of which are dealt with from the viewpoint of the simulator, as compared to that of the mathematician.

Chapter 3 concerns the dynamic characteristics of stagewise types of equipment, based on the concept of the well-stirred tank. In this, the various types of stirred-tank chemical reactor operation are considered, together with allowance for heat effects, non-ideal flow, control and safety. Also included is the modelling of stagewise mass transfer applications, based on liquid-liquid extraction, gas absorption and distillation.

Chapter 4 concerns differential processes, which take place with respect to both time and position and which are normally formulated as partial differential equations. Applications include heterogeneous catalysis, tubular chemical reactors, differential mass transfer, heat exchangers and chromatography. It is shown that such problems can be solved with relative ease, by utilising a finite-differencing solution technique in the simulation approach.

Chapter 5 comprises the computer simulation examples. The exercises are intended to draw the simulators attention to the most important features of each example. Most instructive is to study the influence of important model parameters, using the interactive and graphical features of MADONNA. Interesting features include the possibility of making "parametric runs" to investigate the influence of one parameter on the steady state values. When working with arrays to solve multistage or diffusion problems, the variables can be plotted versus the array number, thus achieving output plots as a function of a distance measure.

Working through a particular example will often suggest an interesting variation, such as a control loop, which can then be inserted into the model. In running our courses, the exercises have proven to be very open-ended and in tackling them, we hope you will share our conviction that computer simulation is fun, as well as being useful and informative. An Appendix provides an instructional guide to the MADONNA software, which is sufficient for work with the simulation examples.

In this edition some of our favourite examples from our previous book "Environmental Bioprocesses" have been added in a new section of Chapter 5. Also the exercises from some examples have been expanded, according to our teaching experience in the area of reactor safety and control.

We are confident that the book will be useful to all who wish to obtain a better understanding of chemical engineering dynamics and to those who have an interest in sharpening their modelling skills. We hope that teachers with an interest in modelling will find this to be a useful textbook for chemical engineering and applied chemistry courses, at both undergraduate and postgraduate levels.

Acknowledgements

We gladly acknowledge all who have worked previously in this field for the stimulation they have provided to us in the course of development of this book and our post-experience teaching. We are very fortunate in having the use of efficient PC and Macintosh based software, which was not available to those who were the major pioneers in the area of digital simulation. The modeller is now free to concentrate on the prime task of developing a realistic process model and to use this then in practical application, as was originally suggested by Franks (1967, 1972).

We are very grateful to all our past post-experience course participants and university students who have helped us to develop and improve some of the examples. In addition, we would like to thank Tim Zahnley, one of the developers of BERKELEY-MADONNA, for his help with software questions. Members of Wiley-VCH helped us in the editing and printing of this third edition, and for this we are grateful.

Nomenclature for Chapters 1–4

Symbols		Units
A	Area	m^2
A	Magnitude of controller input signal	various
a	Specific interfacial area	m^2/m^3 and cm^2/cm^3
a	Various parameters	various
B	Magnitude of controller output signal	various
b	Various parameters	various
C	Concentration	kg/m^3, $kmol/m^3$
c_P	Heat capacity at constant pressure	kJ/kg K, kJ/mol K
c_V	Heat capacity at constant volume	kJ/kg K, kJ/mol K
D	Diffusivity	m^2/s
d	Differential operator	–
d, D	Diameter	m
E	Energy	kJ or kJ/kg
E	Activation energy	kJ/mol
E	Residence time distribution	–
F	Residence time distribution	–
F	Volumetric flow rate	m^3/s
f	Frequency in the ultimate gain method	1/s
G	Gas or light liquid flow rate	m^3/s
g	Gravitational acceleration	m/s^2
G'	Superficial light phase velocity	m/s
H	Enthalpy	kJ/mol, kJ/kg
ΔH	Enthalpy change	kJ/mol, kJ/kg
H	Height	m
H	Henry's law constant	bar m^3/kg
H_G	Rate of heat gain	kJ/s
H_L	Rate of heat loss	kJ/s
h	Fractional holdup	–
h_i	Partial molar enthalpy	kJ/mol
J	Total mass flux	kg/s, kmol/s
j	Mass flux	kg/m^2 s, mol/m^2 s

Chemical Engineering Dynamics: An Introduction to Modelling and Computer Simulation, Third Edition
J. Ingham, I.J. Dunn, E. Heinzle, J.E. Prenosil, J.B. Snape
Copyright © 2007 WILEY-VCH Verlag GmbH & Co. KGaA, Weinheim
ISBN: 978-3-527-31678-6

Nomenclature for Chapters 1–4

K	Constant in Cohen-Coon method	various
K	Mass transfer coefficient	m/s
K	kinetic growth constant	s^{-1}
k	Constant	various
k_d	specific death rate coefficient	s^{-1}
$K_G a$	Gas-liquid mass transfer coefficient referring to concentration in G-phase	1/s
$k_G a$	Gas film mass transfer coefficient	1/s
$K_L a$	Gas-liquid mass transfer coefficient referring to concentration in L-phase	1/s
$k_L a$	Liquid film mass transfer coefficient	1/s
$K_{LX} a$	Overall mass transfer capacity coefficient based on the aqueous phase mole ratio X	$kmol/m^3\ s$
K_p	Proportional controller gain constant	various
K_s	saturation constant	$kg\ m^{-3}$
L	Length	m
L	Liquid or heavy phase flow rate	m^3/s, mol/s
L'	Superficial heavy phase velocity	m/s
M	Mass	kg, mol
$\dot{M}$	Mass flow rate	kg/s
m	maintenance factor	kg S/kg X
N	Mass flux	$kg/m^2\ s$
N	Molar flow rate	mol/s
n	Number of moles	–
n	Reaction order	–
P	Controller output signal	various
P	Total pressure or pure component vapour pressure	bar
p	Partial pressure	bar
Pe	Peclet number (L v/D)	–
Q	Heat transfer rate	kJ/s
Q	Total transfer rate	kg/s, mol/s
q	Heat flux	$kJ/m^2\ s$
R	Ideal gas constant	$bar\ m^3/K\ mol$
R	Reaction rate	kg/s, kmol/s
R	Number of reactions	–
r	Reaction rate	$kg/m^3\ s$, $kmol/m^3\ s$
r_{Ads}	Adsorption rate of the sorbate	$g/cm^3\ s$
r_d	death rate	$kg\ m^{-3}\ s^{-1}$
r_i	Reaction rate of component i	$kg\ i/m^3\ s$, $kmol/m^3\ s$
r_Q	Heat production rate	$kJ/m^3\ s$
r_S	Rate of substrate uptake	$kg\ S\ m^{-3}\ s^{-1}$
r_X	Growth rate	$kg\ biomass/m^3\ h$
S	Slope of process reaction curve/A	various
S	Selectivity	–

Symbol	Description	Units
S	Number of compounds	–
S	Concentration of substrate	kg/m³
s	Laplace operator	–
T	Temperature	°C, K
t	Time	h, min, s
Tr$_A$	Transfer rate of sorbate	g/s
U	Heat transfer coefficient	kJ/m² K s
U	Internal energy	kJ/mol
V	Vapour flow rate	mol/s
V	Volume	m³
v	Flow velocity	m/s
W	Rate of work	kJ/s
W	Mass flow rate	kg/s
X	Concentration in heavy phase	kg/m³, mol/m³
X	Mole ratio in the heavy phase	–
X	Conversion	–
X	Biomass concentration	kg/m³
x	Mole fraction in heavy phase	–
x	Input variable	various
Y	Fractional yield	–
Y	Concentration in light phase	kg/m³, mol/m³
Y	Mole ratio in the light phase	–
Y	Yield coefficient	kg/kg
Y$_{i/j}$	Yield of i from j	kg i/kg j
y	Mole fraction in light phase	–
y	Output variable	various
Z	Arrhenius constant	various
Z	Length variable	m
z	Length variable	m

Greek

Symbol	Description	Units
Δ	Difference operator	–
Φ	Thiele modulus	–
Θ	Dimensionless time	–
Σ	Summation operator	–
α	Backmixing factor	–
α	Relative volatility	–
α, β	Reaction order	–
ε	Controller error	various
η	Effectiveness factor	–
η	Plate efficiency	–
μ	Dynamic viscosity	kg m/s
μ	Specific growth rate	1/h

μ_m	Maximum growth rate	1/h
ν	Stoichiometric coefficient	–
θ	Dimensionless temperature	–
ρ	Density	kg/m^3
τ	Controller time constant	s
τ	Residence time	h and s
τ	Shear stress	kg m/s^2
τ	Time constant	h, min, s
τ_L	Time lag	h, min, s
∂	Partial differential operator	–

Indices

0	Refers to initial, inlet, external, or zero order
1	Refers to outlet or first order
1, 2,…, n	Refers to segment, stage, stream, tank or volume element
A	Refers to component A
a	Refers to ambient
abs	Refers to absorption
agit	Refers to agitation
app	Refers to apparent
avg	Refers to average
B	Refers to component B, base, backmixing, surface position or boiler
C	Refers to component C or combustion
c	Refers to cross-sectional or cold
D	Refers to derivative control, component D, delay or drum
E	Refers to electrode
eq	Refers to equilibrium
F	Refers to formation or feed
f	Refers to final or feed plate
G	Refers to gas or light liquid phase or generation
h	Refers to hot
ht	Refers to heat transfer
I	Refers to integral control
i	Refers to component i or to interface
inert	Refers to inert component
j	Refers to reaction j or to jacket
L	Refers to liquid phase, heavy liquid phase or lag
m	Refers to metal wall or mixer
max	Refers to maximum
mix	Refers to mixer
mt	Refers to mass transfer
n	Refers to tank, section, segment or plate number
p	Refers to plug flow, pocket and particle

Q	Refers to heat
R	Refers to recycle stream
r	Refers to reactor
S	Refers to settler, steam, solid or surroundings
s	Refers to surface, settler or shell side
SL	Refers to liquid film at solid interface
ss	Refers to steady state
St	Refers to standard
t	Refers to tube
tot	Refers to total
V	Refers to vapour
w	Refers to water or wall
–	Bar above symbol refers to dimensionless variable
′	Refers to perturbation variable, superficial velocity or stripping section
*	Refers to equilibrium concentration

1
Basic Concepts

1.1
Modelling Fundamentals

Models are an integral part of any kind of human activity. However, we are mostly unaware of this. Most models are qualitative in nature and are not formulated explicitly. Such models are not reproducible and cannot easily be verified or proven to be false. Models guide our activities, and throughout our entire life we are constantly modifying those models that affect our everyday behaviour. The most scientific and technically useful types of models are expressed in mathematical terms. This book focuses on the use of dynamic mathematical models in the field of chemical engineering.

1.1.1
Chemical Engineering Modelling

The use of models in chemical engineering is well established, but the use of dynamic models, as opposed to the more traditional use of steady-state models for chemical plant analysis, is much more recent. This is reflected in the development of new powerful commercial software packages for dynamic simulation, which has arisen owing to the increasing pressure for design validation, process integrity and operation studies for which a dynamic simulator is an essential tool. Indeed it is possible to envisage dynamic simulation becoming a mandatory condition in the safety assessment of plant, with consideration of such factors as start up, shutdown, abnormal operation, and relief situations assuming an increasing importance. Dynamic simulation can thus be seen to be an essential part of any hazard or operability study, both in assessing the consequences of plant failure and in the mitigation of possible effects. Dynamic simulation is thus of equal importance in large scale continuous process operations, as in other inherently dynamic operations such as batch, semi-batch and cyclic manufacturing processes. Dynamic simulation also aids in a very positive sense in enabling a better understanding of process performance and is a powerful tool for plant optimisation, both at the operational and at the design stage. Furthermore steady-state operation is then seen in its rightful place as the end result of a dynamic process for which rates of change have become eventually zero.

Chemical Engineering Dynamics: An Introduction to Modelling and Computer Simulation, Third Edition
J. Ingham, I. J. Dunn, E. Heinzle, J. E. Prenosil, J. B. Snape
Copyright © 2007 WILEY-VCH Verlag GmbH & Co. KGaA, Weinheim
ISBN: 978-3-527-31678-6

The approach in this book is to concentrate on a simplified approach to dynamic modelling and simulation. Large scale commercial software packages for chemical engineering dynamic simulation are now very powerful and contain highly sophisticated mathematical procedures, which can solve both for the initial steady-state condition as well as for the following dynamic changes. They also contain extensive standard model libraries and the means of synthesising a complete process model by combining standard library models. Other important aspects are the provision for external data interfaces and built-in model identification and optimisation routines, together with access to a physical property data package. The complexity of the software, however, is such that the packages are often non-user friendly and the simplicity of the basic modelling approach can be lost in the detail of the solution procedures. The correct use of such design software requires a basic understanding of the sub-model blocks and hence of the methodology of modelling. Our simplified approach to dynamic modelling and simulation incorporates no large model library, no attached database and no relevant physical property package. Nevertheless quite realistic process phenomena can be demonstrated, using a very simple approach. Also, this can be very useful in clarifying preliminary ideas before going to the large scale commercial package, as we have found several times in our research. Again this follows our general philosophy of starting simple and building in complications as the work and as a full understanding of the process model progresses. This allows the use of models to be an explicit integral part of all our work.

Kapur (1988) has listed thirty-six characteristics or principles of mathematical modelling. Mostly a matter of common sense, it is very important to have them restated, as it is often very easy to lose sight of the principles during the active involvement of modelling. They can be summarised as follows:

1. The mathematical model can only be an approximation of real-life processes, which are often extremely complex and often only partially understood. Thus models are themselves neither good nor bad but should satisfy a previously well defined aim.
2. Modelling is a process of continuous development, in which it is generally advisable to start off with the simplest conceptual representation of the process and to build in more and more complexities, as the model develops. Starting off with the process in its most complex form often leads to confusion.
3. Modelling is an art but also a very important learning process. In addition to a mastery of the relevant theory, considerable insight into the actual functioning of the process is required. One of the most important factors in modelling is to understand the basic cause and effect sequence of individual processes.
4. Models must be both realistic and robust. A model predicting effects, which are quite contrary to common sense or to normal experience, is unlikely to be met with confidence.

1.1.2
General Aspects of the Modelling Approach

An essential stage in the development of any model is the formulation of the appropriate mass and energy balance equations. To these must be added appropriate kinetic equations for rates of chemical reaction, rates of heat and mass transfer and equations representing system property changes, phase equilibrium, and applied control. The combination of these relationships provides a basis for the quantitative description of the process and comprises the basic mathematical model. The resulting model can range from a simple case of relatively few equations to models of great complexity. The greater the complexity of the model, however, the greater is then the difficulty in identifying the increased number of parameter values. One of the skills of modelling is thus to derive the simplest possible model, capable of a realistic representation of the process.

The application of a combined modelling and simulation approach leads to the following advantages:
1. Modelling improves understanding.
2. Models help in experimental design.
3. Models may be used predictively for design and control.
4. Models may be used in training and education.
5. Models may be used for process optimisation.

1.1.3
General Modelling Procedure

One of the more important features of modelling is the frequent need to reassess both the basic theory (physical model), and the mathematical equations, representing the physical model (mathematical model), in order to achieve agreement, between the model prediction and actual process behaviour (experimental data).

As shown in Fig. 1.1, the following stages in the modelling procedure can be identified:
(1) The first involves the proper definition of the problem and hence the goals and objectives of the study.
(2) All the available knowledge concerning the understanding of the problem must be assessed in combination with any practical experience, and perhaps alternative physical models may need to be developed and examined.
(3) The problem description must then be formulated in mathematical terms and the mathematical model solved by computer simulation.
(4) The validity of the computer prediction must be checked. After agreeing sufficiently well with available knowledge, experiments must then be designed to further check its validity and to estimate parameter values. Steps (1) to (4) will often need to be revised at frequent intervals.
(5) The model may now be used at the defined depth of development for design, control and for other purposes.

1 Basic Concepts

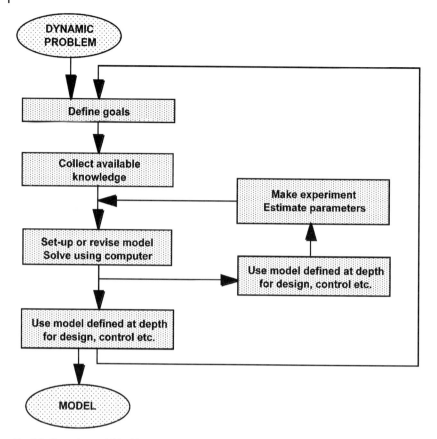

Fig. 1.1 Steps in model building.

1.2
Formulation of Dynamic Models

1.2.1
Material Balance Equations

Steady-State Balances

One of the basic principles of modelling is that of the conservation of mass or matter. For a steady-state flow process, this can be expressed by the statement:

$$\begin{pmatrix} \text{Rate of mass flow} \\ \text{into the system} \end{pmatrix} = \begin{pmatrix} \text{Rate of mass flow} \\ \text{out of the system} \end{pmatrix}$$

Dynamic Total Material Balances

Most real situations are, however, such that conditions change with respect to time. Under these circumstances, a steady-state material balance is inappropriate and must be replaced by a dynamic or unsteady-state material balance, expressed as

$$\begin{pmatrix} \text{Rate of accumulation of} \\ \text{mass in the system} \end{pmatrix} = \begin{pmatrix} \text{Rate of} \\ \text{mass flow in} \end{pmatrix} - \begin{pmatrix} \text{Rate of} \\ \text{mass flow out} \end{pmatrix}$$

Here the rate of accumulation term represents the rate of change in the total mass of the system, with respect to time, and at steady state, this is equal to zero. Thus, the steady-state material balance is seen to be a simplification of the more general dynamic balance.

At steady state

$$\begin{pmatrix} \text{Rate of} \\ \text{accumulation of mass} \end{pmatrix} = 0 = (\text{Mass flow in}) - (\text{Mass flow out})$$

hence, when steady state is reached

$$(\text{Mass flow in}) = (\text{Mass flow out})$$

Component Balances

The previous discussion has been in terms of the total mass of the system, but most process streams, encountered in practice, contain more than one chemical species. Provided no chemical change occurs, the generalised dynamic equation for the conservation of mass can also be applied to each chemical component of the system. Thus for any particular component

$$\begin{pmatrix} \text{Rate of} \\ \text{accumulation of mass} \\ \text{of component} \\ \text{in the system} \end{pmatrix} = \begin{pmatrix} \text{Mass flow of} \\ \text{the component} \\ \text{into the system} \end{pmatrix} - \begin{pmatrix} \text{Mass flow of} \\ \text{the component out} \\ \text{of the system} \end{pmatrix}$$

Component Balances with Reaction

Where a chemical reaction occurs, the change, due to reaction, can be taken into account by the addition of a reaction rate term into the component balance equation. Thus in the case of material produced by the reaction

$$\begin{pmatrix} \text{Rate of} \\ \text{accumulation} \\ \text{of mass} \\ \text{of component} \\ \text{in the system} \end{pmatrix} = \begin{pmatrix} \text{Mass flow} \\ \text{of the} \\ \text{component} \\ \text{into} \\ \text{the system} \end{pmatrix} - \begin{pmatrix} \text{Mass flow} \\ \text{of the} \\ \text{component} \\ \text{out of} \\ \text{the system} \end{pmatrix} + \begin{pmatrix} \text{Rate of} \\ \text{production} \\ \text{of the} \\ \text{component} \\ \text{by the reaction} \end{pmatrix}$$

The principle of the component material balance can also be extended to the atomic level and can also be applied to particular elements.

Thus for the case of carbon, in say a fuel combustion process

$$\begin{pmatrix} \text{Rate of} \\ \text{accumulation} \\ \text{of carbon mass} \\ \text{in the system} \end{pmatrix} = \begin{pmatrix} \text{Mass flow} \\ \text{rate of} \\ \text{carbon into} \\ \text{the system} \end{pmatrix} - \begin{pmatrix} \text{Mass flow} \\ \text{rate of} \\ \text{carbon out} \\ \text{of the system} \end{pmatrix}$$

Note that the elemental balances do not involve additional reaction rate terms since the elements are unchanged by chemical reaction.

While the principle of the material balance is very simple, its application can often be quite difficult. It is important therefore to have a clear understanding of the nature of the system (physical model) which is to be modelled by the material balance equations and also of the methodology of modelling.

1.2.2
Balancing Procedures

The methodology described below outlines five steps **I** through **V** to establish the model balances. The first task is to define the system by choosing the balance or control region. This is done using the following procedure:

I. Choose the Balance Region Such That the Variables Are Constant or Change Little Within the System. Draw Boundaries Around the Balance Region

The balance region can vary substantially, depending upon the particular area of interest of the model, ranging from say the total reactor, a region of a reactor, a single phase within a reactor, to a single gas bubble or a droplet of liquid. The actual choice, however, will always be based on a region of assumed uniform composition, or on another property as in the case of population balances. Generally, the modelling exercises will involve some prior simplification of the real system. Often the system being modelled will be considered in terms of a representation, based on systems of tanks (stagewise or lumped parameter systems) or systems of tubes (differential systems), or even combinations of tanks and tubes.

1.2 Formulation of Dynamic Models

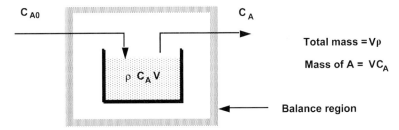

Fig. 1.2 The balance region around the continuous reactor.

1.2.2.1 Case A: Continuous Stirred-Tank Reactor

If the tank is well-mixed, the concentrations and density of the tank contents are uniform throughout. This means that the outlet stream properties are identical with the tank properties, in this case concentration C_A and density ρ. The balance region can therefore be taken around the whole tank (Fig. 1.2).

The total mass in the system is given by the product of the volume of the tank contents V (m^3) multiplied by the density ρ (kg/m^3), thus Vρ (kg). The mass of any component A in the tank is given in terms of actual mass or number of moles by the product of volume V times the concentration of A, C_A (kg of A/m^3 or kmol of A/m^3), thus giving VC$_A$ in kg or kmol.

1.2.2.2 Case B: Tubular Reactor

In the case of tubular reactors, the concentrations of the products and reactants will vary continuously along the length of the reactor, even when the reactor is operating at steady state. This variation can be regarded as being equivalent to that of the time of passage of material as it flows along the reactor and is equivalent to the time available for reaction to occur. Under steady-state conditions the concentration at any position along the reactor will be constant with respect to time, though not with position. This type of behaviour, obtained with tubular reactors, can be approximated by choosing the incremental volume of

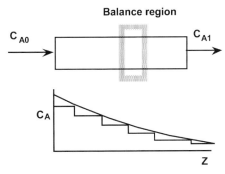

Fig. 1.3 The tubular reactor concentration gradients.

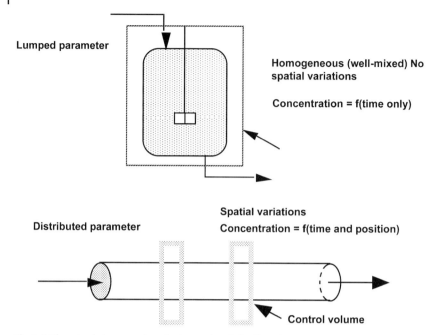

Fig. 1.4 Choosing balance regions for lumped and distributed parameter systems.

the balance regions sufficiently small so that the concentration of any component within the region can be assumed approximately uniform. Thus in this case, many uniform property sub-systems (well-stirred tanks or increments of different volume but all of uniform concentration) comprise the total reactor volume. This situation is illustrated in Fig. 1.3.

The basic concepts of the above lumped parameter and distributed parameter systems are shown in Fig. 1.4.

1.2.2.3 Case C: Coffee Percolator

A coffee percolator operates by circulating a stream of boiling coffee solution from the reservoir in the base of the coffee pot up through a central rise-pipe to the top of a bed of coffee granules, through which the solution then percolates, before returning in a more concentrated state to the base reservoir, as shown in Fig. 1.5.

The above system can be thought of as consisting of two parts with 1) the base reservoir acting effectively as a single well-stirred tank and 2) a fixed bed system of coffee granules irrigated by the flowing liquid stream. Solute coffee is removed from the granules by mass transfer under the action of a concentration driving force and is extracted into the liquid.

The concentrations of the coffee, both in the granules and in the liquid flowing through the bed, will vary continuously both with distance and with time. The behaviour of the packed bed is therefore best approximated by a series of

1.2 Formulation of Dynamic Models | 9

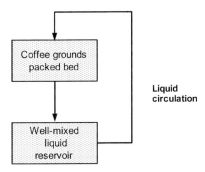

Fig. 1.5 Conceptual of coffee percolator.

many uniform property subsystems. Each segment of solid is related to its appropriate segment of liquid by interfacial mass transfer, as shown in Fig. 1.6.

The resulting model would therefore consist of component balance equations for the soluble component written over each of the many solid and liquid subsystems of the packed bed, combined with the component balance equation for the coffee reservoir. The magnitude of the recirculating liquid flow will depend on the relative values of the pressure driving force generated by the boiling liquid and the fluid flow characteristics of the system.

The concept of modelling a coffee percolator as a dynamic process comes from a problem first suggested by Smith et al. (1970).

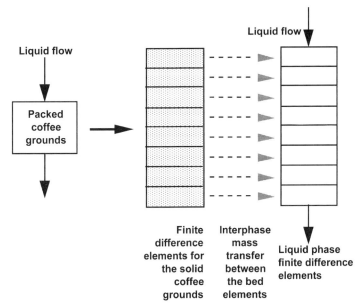

Fig. 1.6 Modelling concepts for the packed bed solid-liquid extraction process of coffee percolation.

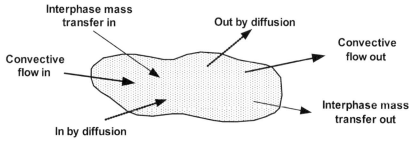

Fig. 1.7 Balance region showing convective and diffusive flows as well as interphase mass transfer in and out.

II. Identify the Transport Streams Which Flow Across the System Boundary

Having defined the balance regions, the next task is to identify all the relevant inputs and outputs to the system (Fig. 1.7). These may be well-defined physical flow rates (convective streams), diffusive fluxes, but may also include interphase transfer rates.

It is important to assume transfer to occur in a particular direction and to specify this by means of an arrow. This direction may reverse itself, but the change will be accommodated by a reversal in sign of the transfer rate term.

III. Write the Material Balance in Word Form

This is an important step because it helps to ensure that the resulting mathematical equation will have an understandable physical meaning. Just starting off by writing down equations is often liable to lead to fundamental errors, at least on the part of the beginner. All balance equations have a basic logic, as expressed by the generalised statement of the component balance given below, and it is very important that the model equations also retain this. Thus

$$\begin{pmatrix} \text{Rate of} \\ \text{accumulation} \\ \text{of mass} \\ \text{of component} \\ \text{in the system} \end{pmatrix} = \begin{pmatrix} \text{Mass flow} \\ \text{of the} \\ \text{component} \\ \text{into} \\ \text{the system} \end{pmatrix} - \begin{pmatrix} \text{Mass flow} \\ \text{of the} \\ \text{component} \\ \text{out of} \\ \text{the system} \end{pmatrix} + \begin{pmatrix} \text{Rate of} \\ \text{production} \\ \text{of the} \\ \text{component by} \\ \text{the reaction} \end{pmatrix}$$

This can be abbreviated as

$$(\text{Accumulation}) = (\text{In}) - (\text{Out}) + (\text{Production})$$

IV. Express Each Balance Term in Mathematical Form with Measurable Variables

A. Rate of Accumulation Term

This is given by the rate of change of the mass of the system, or the mass of some component within the system, with changing time and is expressed as the derivative of the mass with respect to time. Hence

$$\begin{pmatrix} \text{Rate of accumulation of mass} \\ \text{of component i within the system} \end{pmatrix} = \left(\frac{dM_i}{dt}\right)$$

where M is in kg or mol and time is in h, min or s.

Volume, concentration and, in the case of gaseous systems, partial pressure are usually the measured variables. Thus for any component i

$$\frac{dM_i}{dt} = \frac{d(VC_i)}{dt}$$

where C_i is the concentration of component i (kg/m³). In the case of gases, the Ideal Gas Law can be used to relate concentration to partial pressure and mol fraction. Thus,

$$p_i V = n_i RT$$

where p_i is the partial pressure of component i, within the gas phase system, and R is the Ideal Gas Constant, in units compatible with p, V, n and T.

In terms of concentration,

$$C_i = \frac{n_i}{V} = \frac{p_i}{RT} = \frac{y_i P}{RT}$$

where y_i is the mol fraction of the component in the gas phase and P is the total pressure of the system.

The accumulation term for the gas phase can be therefore written in terms of number of moles as

$$\frac{dn_i}{dt} = \frac{d(VC_i)}{dt} = \frac{d\left(\frac{p_i V}{RT}\right)}{dt} = \frac{d\left(\frac{y_i PV}{RT}\right)}{dt}$$

For the total mass of the system

$$\frac{dM}{dt} = \frac{d(V\rho)}{dt}$$

with units

$$\frac{kg}{s} = \frac{kg}{m^3} \frac{m^3}{s}$$

B. Convective Flow Terms

Total mass flow rates are given by the product of volumetric flow multiplied by density. Component mass flows are given by the product of volumetric flow rates multiplied by concentration.

$$(\text{Convective mass flow rate}) = (\text{Volumetric flow rate})\left(\frac{\text{Mass}}{\text{Volume}}\right)$$

for the total mass flow

$$\dot{M} = \frac{dM}{dt} = F\rho$$

and for the component mass flow

$$\dot{M}_i = \frac{dM_i}{dt} = FC_i$$

with units

$$\frac{kg}{s} = \frac{m^3}{s} \frac{kg}{m^3}$$

A stream leaving a well-mixed region, such as a well-stirred tank, has the identical properties as in the system, since for perfect mixing the contents of the tank will have spatially uniform properties, which must then be identical to the properties of the fluid leaving at the outlet. Thus, the concentrations of component i both within the tank and in the tank effluent are the same and equal to C_{i1}, as shown in Fig. 1.8.

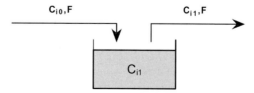

Fig. 1.8 Convective flow terms for a well-mixed tank reactor.

C. Diffusion of Components

As shown in Fig. 1.9, diffusional flow contributions in engineering situations are usually expressed by Fick's Law for molecular diffusion

$$j_i = -D_i \frac{dC_i}{dZ}$$

where j_i is the flux of any component i flowing across an interface (kmol/m² s or kg/m² s) and dC_i/dZ (kmol/m³ m) is the concentration gradient and D_i is the diffusion coefficient of component i (m²/s) for the material.

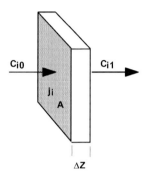

Fig. 1.9 Diffusion flux j_i driven by concentration gradient $(C_{i0}-C_{i1})/\Delta Z$ through surface area A.

In accordance with Fick's Law, diffusive flow always occurs in the direction of decreasing concentration and at a rate, which is proportional to the magnitude of the concentration gradient. Under true conditions of molecular diffusion, the constant of proportionality is equal to the molecular diffusivity of the component i in the system, D_i (m²/s). For other cases, such as diffusion in porous matrices and for turbulent diffusion applications, an effective diffusivity value is used, which must be determined experimentally.

The concentration gradient may have to be approximated in finite difference terms (finite differencing techniques are described in more detail in Sections 4.2 to 4.4). Calculating the mass diffusion rate requires knowledge of the area, through which the diffusive transfer occurs, since

$$\begin{pmatrix} \text{Mass rate} \\ \text{of} \\ \text{component i} \end{pmatrix} = - \begin{pmatrix} \text{Diffusivity} \\ \text{of} \\ \text{component i} \end{pmatrix} \begin{pmatrix} \text{Concentration} \\ \text{gradient} \\ \text{of i} \end{pmatrix} \begin{pmatrix} \text{Area} \\ \text{perpendicular} \\ \text{to transport} \end{pmatrix}$$

$$j_i A = -D_i \left(\frac{dC_i}{dZ}\right) A$$

The concentration gradient can often be approximated by difference quantities, where

$$j_i A = -D_i \left(\frac{\Delta C_i}{\Delta Z}\right) A$$

with units

$$\frac{kg}{s\ m^2} m^2 = \frac{m^2}{s} \frac{kg}{m^3} \frac{1}{m} m^2$$

D. Interphase Transport

Interphase mass transport also represents a possible input to or output from the system. In Fig. 1.10, transfer of a soluble component takes place across the interface which separates the two phases. Shown here is the transfer from phase G to phase L, where the separate phases may be gas, liquid or solid.

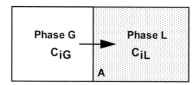

Fig. 1.10 Transfer across an interface of area A from phase G to phase L.

When there is transfer from one phase to another, the component balance equations must consider this. Thus taking a balance for component i around the well-mixed phase G, with transfer of i from phase G to phase L, gives

$$\begin{pmatrix} \text{Rate of} \\ \text{accumulation} \\ \text{of i} \\ \text{in phase G} \end{pmatrix} = - \begin{pmatrix} \text{Rate of interfacial} \\ \text{mass transfer of i} \\ \text{from phase G} \\ \text{into phase L} \end{pmatrix}$$

This form of the transfer rate equation will be examined in more detail in Section 1.4. Suffice it to say here that the rate of transfer can be expressed in the form shown below

$$\begin{pmatrix} \text{Rate of} \\ \text{mass transfer} \end{pmatrix} = \begin{pmatrix} \text{Mass transfer} \\ \text{coefficient} \end{pmatrix} \begin{pmatrix} \text{Area of} \\ \text{the interface} \end{pmatrix} \begin{pmatrix} \text{Concentration} \\ \text{driving force} \end{pmatrix}$$

$$Q = K A \Delta C$$

The units of the transfer rate equation (with appropriate molar quantities) are

$$\frac{kmol}{s} = \frac{m}{s} m^2 \frac{kmol}{m^3}$$

where Q is the total mass transfer rate, A is the total interfacial area for mass transfer (m²), ΔC is the concentration driving force (kmol/m³), and K is the overall mass transfer coefficient (m/s). It is important to note that the concentration driving force is represented as a difference between the actual concentration and the corresponding equilibrium value and is not a simple difference between actual phase concentrations. Mass transfer rates can be converted to mass flows (kg/s), by multiplying by the molar mass of the component.

E. Production Rate

The production rate term allows for the production or consumption of material by chemical reaction and can be incorporated into the component balance equation. Thus,

$$\begin{pmatrix} \text{Rate of} \\ \text{accumulation} \\ \text{of mass} \\ \text{of component} \\ \text{in the system} \end{pmatrix} = \begin{pmatrix} \text{Mass flow} \\ \text{of the} \\ \text{component} \\ \text{into} \\ \text{the system} \end{pmatrix} - \begin{pmatrix} \text{Mass flow} \\ \text{of the} \\ \text{component} \\ \text{out of} \\ \text{the system} \end{pmatrix} + \begin{pmatrix} \text{Rate of} \\ \text{production} \\ \text{of the} \\ \text{component by} \\ \text{the reaction} \end{pmatrix}$$

Chemical production rates are often expressed on a molar basis but can be easily converted to mass flow quantities (kg/s). The material balance equation can then be expressed as

$$\begin{pmatrix} \text{Mass rate} \\ \text{of production} \\ \text{of component A} \end{pmatrix} = \begin{pmatrix} \text{Reaction} \\ \text{rate per} \\ \text{unit volume} \end{pmatrix} \begin{pmatrix} \text{Volume} \\ \text{of the} \\ \text{system} \end{pmatrix}$$

$$R_A = r_A V$$

where R_A is the total reaction rate. The units are

$$\frac{\text{kg}}{\text{s}} = \frac{\text{kg}}{\text{s m}^3} \text{m}^3$$

Equivalent molar quantities may also be used. The quantity r_A is positive when A is formed as product, and r_A is negative when reactant A is consumed.

V. Introduce Other Relationships and Balances Such That the Number of Equations Equals the Number of Dependent Variables

The system material balance equations are often the most important elements of any modelling exercise, but are themselves rarely sufficient to completely formulate the model. Thus other relationships are needed to complete the model in terms of other important aspects of behaviour in order to satisfy the mathematical rigour of the modelling, such that the number of unknown variables must be equal to the number of defining equations.

Examples of this type of relationships, which are not based on balances, but which nevertheless form a very important part of any model are:
- Reaction stoichiometry.
- Reaction rates as functions of concentration and temperature.
- Equations of state or Ideal Gas Law behaviour.
- Physical property correlations as functions of concentration, temperature, etc.
- Hydraulic flow equations.
- Pressure variations as a function of flow rate.
- Equilibrium relationships (e.g., Henry's law, relative volatilities, etc.).
- Activity coefficients.
- Dynamics of measurement instruments, as a function of instrument response time.
- Controller equations with an input variable dependent on a measured variable.
- Correlations for mass transfer coefficients, gas holdup volume, and interfacial area, as functions of system physical properties and agitation rate or flow velocity, etc.

How these and other relationships are incorporated within the development of particular modelling instances is illustrated, throughout the text and in the simulation examples.

1.2.3
Total Material Balances

In this section, the application of the total material balance principle is presented. Consider some arbitrary balance region, as shown in Fig. 1.11 by the shaded area. Mass accumulates within the system at a rate dM/dt, owing to the competing effects of a convective flow input (mass flow rate in) and an output stream (mass flow rate out).

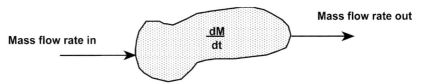

Fig. 1.11 Balancing the total mass of an arbitrary system.

The total material balance is expressed by

$$\begin{pmatrix} \text{Rate of} \\ \text{accumulation of mass} \\ \text{in the system} \end{pmatrix} = \begin{pmatrix} \text{Mass flow} \\ \text{into the system} \end{pmatrix} - \begin{pmatrix} \text{Mass flow out} \\ \text{of the system} \end{pmatrix}$$

or in terms of volumetric flow rates F, densities ρ, and volume V,

$$\frac{dM}{dt} = \frac{d(\rho_1 V)}{dt} = F_0 \rho_0 - F_1 \rho_1$$

When densities are equal, as in the case of water flowing in and out of a tank

$$\frac{dV}{dt} = F_0 - F_1$$

The steady-state condition of constant volume in the tank (dV/dt=0) occurs when the volumetric flow in, F_0, is exactly balanced by the volumetric flow out, F_1. Total material balances therefore are mostly important for those modelling situations in which volumes are subject to change, as in simulation examples CONFLO, TANKBLD, TANKDIS and TANKHYD.

1.2.3.1 Case A: Tank Drainage

A tank of diameter D, containing liquid of depth H, discharges via a short base connection of diameter d, as shown in Fig. 1.12 (Brodkey and Hershey, 1988).

In this case, the problem involves a combination of the total material balance with a hydraulic relationship, representing the rate of drainage.

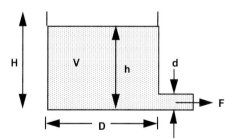

Fig. 1.12 Liquid discharge from the bottom of a tank.

For zero flow of liquid into the tank and assuming constant density conditions, the total material balance equation becomes

$$\frac{dV}{dt} = -F$$

Assuming the absence of any frictional flow effects, the outlet flow velocity, v, is related to the instantaneous depth of liquid within the tank, by the relationship

where

$$v = (2gh)^{1/2}$$

and

$$F = \frac{\pi d^2}{4} v$$

$$V = \frac{\pi D^2}{4} h$$

v is the discharge pipe velocity, V is the volume of liquid in the tank, h is the depth of liquid in the tank and g is the constant of gravitational acceleration.

The above equations are then sufficient to define the model which has the following simple analytical solution

$$h = \left(\sqrt{H} - \frac{d^2}{D^2} \sqrt{\frac{g}{2}} t \right)^2$$

where H is the liquid depth at time t=0.

However, with a time variant flow of liquid into the tank, then analytical solution is not so simple. The problem is treated in more detail in simulation example TANKDIS.

1.2.4
Component Balances

Each chemical species, in the system, can be described by means of a component balance around an arbitrary, well-mixed, balance region, as shown in Fig. 1.13.

In the case of chemical reaction, the balance equation is represented by

$$\begin{pmatrix} \text{Rate of} \\ \text{accumulation} \\ \text{of mass} \\ \text{of component i} \\ \text{in the system} \end{pmatrix} = \begin{pmatrix} \text{Mass flow of} \\ \text{component i} \\ \text{into} \\ \text{the system} \end{pmatrix} - \begin{pmatrix} \text{Mass flow of} \\ \text{component i} \\ \text{out of} \\ \text{the system} \end{pmatrix} \begin{pmatrix} \text{Rate of} \\ \text{production of} \\ \text{component i} \\ \text{by reaction} \end{pmatrix}$$

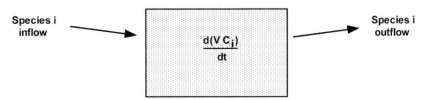

Fig. 1.13 Component balancing for species i.

Expressed in terms of volume, volumetric flow rate and concentration, this is equivalent to

$$\frac{d(VC_i)}{dt} = (F_0 C_{i0}) - (F_1 C_{i1}) + (r_i V)$$

with dimensions of mass/time

$$m^3 \frac{\frac{kg}{m^3}}{s} = \frac{m^3}{s}\frac{kg}{m^3} - \frac{m^3}{s}\frac{kg}{m^3} + \frac{kg}{m^3 s}m^3 = \frac{kg}{s}$$

In the case of an input of component i to the system by interfacial mass transfer, the balance equation now becomes

$$\begin{pmatrix} \text{Rate of} \\ \text{accumulation} \\ \text{of mass} \\ \text{of component i} \\ \text{in the system} \end{pmatrix} = \begin{pmatrix} \text{Mass flow of} \\ \text{component i} \\ \text{into} \\ \text{the system} \end{pmatrix} - \begin{pmatrix} \text{Mass flow of} \\ \text{component i} \\ \text{out of} \\ \text{the system} \end{pmatrix} + \begin{pmatrix} \text{Rate of} \\ \text{interfacial} \\ \text{transfer} \\ \text{of component i} \\ \text{into the system} \end{pmatrix}$$

$$\frac{d(VC_i)}{dt} = (F_0 C_{i0}) - (F_1 C_{i1}) + Q_i$$

where Q_i, the rate of mass transfer, is given by

$$Q_i = K_i A \Delta C_i$$

1.2.4.1 Case A: Waste Holding Tank

A plant discharges an aqueous effluent at a volumetric flow rate F. Periodically, the effluent is contaminated by an unstable noxious waste, which is known to decompose at a rate proportional to its concentration. The effluent must be diverted to a holding tank, of volume V, prior to final discharge, as in Fig. 1.14 (Bird et al. 1960).

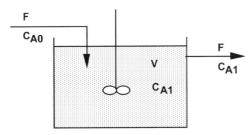

Fig. 1.14 Waste holding tank.

1 Basic Concepts

This situation is one involving both a total and a component material balance, combined with a kinetic equation for the rate of decomposition of the waste component. Neglecting density effects, the total material balance equation is

$$\frac{dV}{dt} = F_0 - F_1$$

The rate of the decomposition reaction is given by

$$r_A = -kC_A$$

and the component balance equation by

$$\frac{d(VC_{A1})}{dt} = (F_0 C_{A0}) - (F_1 C_{A1}) + k_1 C_{A1} V$$

The tank starts empty, so that at time t=0, volume V=0, and the outlet flow from the tank F_1=0. At time t=V/F$_0$ the tank is full; then F_1=F_0=F, and the condition that dV/dt=0 also applies.

The above model equations can be solved analytically. For the conditions that, at time t=0, the initial tank concentration C_A=0, the tank is full and overflowing and that both F and C_{A0} are constant, analytical solution gives

$$C_A = C_{A0}(1 - Z)e^{(-Zt)} + Z$$

where

$$Z = \frac{F}{F + kV}$$

When the flow and inlet concentration vary with time, a solution is best obtained by numerical integration.

1.2.4.2 Case B: Extraction from a Solid by a Solvent

An agitated batch tank is used to dissolve a solid component from a solid matrix into a liquid solvent medium, as in Fig. 1.15.

For a batch system, with no inflow and no outflow, the total mass of the system remains constant. The solution to this problem thus involves a liquid-phase component material balance for the soluble material, combined with an expression for the rate of mass transfer of the solid into the liquid.

The component material balance is then

$$\begin{pmatrix} \text{Rate of accumulation} \\ \text{of the material} \\ \text{in the solvent} \end{pmatrix} = \begin{pmatrix} \text{Rate of transfer} \\ \text{of the solid} \\ \text{to the solvent} \end{pmatrix}$$

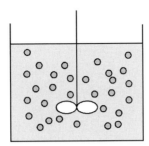

Fig. 1.15 Agitated tank for dissolving solids.

giving

$$V_L \frac{dC_L}{dt} = k_L A (C_L^* - C_L)$$

where V_L is the volume of the liquid, C_L is the concentration of the component in the liquid, k_L is the liquid phase mass transfer coefficient, A is the total interfacial area for mass transfer and C_L^* is the equilibrium value.

The analytical solution to the above equation, assuming constant V_L, k_L, A and equilibrium concentration, C_L^*, is given by

$$\frac{C_L^* - C_L}{C_L^* - C_{L0}} = e^{-k_L A t / V_L}$$

For the case, where the soluble component is leaching from an inert solid carrier, a separate solid phase component balance would be required to establish the solute concentration in the solid phase and hence the time-dependent value of the equilibrium concentration, C_L^*.

If during this extraction the volume and area of the solid remains approximately constant, the balance for the component in the solid phase is

$$V_S \frac{dC_S}{dt} = -k_L A (C_L^* - C_L)$$

where

$$C_L^* = f_{eq}(C_S)$$

and the subscript "eq" refers to the equilibrium condition.

1.2.5
Energy Balancing

Energy balances are needed whenever temperature changes are important, as caused by reaction heating effects or by cooling and heating for temperature control. For example, such a balance is needed when the heat of reaction causes a change in reactor temperature. This change obviously influences the reaction rate and therefore the rate of heat evolution. This is seen in the information flow diagram for a non-isothermal continuous reactor as shown in Fig. 1.16.

Energy balances are formulated by following the same set of guidelines as those given in Section 1.2.2 for material balances. Energy balances are however considerably more complex, because of the many different forms energy occurs in chemical systems. The treatment considered here is somewhat simplified, but is adequate to understand the non-isothermal simulation examples. The various texts cited in the reference section provide additional advanced reading in this subject.

Based on the law of conservation of energy, energy balances are a statement of the first law of thermodynamics. The internal energy depends not only on temperature, but also on the mass of the system and its composition. For that reason, material balances are almost always a necessary part of energy balancing.

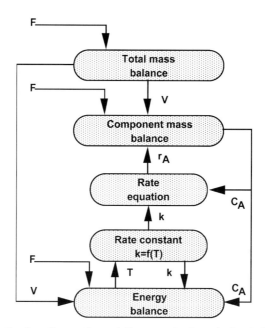

Fig. 1.16 Information flow diagram for modelling a non-isothermal, chemical reactor, with simultaneous mass and energy balances.

For an open system with energy exchange across its boundaries, as shown in Fig. 1.17, the energy balance can be written as

$$\begin{pmatrix} \text{Rate of} \\ \text{accumulation} \\ \text{of energy} \end{pmatrix} = \begin{pmatrix} \text{Rate of} \\ \text{energy} \\ \text{input due} \\ \text{to flow} \end{pmatrix} - \begin{pmatrix} \text{Rate of} \\ \text{energy} \\ \text{output due} \\ \text{to flow} \end{pmatrix} + \begin{pmatrix} \text{Rate of} \\ \text{energy} \\ \text{input due} \\ \text{to transfer} \end{pmatrix}$$

$$- \begin{pmatrix} \text{Rate of work} \\ \text{done by the} \\ \text{system on the} \\ \text{surroundings} \end{pmatrix}$$

$$\frac{dE}{dt} = \sum_{i=1}^{S} N_{i0} E_{i0} - \sum_{i=1}^{S} N_{i1} E_{i1} + Q - W$$

Here, E is the total energy of the system, E_i is the energy per mole of component i, N_i is the molar flow rate of component i, Q is the rate of energy input to the system due transfer and S is the total number of components (reactants and inerts).

The work term can be separated into flow work and other work W_S, according to

$$W = -\sum_{i=1}^{S} N_{i0} P V_{i0} + \sum_{i=1}^{S} N_{i1} P V_{i1} + W_S$$

V_i is the molar volume of component i. The energy E_i is the sum of the internal energy U_i, the kinetic energy, the potential energy and any other forms of energy. Of these various forms of energy, changes of internal energy are usually dominant in chemical systems. The other terms are usually neglected.

Internal energy, U, can be expressed in terms of enthalpy, H

$$E = U = H - (PV)$$

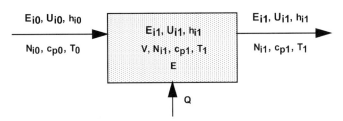

Fig. 1.17 A continuous reactor showing only the energy-related variables.

The accumulation term is given only in terms of enthalpies, since the d(PV) term is usually negligible in chemical reactors.

$$\frac{dE}{dt} = \frac{d}{dt}\sum_{i=1}^{S} n_i h_{i1}$$

h_i is the partial molar enthalpy of component i. Combining these equations and assuming the other work $W_S = 0$, yields the energy balance equation

$$\frac{d}{dt}\sum_{i=1}^{S} n_i h_{i1} = \sum_{i=1}^{S} N_{i0} h_{i0} - \sum_{i=1}^{S} N_{i1} h_{i1} + Q$$

Here, n_i the number of moles of component i, h_i the partial molar enthalpy and Q is the rate of energy input from the environment. Enthalpies are generally dependent on temperature where:

$$h_i = h_{i0} + \int_{T_0}^{T} c_{pi} dT$$

The temperature dependency for heat capacity can usually be described by a polynomial expression, e.g.,

$$c_p = a + bT + cT^2$$

where a, b and c are empirical constants.

A detailed derivation of the energy balance is given in various textbooks (e.g., Aris, 1989 and Fogler, 2005).

With $\frac{\partial h_{i1}}{\partial T} = c_{pi1}$ and $n_{i1} \sum \frac{\partial h_{i1}}{\partial n_k} = 0$, the accumulation term in the energy balance equation can be rewritten as

$$\frac{d}{dt}\sum_{i=1}^{S} n_i h_{i1} = \sum_{i=1}^{S} h_{i1} \frac{dn_{i1}}{dt} + \frac{dT_1}{dt}\sum_{i=1}^{S} n_{i1} c_{pi1}$$

For the solution of the energy balance it is necessary that this is combined with material balance relationships.

Using a general material balance for component i

$$\frac{dn_{i1}}{dt} = N_{i0} - N_{i1} + r_i V$$

multiplying this equation by h_{i1} and summing for all the S components gives

$$\sum_{i=1}^{S} h_{i1} \frac{dn_{i1}}{dt} = \sum_{i=1}^{S} h_{i1} N_{i0} - \sum_{i=1}^{S} h_{i1} N_{i1} + V \sum_{i=1}^{S} r_i h_{i1}$$

Introducing the reaction enthalpy ΔH

$$\sum_{i=1}^{S} r_i h_{i1} = \sum_{j=1}^{R} \frac{r_{ij}}{v_{ij}} (-\Delta H_j(T_1))$$

and allowing for R reactions to occur gives the general energy balance as:

$$\sum_{i=1}^{S} (n_{i1} c_{pi1}) \frac{dT_1}{dt} = \sum_{i=1}^{S} N_{i0}(h_{i0} - h_{i1}) + \sum_{j=1}^{R} \frac{r_{ij}}{v_{ij}} (-\Delta H_j(T_1)) + Q$$

where c_{pi} is the partial molar heat capacity of component i, R_{ij} the reaction rate of component i in reaction j, v_{ij} the stoichiometric coefficient of component i in reaction j and $\Delta H_j(T_1)$ is the reaction enthalpy of reaction j at temperature T_1.

The heat of reaction ΔH is defined by

$$\Delta H = \sum_{i=1}^{n} v_i h_i \approx \sum_{i=1}^{n} v_i \Delta H_{Fi}$$

where ΔH_{Fi} is the heat of formation of component i.

Considering the above temperature dependencies, the complete heat balance can then be written in the following form

$$\sum_{i=1}^{S} (n_{i1} c_{pi1}) \frac{dT_1}{dt} = -\sum_{i=0}^{S} N_{i0} \int_{T_0}^{T_1} c_{pi} dT_1 + \sum_{j=1}^{R} \frac{r_{ij}}{v_{ij}} (-\Delta H_j(T_1)) + Q$$

This equation can be used directly for any well-mixed, batch, semi-batch or continuous volume element. The term on the left-hand side represents the rate of energy accumulation. The first term on the right-hand side depicts the energy needed to raise the temperature of the incoming reactants, including inert material, to the reactor temperature. The second term describes the heat released by the chemical reactions. Since ΔH is a function of state, the energy balance could also be formulated such that the reaction is considered to take place at the inlet temperature T_0, followed by heating the reactor contents to temperature T_1. The above general energy balance equation is applied in simulation example REVTEMP and in Case C (Section 1.2.5.3).

The general heat balance can often be simplified for special situations.

Accumulation Term

At moderate temperature changes, c_{pi} may be assumed to be independent of temperature and therefore

$$\sum_{i=1}^{S} n_{i1} c_{pi1} \approx V \rho c_p$$

The total heat capacity in the accumulation term must also include the reactor parts. c_p is the heat capacity per unit mass (J/kgk). Then

$$\sum_{i=1}^{S} n_{i1} c_{pi1} \frac{dT}{dt} \approx V \rho c_p \frac{dT}{dt}$$

with units

$$\text{mol} \frac{J}{\text{mol K}} \frac{K}{s} = m^3 \frac{kg}{m^3} \frac{J}{kg\,K} \frac{K}{s} = \frac{J}{s}$$

Thus the accumulation term has the units of (energy)/(time), for example J/s.

Flow Term

At moderate temperature changes, c_{pi} is again assumed constant and therefore the flow term is

$$-\sum_{i=0}^{S} N_{i0} \int_{T_0}^{T_1} c_{pi} dT_1 \approx F_0 \rho c_p (T_0 - T_1)$$

with the units

$$\frac{\text{mol}}{s} \frac{J}{\text{mol K}} K = \frac{m^3}{s} \frac{kg}{m^3} \frac{J}{kg\,K} K = \frac{J}{s}$$

This term actually describes the heating of the stream entering the system with temperature T_0 to the reaction temperature T_1, and is therefore only needed if streams are entering the system.

Heat Transfer Term

The important quantities in this term are the heat transfer area A, the temperature driving force or difference $(T_a - T_1)$, where T_a is the temperature of the heating or cooling source, and the overall heat transfer coefficient U. The heat transfer coefficient, U, has units of (energy)/(time)(area)(degree), e.g., J/s m² K.

$$(\text{heat transfer rate}) = UA(T_a - T_1)$$

The units for $U A \Delta T$ are thus

$$\frac{J}{s} = \frac{J}{m^2 s K} m^2 K$$

The sign of the temperature difference determines the direction of heat flow. Here if $T_a > T_1$, heat flows into the reactor.

Reaction Heat Term
For exothermic reactions, the value of ΔH is by convention negative and for endothermic reactions positive. For a set of R individual reactions, the total rate of heat production by reaction is given by

$$r_Q V = \sum_{j=1}^{R} \frac{r_{ij}}{v_{ij}} (-\Delta H_j)$$

with the units

$$\frac{J}{m^3 s} m^3 = \frac{mol}{s} \frac{J}{mol} = \frac{J}{s}$$

Other Heat Terms
The heat of agitation may be important, depending on the relative magnitudes of the other heat terms in the general balance equation and especially with regard to highly viscous reaction mixtures. Other terms, such as heat losses from the reactor, by radiation or by mixing, can also be important in the overall energy balance equation.

Simplified Energy Balance
If specific heat capacities can be assumed constant and for zero mechanical work done to the system, the energy balance equation simplifies to

$$V \rho c_p \frac{dT_1}{dt} = F_0 \rho c_p (T_0 - T_1) + V r_Q + UA(T_a - T_1)$$

where the units of each term of the balance equation are energy per unit time (kJ/s). This equation can be applied to batch, semi-batch and continuous reactors. Most of the non-isothermal simulation examples in Chapter 5 use the above form of the energy balance.

1.2.5.1 Case A: Continuous Heating in an Agitated Tank
Liquid is fed continuously to a stirred tank, which is heated by internal steam coils (Fig. 1.18). The tank operates at constant volume conditions. The system is

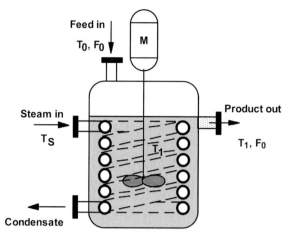

Fig. 1.18 Continuous stirred tank heated by internal steam coil.

therefore modelled by means of a dynamic heat balance equation, combined with an expression for the rate of heat transfer from the coils to the tank liquid.

With no heat of reaction and neglecting any heat input from the agitator, the heat balance equation becomes

$$V\rho c_p \frac{dT_1}{dt} = F_0 \rho c_p (T_0 - T_1) + UA(T_S - T_1)$$

where T_S is the steam temperature.

1.2.5.2 Case B: Heating in a Filling Tank

The situation is the same as in Fig. 1.18 but without material leaving the reactor. Liquid flows continuously into an initially empty tank, containing a full-depth heating coil. As the tank fills, an increasing proportion of the coil is covered by liquid. Once the tank is full, the liquid starts to overflow, but heating is maintained. A total material balance is required to model the changing liquid volume and this is combined with a dynamic heat balance equation.

Assuming constant density, the material balance equation is

$$\frac{dV}{dt} = F_0 - F_1$$

where for time t less than the filling time, V/F_0, the outlet flow, F_1, equals zero, and for time t greater than V/F_0, F_1 equals F_0.

The heat balance is expressed by

$$V\rho c_p \frac{dT}{dt} = F_0 \rho c_p (T_0 - T) + UA(T_S - T)$$

Assuming A_0 is the total heating surface in the full tank, with volume V_0, and assuming a linear variation in heating area with respect to liquid depth, the heat transfer area may vary according to the simple relationship

$$A = A_0 \frac{V}{V_0}$$

More complex relationships can of course be derived, depending on the particular tank geometry concerned.

1.2.5.3 Case C: Parallel Reaction in a Semi-Continuous Reactor with Large Temperature Changes

Let us assume an adiabatic, semi-continuous reactor (see Section 3.2.4) with negligible input of mechanical energy (Fig. 1.19).

Two reactions are assumed to occur in parallel

$$A + B \rightarrow C$$

$$A + 2B \rightarrow D$$

The total energy balance from Section 1.2.5 is given by

$$\sum_{i=1}^{S} (n_i c_{pi}) \frac{dT}{dt} = F_0 \sum_{i=1}^{S} C_{i0} \int_{T}^{T_0} c_{pi} dT + V \sum_{j=1}^{R} \frac{r_{ij}}{\nu_{ij}} (-\Delta H_j)$$

In this case the number of components, S=4 and the number of reactions, R=2.

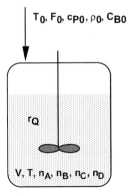

Fig. 1.19 Adiabatic, semi-batch reactor.

The reaction enthalpies at standard temperature, T_{St}, are then

$$\Delta H_{1St} = \Delta H_{FC} - \Delta H_{FA} - \Delta H_{FB}$$

$$\Delta H_{2St} = \Delta H_{FD} - \Delta H_{FA} - 2\Delta H_{FB}$$

All heats of formation, ΔH_{Fi}, are at standard temperature.

Assuming that the temperature dependencies for the specific heats are given by

$$c_{pi} = a_i + b_i T$$

then

$$h_i = h_{iSt} + \int_{T_{St}}^{T} c_{pi} dT = h_{iSt} + a_i(T - T_{St}) + \frac{b_i}{2}(T^2 - T_{St}^2)$$

and the reaction enthalpies, ΔH_1 and ΔH_2, at temperature T are

$$\Delta H_1 = \Delta H_{1St} + (a_C - a_A - a_B)(T - T_{St}) + \frac{b_C - b_A - b_B}{2}(T^2 - T_{St}^2)$$

$$\Delta H_2 = \Delta H_{2St} + (a_C - a_A - 2a_B)(T - T_{St}) + \frac{b_C - b_A - 2b_B}{2}(T^2 - T_{St}^2)$$

With stoichiometric coefficients, $\nu_{A1} = -1$ and $\nu_{A2} = -1$, the total heat of reaction is then

$$r_Q = \sum_{j=1}^{R} \frac{r_{ij}}{\nu_{ij}}(-\Delta H_j) = r_{A1} \Delta H_1 + r_{A2} \Delta H_2$$

The total heat capacity in the accumulation term is

$$V\rho c_p = \sum_{i=1}^{S}(n_i c_{pi}) = n_A(a_A + b_A T) + n_B(a_B + b_B T)$$
$$+ n_C(a_C + b_C T) + n_D(a_D + b_D T)$$

With only component B in the feed, the flow term in the energy balance becomes

$$F_0 \sum_{i=1}^{S} C_{i0} \int_{T}^{T_0} c_{pi} dT = F_0 C_{B0} \left[a_B(T_0 - T) + \frac{b_B}{2}(T_0^2 - T^2) \right]$$

Substitution into the energy balance then gives

$$\frac{dT}{dt} = \frac{F_0 C_{B0} \left[a_B(T_0 - T) + \frac{b_B}{2}(T_0^2 - T^2) \right] + V r_Q}{V\rho c_p}$$

1.2.6
Momentum Balances

Momentum balance equations are of importance in problems involving the flow of fluids. Momentum is defined as the product of mass and velocity and as stated by Newton's second law of motion, force which is defined as mass times acceleration is also equal to the rate of change of momentum. The general balance equation for momentum transfer is expressed by

$$\begin{pmatrix} \text{Rate of change} \\ \text{of momentum} \\ \text{with respect to} \\ \text{time} \end{pmatrix} = \begin{pmatrix} \text{Rate of} \\ \text{momentum} \\ \text{into the} \\ \text{system} \end{pmatrix} - \begin{pmatrix} \text{Rate of} \\ \text{momentum} \\ \text{out of the} \\ \text{system} \end{pmatrix} + \begin{pmatrix} \text{Rate of} \\ \text{generation} \\ \text{of} \\ \text{momentum} \end{pmatrix}$$

Force and velocity are however both vector quantities and in applying the momentum balance equation, the balance should strictly sum all the effects in three dimensional space. This however is outside the scope of this text, and the reader is referred to more standard works in fluid dynamics.

As for the mass and energy balance equations, steady-state conditions are obtained when the rate of change of momentum in the system is zero and

$$\begin{pmatrix} \text{Rate of} \\ \text{momentum} \\ \text{into the} \\ \text{system} \end{pmatrix} - \begin{pmatrix} \text{Rate of} \\ \text{momentum} \\ \text{out of the} \\ \text{system} \end{pmatrix} + \begin{pmatrix} \text{Rate of} \\ \text{generation} \\ \text{of} \\ \text{momentum} \end{pmatrix} = 0$$

Three forms of force, important in chemical engineering flow problems, are pressure forces, shear or viscous forces and gravitational forces (Froment and Bischoff, 1990).

The pressure force is given as the product of pressure and applied area. They are usually taken to be positive when acting on the system surroundings. Shear or viscous forces are also usually taken to be positive when acting on the surroundings and shear force is again the product of shear stress and the applied area. The gravitational forces consist of the force exerted by gravity on the fluid and is equal to the product of the mass of fluid in the control volume times the local acceleration due to gravity. The simulation example TANKHYD utilizes a simple momentum balance to calculate flow rates.

1.2.7
Dimensionless Model Equations

The model mass and energy balance equations will have consistent units, throughout, i.e., kg/s, kmol/s or kJ/s, and corresponding dimensions of mass/time or energy/time. The major system variables, normally concentration or temperature, will also be expressed in terms of particular units, e.g., kmol/m^3 or °C, and the model solution will also usually be expressed in terms of the resultant concentration or temperature profiles, obtained with respect to either time or distance. Each variable will have some maximum value, which is usually possible to establish by simple inspection of the system. Time, as a variable, usually does not have a maximum value, but some characteristic value of time can always be identified. Using these values, a new set of dimensionless variables can be obtained, simply, by dividing all the variables by the appropriate maximum value or by the characteristic time value. This leads to a model composed of dependent variables that vary only between the limits of zero and unity. This means that the solution may, for example, be in terms of the variation of dimensionless concentration versus dimensionless time, and now has a much greater significance, since the particular units of the problem are no longer relevant to the model formulation. The model can now be used much more generally. Furthermore, the various parameters in the original model can be grouped together such that each group of terms also becomes dimensionless. As an added result, the model equations can now be formulated in terms of a fewer number of dimensionless groups than the original number of single parameters. In addition to extending the utility of the mathematical models, the resulting dimensionless groups are especially valuable in correlating experimental data. The above procedure is best illustrated in the simulation examples (BATCHD, TANKD, HOMPOLY, KLADYN, TUBED, TUBDYN, DISRE, DISRET, ENZSPLIT, ENZDYN and BEAD).

1.2.7.1 Case A: Continuous Stirred-Tank Reactor (CSTR)

The material balance for a continuous-flow, stirred-tank reactor with constant volume and first-order reaction is

$$V \frac{dC_{A1}}{dt} = (F_0 C_{A0}) - (F_1 C_{A1}) + k_1 C_{A1} V$$

and this is treated in more detail in Chapter 3. The dimensions for each term in the above equation are those of mass per unit time and the units would normally be kmol/s or kg/s.

Dividing the balance equation by the volume of reactor, V, leads to the equation in the form

$$\frac{dC_{A1}}{dt} = \frac{C_{A0} - C_{A1}}{\tau} - kC_{A1}$$

This equation has two parameters τ, the mean residence time ($\tau = V/F$) with dimensions of time and k, the reaction rate constant with dimensions of reciprocal time, applying for a first-order reaction. The concentration of reactant A in the reactor cannot, under normal circumstances, exceed the inlet feed value, C_{A0}, and thus a new dimensionless concentration, $\bar{C}_{A1}$, can be defined as

$$\bar{C}_{A1} = \frac{C_{A1}}{C_{A0}}$$

such that $\bar{C}_{A1}$ normally varies in the range from zero to one.

The other variable time, t, can vary from zero to some undetermined value, but the system is also represented by the characteristic time, τ. Note that the value of $1/k$ also represents a characteristic time for the process.

A new dimensionless time variable is defined here as

$$t = \tau \bar{t}$$

Alternatively the dimensionless time variable

$$\bar{t} = kt$$

could be employed.

In terms of the dimensionless variables, the original variables are

$$C_{A1} = C_{A0}\bar{C}_{A1}$$
$$dC_{A1} = C_{A0}d\bar{C}_{A1}$$
$$t = \tau$$
$$dt = \tau d\bar{t}$$

When substituted into the model equation, the result is

$$\frac{C_{A0}}{\tau}\frac{d\bar{C}_{A1}}{d\bar{t}} = \frac{C_{A0} - C_{A0}\bar{C}_{A1}}{\tau} - kC_{A0}\bar{C}_{A1}$$

This equation can now be rearranged such that the parameter for the time derivative is unity. Thus dividing by C_{A0} and multiplying by τ gives

$$\frac{d\bar{C}_{A1}}{d\bar{t}} = 1 - \bar{C}_{A1} - (k\tau)\bar{C}_{A1}$$

The parameter term $(k\tau)$, which is called the Damköhler Number Da, is dimensionless and is now the single governing parameter in the model. This results in a considerable model simplification because originally the three parameters, τ, k and C_{A0}, all appeared in the model equation.

The significance of this dimensionless equation form is now that only the parameter $(k\tau)$ is important; and this alone determines the system dynamics and the resultant steady state. Thus, experiments to prove the validity of the model need only consider different values of the combined parameter $(k\tau)$.

For this, the dimensionless reactant concentration, $\bar{C}_{A1}$, should be plotted versus dimensionless time, $\bar{t}$, for various values of the dimensionless system parameter $(k\tau)$. Although, k is not an operating variable and cannot be set independently, this type of plot may be useful estimating a value from experimental data, as illustrated below.

At steady state

$$0 = 1 - \bar{C}_{A1} - (k\tau)\bar{C}_{A1}$$

so that

$$(k\tau) = \frac{1 - \bar{C}_{A1}}{\bar{C}_{A1}}$$

Knowing τ thus permits determination of the value k from experimental data.

Consider an nth-order reaction, the equivalent dimensionless model for the stirred-tank reactor becomes

$$\frac{d\bar{C}_{A1}}{d\bar{t}} = 1 - \bar{C}_{A1} - \left(k\tau C_{A0}^{n-1}\right)\bar{C}_{A1}^{n}$$

The variables are defined as previously. Thus if, for example, experimental data is to be tested for second-order reaction behaviour, then data plotted as $\bar{C}_{A1}$ versus $\bar{t}$ should be examined from experiments, for which $(k\tau C_{A0})$ is kept constant.

1.2.7.2 Case B: Gas-Liquid Mass Transfer to a Continuous Tank Reactor with Chemical Reaction

A second-order reaction takes place in a two-phase continuous system. Reactant A is supplied by gas-liquid transfer, and reactant B is supplied by liquid feed as depicted in Fig. 1.20.

The model equations are

$$V\frac{dC_{A1}}{dt} = K_L a(C_A^* - C_{A1})V - kC_{A1}C_{B1}V - FC_{A1}$$

$$V\frac{dC_{B1}}{dt} = FC_{B0} - FC_{B1} - kC_{A1}C_{B1}V$$

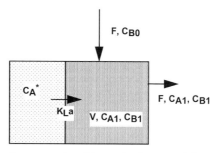

Fig. 1.20 Two phase continuous reactor with supply of reactant A from gas phase.

The gas-liquid saturation value, C_A^*, will be assumed constant.

Dimensionless variables can be defined as

$$\bar{C}_{A1} = \frac{C_{A1}}{C_{B0}}$$

$$\bar{C}_{B1} = \frac{C_{B1}}{C_{B0}}$$

$$\bar{t} = K_L a\, t$$

The equations become

$$\frac{d\bar{C}_{A1}}{d\bar{t}} = \left(\frac{C_A^*}{C_{B0}} - \bar{C}_{A1}\right) - \frac{kC_{B0}}{K_L a}\bar{C}_{A1}\bar{C}_{B1} - \frac{1}{K_L a \tau}\bar{C}_{A1}$$

$$\frac{d\bar{C}_{B1}}{d\bar{t}} = \frac{1}{K_L a \tau}(1 - \bar{C}_{B1}) - \frac{kC_{B0}}{K_L a}\bar{C}_{A1}\bar{C}_{B1}$$

where τ is the residence time ($= V/F$). The number of parameters is reduced, and the equations are in dimensionless form.

An equivalency can be demonstrated between the concept of time constant ratios and the new dimensionless parameters as they appear in the model equations. The concept of time constants is discussed in Section 2.2.

Thus the variables in this example can be interpreted as follows

$$\frac{1}{K_L a \tau} = \frac{\text{Transfer time constant}}{\text{Residence time constant}} = \frac{\text{Convection rate}}{\text{Transfer rate}}$$

and

$$\frac{kC_{B0}}{K_L a} = \frac{\text{Transfer time constant}}{\text{Reaction time constant}} = \frac{\text{Reaction rate}}{\text{Transfer rate}}$$

Further examples of the use of dimensionless terms in dynamic modelling applications are given in Sections 1.2.5.1, 4.3.6.1 and 4.3.7 and in the simulation examples KLADYN, DISRET, DISRE, TANKD and TUBED.

1.3 Chemical Kinetics

1.3.1 Rate of Chemical Reaction

By simplifying the general component balance of Section 1.2.4, the material balance for a batch reactor becomes

$$\begin{pmatrix} \text{Rate of} \\ \text{accumulation} \\ \text{of mass} \\ \text{of component i} \\ \text{in the system} \end{pmatrix} = \begin{pmatrix} \text{Rate of} \\ \text{production of} \\ \text{component i} \\ \text{by reaction} \end{pmatrix}$$

Expressed in terms of volume V and concentration C_i, this is equivalent to

$$\frac{d(VC_i)}{dt} = r_i V$$

with units of moles/time. Here the term r_i is the rate of chemical reaction, expressed as the change in the number of moles of a given reactant or product per unit time and per unit volume of the reaction system. Thus for a batch reactor, the rate of reaction for reactant i can be defined as

$$r_i = \frac{1}{V}\frac{dn_i}{dt} \quad \frac{\text{moles of i}}{\text{volume time}}$$

where $n_i = VC_i$ and it is the number of moles of i present at time t. Alternatively the rate equation may be expressed in terms of mass, kg.

The reactants and products are usually related by a stoichiometric equation which is usually expressed as a molar relationship. For the case of components A and B reacting to form product C it has the form

$$v_A A + v_B B \rightarrow v_C C$$

v_i is the stoichiometric coefficient for species i in the reaction. By convention, the value of v is positive for the products and negative for the reactants. The stoichiometric coefficients relate the simplest ratio of the number of moles of reactant and product species, involved in the reaction.

The individual rates of reaction, for all the differing species of a reaction, are related via their stoichiometric coefficients according to

$$r_i = r_j \left(\frac{v_i}{v_j}\right)$$

The value of r_i is therefore negative for reactants and positive for products.

For the reaction

$$A + 2B \rightarrow 3P$$

the individual reaction rates are therefore

$$-r_A = -\frac{1}{2}r_B = \frac{1}{3}r_P$$

Thus in defining the rate of reaction, it is important to state the particular species.

The temperature and concentration dependencies of reaction rates can usually be expressed as separate functions, for example

$$r_A = k_0 f_1(T) f_2(C_A)$$

The dependency of temperature $f_1(T)$ is usually described as the Arrhenius Equation, as explained in the next section.

The exact functional dependence of the reaction rate with respect to concentration must be found by experiment, and for example may have the general form

$$r_A = -k C_A^\alpha C_B^\beta$$

Here k is the reaction rate constant, which is a function of temperature only; C_A, C_B are the concentrations of the reactants A, B (moles/volume); α is the order of reaction, with respect to A; β is the order of reaction, with respect to B; $(\alpha+\beta)$ is the overall order of the reaction. Whatever reference quantity is used to define specific rates this needs to be stated clearly.

It is important to realize that the reaction rate may represent the overall summation of the effect of many individual elementary reactions, and therefore only rarely represents a particular molecular mechanism. The orders of reaction, α or β, can not be assumed from the stoichiometric equation and must be determined experimentally.

For heterogeneous catalytic reactions, the rate of reaction is often expressed as the number of moles (or kg) of component reacting per unit time, per unit mass of catalyst. For a batch reactor

$$r_A = \frac{1}{M} \frac{dn_A}{dt} \quad \frac{\text{moles}}{\text{mass time}}$$

where M is the mass of catalyst. Sometimes the surface area is used as the reference quantity for solid surface reactions.

In vapour phase reactions, partial pressure units are often used in place of concentration in the rate equation, for example

$$r_A = -k p_A^\alpha p_B^\beta$$

where p_A and p_B are the gas phase partial pressures of reactants A and B. In this case, k would be expressed in terms of pressure units. Detailed treatments of chemical kinetics are found e.g. in Walas (1989), Missen et al. (1999), Levenspiel (1999).

1.3.2
Reaction Rate Constant

The reaction constant, k, is normally an exponential function of the absolute temperature, T, and is described by the Arrhenius equation

$$k = Ze^{-E/RT}$$

The exponential term gives rise to the highly non-linear behaviour in reactor systems which are subject to temperature changes.

The parameters Z and E, the activation energy, are usually determined by measuring k, over a range of temperatures, and plotting ln k versus the reciprocal absolute temperature, 1/T, as shown in Fig. 1.21.

High reaction temperatures can cause numerical overflow problems in the computer calculation of k, owing to the very large values generated by the exponential term. This can often be eliminated by defining a value of the rate constant, k_0, for some given temperature, T_0.

Thus

$$k_0 = Ze^{-E/RT_0}$$

The logarithmic form of the Arrhenius relationship is then

$$\ln \frac{k}{k_0} = -\frac{E}{R}\left(\frac{1}{T} - \frac{1}{T_0}\right)$$

which permits the calculation of k at any temperature T. The above procedure is used in the simulation examples THERM and THERMPLOT.

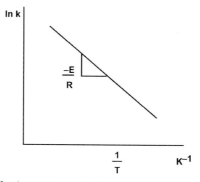

Fig. 1.21 Arrhenius plot for determining activation energy.

1.3.3
Heat of Reaction

The heat of reaction, ΔH, can be calculated from the heats of formation or heats of combustion

$$\Delta H = \sum_{i=1}^{n} v_i \Delta H_{Fi} = \sum_{i=1}^{n} v_i \Delta H_{Ci}$$

where ΔH_{Fi} is the heat of formation of component i, ΔH_{Ci} is the heat of combustion of component i and v_i is the stoichiometric coefficient for component i. If heats of formation are not available, heats of combustion can easily be determined from calorimetric heats of combustion data. The resulting heat of reaction, ΔH, can be calculated and by convention is negative for exothermic reactions and positive for endothermic reactions. The temperature dependence of ΔH is described in Section 1.2.5. For complex reaction systems, the heats of reaction of all individual reactions have to be estimated, and the dynamic heat balance equations must include the heats of all the reactions.

1.3.4
Chemical Equilibrium and Temperature

Chemical equilibrium depends on temperature as described by the van't Hoff equation

$$\frac{d(\ln K)}{dT} = \frac{\Delta H}{RT^2}$$

Here K is the thermodynamic chemical equilibrium constant. If ΔH is constant, direct integration yields an explicit expression. If ΔH is a function of temperature, as described in Section 1.3.3, then its dependency on c_p can be easily included and integration is again straight-forward. A calculation with varying ΔH and c_p as functions of temperature is given in the simulation example REVTEMP developed in Section 1.2.5.3.

1.3.5
Yield, Conversion and Selectivity

The fractional conversion of a given reactant, X_A, is defined for a batch system as

$$X_A = \frac{\text{moles of A reacted}}{\text{moles of A initially present}}$$

giving

$$X_A = \frac{n_{A0} - n_A}{n_{A0}}$$

or

$$n_A = n_{A0}(1 - X_A)$$

and at constant volume

$$X_A = \frac{C_{A0} - C_A}{C_{A0}}$$

where n_{A0} is the initial number of moles of A, n_A is the number of moles of A at fractional conversion X_A and $(n_{A0} - n_A)$ is the number of moles of A reacted. From this it follows for a constant-volume batch system that

$$r_A = -\frac{n_{A0}}{V}\frac{dX_A}{dt}$$

For a well-mixed flow system at steady state, the fractional conversion X_A is the ratio of the number of moles of A converted to the moles A fed to the system

$$X_A = \frac{F_0 C_{A0} - F_1 C_{A1}}{F_0 C_{A0}}$$

where for equal volumetric flow rates at inlet and outlet $(F_0 = F_1)$

$$X_A = \frac{C_{A0} - C_{A1}}{C_{A0}}$$

This definition is identical to that of the batch case.

Fractional yield is defined by

$$Y_{C/A} = \frac{\text{Moles of A transformed into a given product C}}{\text{Total moles of A reacted}}$$

Again it is important that both the particular reactant and product, concerned, should be stated, when defining a fractional yield.

A definition of instantaneous fractional yield is based on the ratio of reaction rates

$$Y_{C/A} = \frac{r_C}{-r_A}$$

where A is the key reactant and C the product.

Multiple reaction selectivity can be defined similarly as the ratio of the rate of formation of the desired product to the formation rate of an undesired product as in a parallel reaction

where

$$S_{B,C} = \frac{r_B}{r_C}$$

Here $S_{B,C}$ is the instantaneous selectivity of the desired product B to unwanted product in this parallel reaction.

1.3.6
Microbial Growth Kinetics

Under ideal batch growth conditions, the quantity of biomass, and therefore the biomass concentration will increase exponentially with respect to time and in accordance with all cells having the same probability to multiply. Thus the overall rate of biomass formation is proportional to the biomass itself where

$$r_X = kX$$

Here r_X is the rate of cell growth (kg cell m^{-3} s^{-1}), X is the cell concentration (kg cell m^{-3}) and k is a kinetic growth constant (s^{-1}). For a batch system, this is equivalent to

$$\frac{dX}{dt} = kX$$

where dX/dt is the rate of change of cell concentration with respect to time (kg cell m^{-3}s^{-1}). The analytical solution of this simple first order differential equation is

$$\frac{X}{X_0} = e^{kt}$$

where X_0 is the initial cell concentration at time $t=0$.

The plot of the logarithm of cell concentration versus time will often yield a straight line over a large portion of the curve, as shown in Fig. 1.22.

The initial time period up to t_1 represents a period of zero growth, which is known as the lag phase. In this period the cells synthesise enzymes and other cellular components appropriate to the particular environmental conditions of the fermentation.

An exponential (or logarithmic) growth phase follows the lag phase, and during this period the cell mass increases exponentially. The growth rate is at a maximum during this phase, and the population of cells are fairly uniform with respect to chemical composition and metabolic composition.

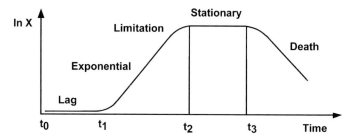

Fig. 1.22 Biomass concentration during batch growth.

The growth of micro-organisms in a batch reactor will eventually end, owing either to the depletion of some essential nutrient or perhaps to the accumulation of some toxic product. The result is that the growth rate gradually slows, and the growth becomes nutrient limited or product inhibited. When the growth rate falls to that of the cell death rate, the cell concentration remains constant, during the stationary phase part of the curve.

Following the stationary phase the rate of cell death exceeds that of cell growth, and the cell number begins to decrease, resulting in the final death rate part of the curve.

The slope of the curve in the limitation region decreases as it approaches t_2. The slope represents the growth rate per unit mass of cells or specific growth rate and is given the symbol μ (s^{-1}), where:

$$\frac{d \ln X}{dt} = \frac{1}{X}\frac{dX}{dt} = \text{specific growth rate} = \mu$$

In many processes, cells may die continuously or may start dying (after time, t_3) because of a lack of nutrients, toxic effects or cell ageing. This process can typically be described by a first order decay relationship:

$$r_d = -k_d X$$

where r_d is the death rate and k_d (s^{-1}) is the specific death rate coefficient.

The exponential and limiting regions of cell growth can be described by a single relation, in which μ is a function of substrate concentration, i.e., the Monod equation

$$\mu = \frac{\mu_{max} S}{K_S + S}$$

Although very simple, the Monod equation frequently describes experimental growth rate data very well. The form of this relation is shown in Fig. 1.23.

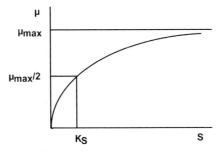

Fig. 1.23 Specific growth rate versus limiting substrate concentration according to the Monod relation.

The important properties of this relationship are as follows:

$$S \to 0, \quad \mu \to \frac{\mu_{max}}{K_S} S$$

$$S \to \infty, \quad \mu \to \mu_{max}$$

$$S = K_S, \quad \mu = \frac{\mu_{max}}{2}$$

Substrate Uptake Kinetics
The rate of uptake of substrate by micro-organisms is generally considered to be related to the rate of growth and to the rate required for maintenance,

$$r_S = \frac{-r_X}{Y_{X/S}} - mX$$

Here r_S is the rate of substrate uptake by the cells (kg substrate m^{-3} s^{-1}). $Y_{X/S}$ (kg/kg) is the stoichiometric factor or yield coefficient, relating the mass of cell produced per unit mass of substrate consumed and the maintenance factor m (kg substrate/kg biomass S), represents the utilisation of substrate by the cells for non-growth related functions.

For further details of microbial kinetics refer to Dunn et al. (2003), Moser (1988), Shuler and Kargi (1999) and Blanch and Clark (1996).

1.4
Mass Transfer Theory

1.4.1
Stagewise and Differential Mass Transfer Contacting

Mass transfer separation processes, e.g., distillation, gas absorption, etc., are normally treated in terms of stagewise or differential procedures. In a stagewise procedure, concentration changes are taken to occur in distinct jumps, as, for

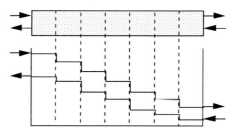

Fig. 1.24 Concentration profiles for countercurrent, stagewise contactor.

example, between the neighbouring plates of a distillation column. In a differential procedure the concentrations are assumed to vary continuously throughout the total length or volume of the contactor, as, in say, a packed bed gas absorption column. The two different types of operation lead to two quite distinct design approaches, namely a stagewise design and a differential design. Both can be handled to advantage by simulation methods.

Figure 1.24 shows a countercurrent stagewise mass transfer cascade and the resulting staged profile of the two streams, owing to the mass transfer between the streams.

In the stagewise simulation method the procedure is based on the assessment of the separation achieved by a given number of equilibrium contacting stages. The concept of the equilibrium stage is illustrated, for a particular stage n of the cascade, as shown in Fig. 1.25.

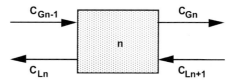

Fig. 1.25 Equilibrium mass-transfer stage.

According to this assumption the two streams are so well mixed that the compositions of each phase within the stage are uniform. Further, the mass transfer is so efficient that the compositions of the streams leaving the stage are in equilibrium.

$$C_{Ln} = f_{eq}(C_{Gn})$$

The actual stage can be a mixing vessel, as in a mixer-settler used for solvent extraction applications, or a plate of a distillation or gas absorption column. In order to allow for non-ideal conditions in which the compositions of the two exit streams do not achieve full equilibrium, an actual number of stages can be related to the number of theoretical stages, via the use of a stage-efficiency factor. Also it will be seen that a rate approach will account for this.

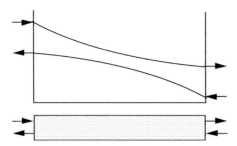

Fig. 1.26 Concentration profiles for countercurrent, differential contactor.

Figure 1.26 shows a differential type of contactor as in, say, a countercurrent flow-packed gas absorption column, together with the resulting approximate continuous concentration profiles.

In this type of apparatus, the two phases do not come to equilibrium, at any point in the contactor and the simulation method is based, therefore, not on a number of equilibrium stages, but rather on a consideration of the relative rates of transport of material through the contactor by flow and the rate of interfacial mass transfer between the phases. For this, a consideration of mass transfer rate theory becomes necessary.

1.4.2
Phase Equilibria

Knowledge of the phase equilibrium is essential for any mass transfer process, since this is, by definition, implicit in the idea of a theoretical stage. It is also important, however, in determining the concentration driving-force term in the mass-transfer rate expression. At phase equilibrium conditions, the driving force for mass transfer is zero and therefore further concentration changes via a mass transfer mechanism become impossible. The equilibrium is therefore also important in determining the maximum extent of the concentration change, possible by mass transfer.

Equilibrium data correlations can be extremely complex, especially when related to non-ideal multicomponent mixtures, and in order to handle such real life complex simulations, a commercial dynamic simulator with access to a physical property data-base often becomes essential. The approach in this text is based, however, on the basic concepts of ideal behaviour, as expressed by Henry's Law for gas absorption, the use of constant relative volatility values for distillation and constant distribution coefficients for solvent extraction. These have the advantage that they normally enable an explicit method of solution and avoid the more cumbersome iterative types of procedure, which would otherwise be required. Simulation examples in which more complex forms of equilibria are employed are STEAM and BUBBLE.

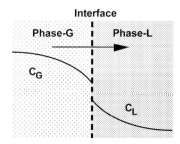

Fig. 1.27 Concentration gradients at a gas-liquid interface.

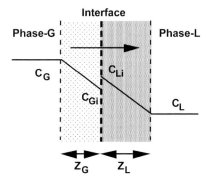

Fig. 1.28 Concentration gradients according to the Whitman Two-Film theory.

1.4.3
Interphase Mass Transfer

Actual concentration profiles (Fig. 1.27) in the very near vicinity of a mass transfer interface are complex, since they result from an interaction between the mass transfer process and the local hydrodynamic conditions, which change gradually from stagnant flow, close to the interface, to more turbulent flow within the bulk phases.

According to the Whitman Two-Film theory, the actual concentration profiles, as shown in Fig. 1.27, are approximated for the steady state with no chemical reaction, by that of Fig. 1.28.

The above theory makes the following assumptions:
1. A thin film of fluid exists on either side of the interface.
2. Each film is in stagnant or laminar flow, such that mass transfer across the films is by a process of molecular diffusion and can therefore be described by Fick's Law.
3. There is zero resistance to mass transfer at the interface, itself, and therefore the concentrations at the interface are in local equilibrium.

1.4 Mass Transfer Theory

4. Each of the bulk phases, outside the films, are in turbulent flow. Concentrations within the bulk phases are therefore uniform and the bulk phases constitute zero resistance to mass transfer.
5. All the resistance to mass transfer therefore occurs within the films.

Fick's Law states that the flux j (mol/s m^2) for molecular diffusion, for any given component is given by

$$j = -D\frac{dC}{dZ}$$

where D is the molecular diffusion coefficient (m^2/s), and dC/dZ is the steady-state concentration gradient (mol/m^{3}m). Thus applying this concept to mass transfer across the two films

$$j_A = D_G\frac{C_G - C_{Gi}}{Z_G} = D_L\frac{C_{Li} - C_L}{Z_L}$$

where D_G and D_L are the effective diffusivities of each film, and Z_G and Z_L are the respective thicknesses of the two films.

The above equations can be expressed as

$$j = k_G(C_G - C_{Gi}) = k_L(C_{Li} - C_L)$$

where k_G and k_L (m/s) are the mass transfer coefficients for the G-phase and L-phase films, respectively.

The total rate of mass transfer, Q (mol/s), is given by

$$Q = jA = j(aV)$$

where A is the total interfacial area for mass transfer; a is defined as the specific area for mass transfer or interfacial area per volume (m^2/m^3) and V is the volume (m^3).

Thus

$$Q = k_G A(C_G - C_{Gi}) = k_L A(C_{Li} - C_L)$$

or in terms of a and V

$$Q = k_G a(C_G - C_{Gi})V = k_L a(C_{Li} - C_L)V$$

Since the mass transfer coefficient, k, and the specific interfacial area, a, vary in a similar manner, dependent upon the hydrodynamic conditions and system physical properties, they are frequently combined and referred to as a "ka" value or more properly as a mass transfer capacity coefficient.

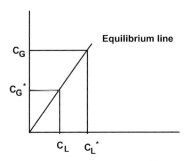

Fig. 1.29 The bulk phase concentrations determine the equilibrium concentrations.

In the above theory, the interfacial concentrations C_{Gi} and C_{Li} are not measurable directly and are therefore of relatively little immediate use. In order to overcome this apparent difficulty, overall mass transfer rate equations are defined by analogy to the film equations. These are based on overall coefficients of mass transfer, K_G and K_L, and overall concentration driving forces, where

$$Q = K_G A (C_G - C_G^*) = K_L A (C_L^* - C_L)$$

and C_G^* and C_L^* are the respective equilibrium concentrations, corresponding to the bulk phase concentrations, C_L and C_G, respectively, as shown in Fig. 1.29.

Simple algebra, based on a combination of the film and overall mass transfer rate equations, lead to the following equations, relating the respective overall mass transfer coefficients and the coefficients for the two films

$$\frac{1}{K_G} = \frac{1}{k_G} + \frac{m}{k_L}$$

and

$$\frac{1}{K_L} = \frac{1}{k_L} + \frac{1}{m k_G}$$

where m is the local slope of the equilibrium line

$$m = \frac{dC_G}{dC_L^*}$$

For a non-linear equilibrium relationship, in which the slope of the equilibrium curve varies with concentration, the magnitudes of the overall mass transfer coefficients will also vary with concentration, even when the film coefficients themselves remain constant. The use of overall mass transfer coefficients in mass transfer rate equations should therefore be limited to the case of linear equilibrium or to situations in which the mass transfer coefficient is known to be relatively insensitive to concentration changes. Design equations based on the use of film mass transfer coefficients and film concentration driving forces make use of the identity that:

$$k_G(C_G - C_{Gi}) = k_L(C_{Li} - C_L)$$

and that the interfacial concentrations C_{Gi} and C_{Li} are in local equilibrium. Examples with mass transfer are OXIDAT, KLADYN and all examples in Section 5.8.

2
Process Dynamics Fundamentals

2.1
Signal and Process Dynamics

2.1.1
Measurement and Process Response

The aim of dynamic simulation is to be able to relate the dynamic output response of a system to the form of the input disturbance, in such a way that an improved knowledge and understanding of the dynamic characteristics of the system are obtained. Figure 2.1 depicts the relation of a process input disturbance to a process output response.

In testing process systems, standard input disturbances, such as the unit-step change, unit pulse, unit impulse, unit ramp, sinusoidal, and various randomised signals, can be employed.

All the above changes are easily implementable in dynamic simulations, using MADONNA and other digital simulation languages. The forms of response obtained differ in form, depending upon the system characteristics and can be demonstrated in the various MADONNA simulation examples. The response characteristics of real systems are, however, more complex. In order to be able to explain such phenomena, it is necessary to first examine the responses of simple systems, using the concept of the simple, step-change disturbance.

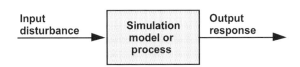

Fig. 2.1 Relation of process input to process output.

2.1.1.1 First-Order Response to an Input Step-Change Disturbance
The simplest response of a linear system is described mathematically by the following standard form of first-order differential equation

$$\tau \frac{dy}{dt} + y = y_0$$

$$\frac{dy}{dt} = \frac{y_0 - y}{\tau}$$

where y is the measured or process response value, t is time, and τ is the equation time constant. In its second form, the equation is often described as a first-order lag equation, in that the response, y, lags behind the input value y_0, imposed on the system at time t = 0.

For the step-change condition, shown in Fig. 2.2, the initial conditions are given by y = 0, when t = 0. The solution to the differential equation, with the above boundary conditions, is given by

$$y = y_0(1 - e^{-t/\tau})$$

and is shown in Fig. 2.2

Substituting the value $t = \tau$, gives

$$y_\tau = y_0(1 - e^{-1}) = 0.632 y_0$$

Hence the value of the equation time constant τ is simply determined as the time at which the response achieves sixty three per cent of its eventual steady-state value, when following a step change disturbance to the system.

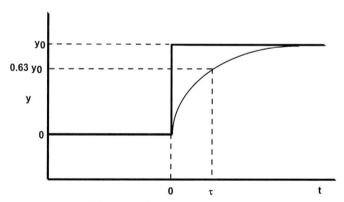

Fig. 2.2 First-order exponential response to an imposed step-change disturbance.

2.1.1.2 Case A: Concentration Response of a Continuous Flow, Stirred Tank

Liquid flows through a tank of constant volume V, with volumetric flow rate F and feed concentration C_0, as shown in Fig. 2.3.

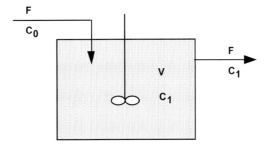

Fig. 2.3 A continuous stirred-tank reactor.

Assuming well-mixed conditions, the component balance equation is given by

$$V\frac{dC_1}{dt} = FC_0 - FC_1$$

This can be expressed as

$$\frac{dC_1}{dt} = \frac{C_0 - C_1}{\tau}$$

where τ is V/F.

Note that the equation has the general form

$$\left(\begin{array}{c}\text{Rate of change}\\ \text{output variable}\end{array}\right) = \left(\frac{\text{Input variable} - \text{Output variable}}{\text{Process time constant}}\right)$$

Thus the time constant for the process is equal to the mean residence or hold-up time in the tank and has units of time (volume/volumetric flow rate).

Integrating with $C_1 = 0$ when $t = 0$ gives

$$C_1 = C_0(1 - e^{-t/\tau})$$

The response to a step change in feed concentration is thus given by

$$C_1 = C_0(1 - e^{-t/\tau}) = C_0(1 - e^{-Ft/V})$$

and follows the same form as that shown in Fig. 2.4.

Note that when time, t, tends to infinity, the value C_1 approaches C_0, and when time, t, is numerically equal to τ

$$C_1 = C_0(1 - e^{-1}) = 0.632 C_0$$

Thus the response is 63.2% complete when the time passed is equal to the time constant τ. Further, for time t equal to four times the value of time constant τ,

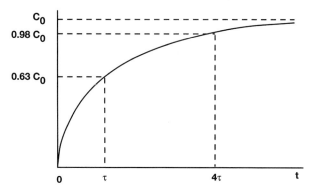

Fig. 2.4 Step response of a first-order system.

$$C_1 = C_0(1 - e^{-4}) = 0.98 C_0$$

then the response is 98% complete.

2.1.1.3 Case B: Concentration Response in a Continuous Stirred Tank with Chemical Reaction

Assuming a chemical reaction in the tank, in which the rate of reaction is proportional to concentration, the component balance equation now becomes

$$V \frac{dC_1}{dt} = FC_0 - FC_1 - kC_1 V$$

where k is the chemical rate coefficient (1/s). This can be rewritten as

$$\frac{V}{F + kV} \frac{dC_1}{dt} + C_1 = \frac{F}{F + kV} C_0$$

and the system time constant now has the value

$$\tau = \frac{V}{F + kV}$$

with high values of k acting to reduce the magnitude of τ.

The above equation now becomes

$$\frac{dC_1}{dt} = \frac{C_{1\infty} - C_1}{\tau}$$

where $C_{1\infty}$ is the final steady-state value at $t = \infty$, and is given by

$$C_{1\infty} = \frac{FC_0}{F + kV}$$

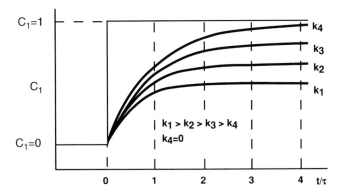

Fig. 2.5 Step response of a stirred tank with first-order chemical reaction ($V = 1$, $F = 1$, $k_4 = 0$, $k_3 = 0.2$, $k_2 = 0.5$, $k_1 = 1$).

Figure 2.5 illustrates the effect on the process response of increasing values of k. This shows that increasing the value of k will decrease the response time of the system and that the final effluent concentration leaving the tank will be reduced in magnitude. Increasing k has, however, very little influence on the initial rate of response.

Note that for $k = 0$, with zero reaction, the final steady-state response is given by C_0, and the response is identical to that of Case A.

2.1.1.4 Case C: Response of a Temperature Measuring Element

Instrument measurement response can often be important in the overall system response. The thermal response of a simple thermometer bulb, immersed in fluid, as shown in Fig. 2.6, is the result of a simple heat balance in which

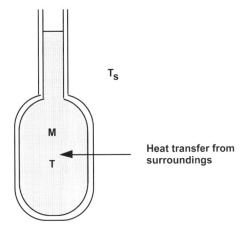

Fig. 2.6 Temperature measuring element.

$$\begin{pmatrix} \text{Rate of accumulation} \\ \text{of heat by the bulb} \end{pmatrix} = \begin{pmatrix} \text{Rate of heat transfer} \\ \text{to the bulb from the fluid} \end{pmatrix}$$

i.e.,

$$Mc_p \frac{dT}{dt} = UA(T_S - T)$$

where M is the mass of the thermometer bulb contents, c_p is the specific heat capacity of the bulb contents, U is the film heat transfer coefficient between the fluid and bulb wall, A is the heat transfer surface area and T_S is the temperature of the surrounding fluid. Note that the stirred-tank or lumped-parameter concept has again been adopted in the modelling approach and that the temperature of the fluid within the bulb is assumed to be uniform.

The simple balance equation can be reformulated as

$$\frac{M c_p}{U A} \frac{dT}{dt} + T = T_S$$

showing that the measurement time constant is

$$\tau = \frac{M c_p}{U A}$$

Often an instrument response measurement can be fitted empirically to a first-order lag model, especially if the pure instrument response to a step change disturbance has the general shape of a first-order exponential.

As shown in Section 2.1.1.1, the time constant for the instrument is then given as the time at which 63% of the final response is achieved and the instrument response may be described by the simple relationship

$$\frac{dT_{meas}}{dt} = \frac{T_{system} - T_{meas}}{\tau}$$

where T_{system}, the actual temperature, and T_{meas}, the measured temperature, are related by the measurement dynamics, as shown in Fig. 2.7, and τ is the experimentally obtained instrument time constant.

The ratio of the time constants, τ_{meas} and τ_{system}, determines whether or not the system value is significantly different from the measurement value when conditions are changing, since the measured value will tend to lag behind that of the system.

For a thermometer to react rapidly to changes in the surrounding temperature, the magnitude of the time constant should be small. This involves a high

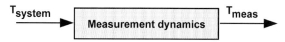

Fig. 2.7 Influence of instrument response on the measured temperature.

surface area to liquid mass ratio, a high heat transfer coefficient and a low specific heat capacity for the bulb liquid. With a large time constant, the instrument will respond slowly and may result in a dynamic measurement error.

2.1.1.5 Case D: Measurement Lag for Concentration in a Batch Reactor

The measurement lag for concentration in a reactor is depicted in Fig. 2.8. The actual reactant concentration in the reactor at any time t is given by C_r, but owing to the slow response of the measuring instrument, the measured concentration, shown by the instrument, C_m, lags behind C_r, as indicated in Fig. 2.9.

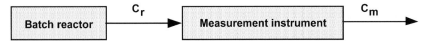

Fig. 2.8 Concentration measurement lag.

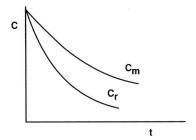

Fig. 2.9 Measured (C_m) and actual (C_r) concentration responses.

The dynamic error existing between C_m and C_r depends on the relative magnitudes of the respective time constants. For the reactor, assuming a first-order constant volume reaction

$$\frac{dC_r}{dt} = -k\,C_r$$

gives a time constant for the reaction

$$\tau_r = \frac{1}{k}$$

Assuming that the instrument response is first order, then as shown in Section 2.1.1.1, the instrument time constant τ_m is given by the value of time at the 63% point (response to a step-change disturbance), where

$$\frac{dC_m}{dt} = \frac{C_r - C_m}{\tau_m}$$

2 Process Dynamics Fundamentals

The ratio of the time constants, τ_r/τ_m, which for this case equals $(k\tau_m)$ will determine whether C_r is significantly different from C_m. When this ratio is less than 1.0 the measurement lag will be important. If $\tau_r/\tau_m > 10$, then $C_m \approx C_r$ and the measurement dynamics become unimportant.

The effects of measurement dynamics are demonstrated in the simulation examples KLADYN, TEMPCONT and CONTUN.

2.1.2
Higher Order Responses

Actual response curves often follow a sigmoidal curve as shown in Fig. 2.10. This is characteristic of systems having a series of multiple lags and hence of systems which are characterised by several time constants.

Examples of higher order response curves are shown by the following case studies.

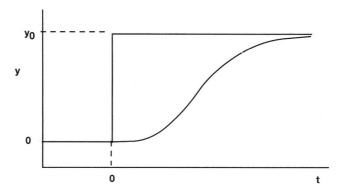

Fig. 2.10 Higher order step response.

2.1.2.1 Case A: Multiple Tanks in Series

Consider the case of three, constant-volume tanks in series, as represented in Fig. 2.11, in which the tanks have differing volumes V_1, V_2, V_3, respectively. Assuming well-mixed tanks, the component balance equations are

for tank 1
$$V_1 \frac{dC_1}{dt} = F\,C_0 - F\,C_1$$

for tank 2
$$V_2 \frac{dC_2}{dt} = F\,C_1 - F\,C_2$$

for tank 3
$$V_3 \frac{dC_3}{dt} = F\,C_2 - F\,C_3$$

2.1 Signal and Process Dynamics

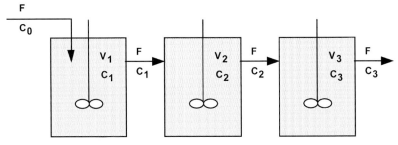

Fig. 2.11 Stirred tanks in series.

The above balance equations may be expressed as

$$\tau_1 \frac{dC_1}{dt} + C_1 = C_0 \quad \tau_1 = \frac{V_1}{F}$$

$$\tau_2 \frac{dC_2}{dt} + C_2 = C_1 \quad \tau_2 = \frac{V_2}{F}$$

and

$$\tau_3 \frac{dC_3}{dt} + C_3 = C_2 \quad \tau_3 = \frac{V_3}{F}$$

or as

$$\frac{dC_1}{dt} = \frac{C_0 - C_1}{\tau_1}$$

$$\frac{dC_2}{dt} = \frac{C_1 - C_2}{\tau_2}$$

$$\frac{dC_3}{dt} = \frac{C_2 - C_3}{\tau_3}$$

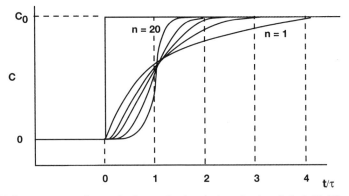

Fig. 2.12 Step response of n equal volume stirred tanks in series (n = 1, 2, 5, 10, 20).

In this case, three time constants in series, τ_1, τ_2 and τ_3, determine the form of the final outlet response C_3. As the number of tanks is increased, the response curve increasingly approximates the original step-change input signal, as shown in Fig. 2.12. The response curves for three stirred tanks in series, combined with chemical reaction are shown in the simulation example CSTRPULSE.

2.1.2.2 Case B: Response of a Second-Order Temperature Measuring Element

The temperature response of the measurement element shown in Fig. 2.13 is strictly determined by four time constants, describing (a) the response of the bulk liquid, (b) the response of the thermometer pocket, (c) the response of the heat conducting liquid between the wall of the bulb and the wall of the pocket and (d) the response of the wall material of the actual thermometer bulb. The time constants (c) and (d) are usually very small and can be neglected. A realistic model should, however, take into account the thermal capacity of the pocket, which can sometimes be significant.

Assuming the pocket to have a uniform temperature T_P, the heat balance for the bulb is now

$$\begin{pmatrix} \text{Rate of accumulation} \\ \text{of heat by the bulb} \end{pmatrix} = \begin{pmatrix} \text{Rate of heat transfer} \\ \text{to the bulb from the pocket} \end{pmatrix}$$

$$M\, c_p \frac{dT}{dt} = U_1 A_1 (T_P - T)$$

where U_1 is the heat transfer coefficient from the pocket to the bulb and A_1 is the heat transfer surface between the fluid and the bulb.

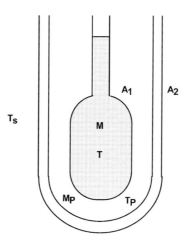

Fig. 2.13 Thermometer enclosed within a pocket.

Since the pocket temperature, T_P, is now a variable in the system, an additional heat balance equation is required for the pocket. This is of the same form as for the bulb, except that heat is now transferred both to the pocket from the surrounding and from the pocket to the bulb. Thus

$$\begin{pmatrix} \text{Rate of} \\ \text{accumulation} \\ \text{of heat} \\ \text{by the pocket} \end{pmatrix} = \begin{pmatrix} \text{Rate of heat transfer} \\ \text{to the pocket} \\ \text{from the fluid} \end{pmatrix} - \begin{pmatrix} \text{Rate of heat transfer} \\ \text{from the pocket} \\ \text{to the bulb} \end{pmatrix}$$

giving

$$M_P c_{pP} \frac{dT_P}{dt} = U_2 A_2 (T_S - T_P) - U_1 A_1 (T_P - T)$$

where U_2 is the heat transfer coefficient between the fluid and the pocket, A_2 is the heat transfer surface between the fluid and the pocket, M_P is the mass of the pocket, and c_{pP} is the specific heat of the pocket.

The overall instrument response is thus now determined by the relative magnitudes of the two major time constants, where for the liquid in the bulk

$$\tau_1 = \frac{M c_P}{U_1 A_1}$$

and for the pocket,

$$\tau_2 = \frac{M_P c_{pP}}{U_1 A_1 + U_2 A_2}$$

For accurate dynamic measurement of process temperature, both τ_2 and τ_1 should be small compared with the time constant of the actual process.

2.1.3
Pure Time Delay

In contrast to the prior lag type response signals, time delays give no immediate response until the elapse of a given period of dead time or delay. In flow processes, this is equivalent to the time required for the system to pass through the signal in an otherwise unchanged state. An example would be the time taken to pump a sample from the process to a measuring instrument. In this case the magnitude of the time delay would simply be the time taken for the sample to travel along the pipe or the volume of the sample pipe divided by the sample flow rate, and thus equal to the mean residence time of the sample system.

As shown in Fig. 2.14, the input signal from the process is transmitted through the sample pipe until it arrives at the measuring instrument at a delay

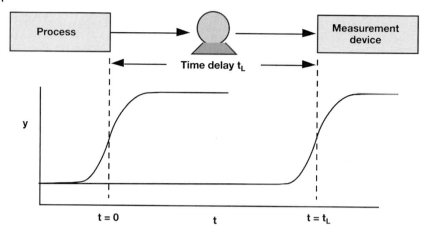

Fig. 2.14 Schematic drawing of a process with immediate and time-delayed responses to a step change of an input signal.

time t_L. In all other respects, however, the signal arriving at the measurement point is identical to the response of the actual system.

Most simulation languages include a standard time delay function, which is pre-programmed into the language structure. This facility is also available in MADONNA and is implemented in several of the simulation examples.

2.1.4
Transfer Function Representation

Complex systems can often be represented by linear time-dependent differential equations. These can conveniently be converted to algebraic form using Laplace transformation and have found use in the analysis of dynamic systems (e.g., Coughanowr and Koppel, 1965; Stephanopoulos, 1984; Luyben, 1990).

Thus, as shown in Fig. 2.15, the input-output transfer-function relationship, G(s), is algebraic, whereas the time domain is governed by the differential equation.

Model representations in Laplace transform form are mainly used in control theory. This approach is limited to linear differential equation systems or their

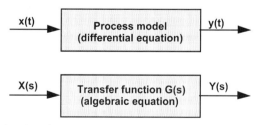

Fig. 2.15 Time and Laplace domain representations.

linearized approximations and is achieved by a combination of first order lag function and time delays. This limitation together with additional complications of modelling procedures are the main reasons for not using this method here. Specialized books in control theory as mentioned above use this approach and are available to the interested reader.

Dynamic problems expressed in transfer function form are often very easily reformulated back into sets of differential equation and associated time delay functions. An example of this is shown in the simulation example TRANSIM.

2.2 Time Constants

As shown in the preceding sections, the magnitude of various process time constants can be used to characterise the rate of response of a process resulting from an input disturbance. A fast process is characterised by a small value of the time constant and a slow process by large time constants. Time constants can therefore be used to compare rates of change and thus also to compare the relative importance of differing rate processes.

The term *time constant* is more or less equivalent to process time, characteristic time and relaxation time. Relaxation time is often used in physics, but is applied only to first-order processes and refers to the time for a process to reach a certain fraction of completion. This fraction is given by $(1 - 1/e) = 0.63$, which for a first-order process, as shown previously, is reached at a time $t = \tau$. Time constants also may be used to describe higher order processes and also non-linear processes. In these cases the time constant is defined as the time in which the process proceeds to a specified fraction of the resultant steady state. Higher order processes are often more elegantly described by a series of time constants.

A knowledge of the relative magnitude of the time constants involved in dynamic processes is often very useful in the analysis of a given problem, since this can be used to
- discover whether a change of regime occurs during scale up
- reduce the complexity of mathematical models
- determine whether the overall rate of a process is limited by a particular rate process, e.g., kinetic limitation or by diffusion, mixing, etc.
- check the controllability of a process
- check the difficulty of numerical solution due to equation stiffness

If the differing time constants for a chemical process are plotted as a function of the system variables, it can often be seen which rate process may be limiting. Many dimensionless groups can be considered as a ratio of the time constants for differing processes, and can give a clearer view on the physical meaning of the group. Such factors are discussed in much greater detail in other texts (Sweere et al., 1987), but here the intention is simply to draw attention to the importance of process time constants in the general field of dynamic simulation.

Table 2.1 Time constants defined by capacity and flow.

Capacity symbol	Dimension	Rate symbol	Dimension	Time constant
L	m	v	m/s	Travelling time τ
V	m^3	F	m^3/s	Residence time τ
V C	kmol	V $k_L a$ C	kmol/s	Mass transfer time τ_{mt}
V C	kmol	V r_C	kmol/s	Reaction time τ_r
V ρc_p dT	J	U A dT	J/s	Heat transfer time τ_{ht}
V ρc_p dT	J	V r_h	J/s	Heat production time τ_{hp}
V C	kmol	A D C/L	kmol/s	Diffusion time τ_d

As shown previously, the general form of equation serving to define time constants is as follows

$$\left(\begin{array}{c}\text{The rate of change}\\ \text{of the variable}\end{array}\right) = \left(\frac{\text{Final value} - \text{Instantaneous value}}{\text{Time constant}}\right)$$

However, a more general way to define time constants is

$$\text{Time constant} = \frac{\text{Capacity}}{\text{Rate}}$$

In this definition both "capacity" and "rate" have to be used in a rather general way. Some examples are presented in Table 2.1. The symbols are defined in the Nomenclature.

The choice of capacity is sometimes a problem, and may change according to the particular circumstance. Sometimes using a definition of time constant, based on the above equations, is not very helpful and other means must be employed. For example, mixing time is a very important time constant relating to liquid mixing, and this is best obtained directly from empirical correlations of experimental data.

2.2.1
Common Time Constants

2.2.1.1 Flow Phenomena
Some common time constants, relating to particular chemical engineering flow applications, are

$$\tau = \frac{\text{Capacity}}{\text{Rate}} = \frac{\text{Length}}{\text{Velocity}} = \text{Travelling time}$$

$$\tau = \frac{\text{Capacity}}{\text{Rate}} = \frac{\text{Volume}}{\text{Volumetric flow}} = \text{Residence time}$$

$$\tau_{circ} = \text{Circulation time}$$

Various empirical equations are available for the circulation time constant, τ_{circ}, in stirred vessels, columns, etc. Usually the value of the time constant, however, will represent a mean value, owing to the stochastic nature of flow.

Mixing time constants, τ_{mix}, are also available based on an empirical correlation and are usually closely related to the value of τ_{circ} (Joshi et al., 1982). A value of $\tau_{mix} = 4\tau_{circ}$ is often used for stirred vessels and a value of $\tau_{mix} = 2$ to $4\tau_{circ}$ for columns. The exact value strongly depends on the degree of mixing obtained.

2.2.1.2 Diffusion–Dispersion
Diffusion and dispersion processes can be characterised by a time constant for the process, given by

$$\tau_D = \frac{\text{Capacity}}{\text{Rate}} = \frac{L^2}{D}$$

where L is the characteristic diffusion or dispersion length and D is the diffusion or dispersion coefficient.

2.2.1.3 Chemical Reaction
Chemical reaction rate processes can be described by time constants.
In general

$$\tau_r = \frac{\text{Capacity}}{\text{Rate}} = \frac{VC}{Vr} = \frac{C}{r}$$

where C is concentration and r is the reaction rate. Hence
- for a zero-order process $r = k$ $\tau_r = C/k$
- for a first-order process $r = kC$ $\tau_r = 1/k$
- for a second-order process $r = kC^2$ $\tau_r = 1/kC$

2.2.1.4 Mass Transfer
Transfer rate processes can also be characterized by time constants formulated as

$$\tau_{mt} = \frac{\text{Capacity}}{\text{Rate}} = \frac{VC}{V\,Ka\,C} = \frac{1}{Ka}$$

where Ka is the mass transfer capacity coefficient, with units (1/s).

For a first-order process the time constant can be found from the defining differential equation as shown in Section 2.1.1.1. For the case of the aeration of a liquid, using a stirred tank, the following component balance equation applies

$$V_L \frac{dC_L}{dt} = k_L a (C_L^* - C_L) V_L$$

where C_L is the concentration of oxygen in the liquid phase (kg/m³), t is time (s), $k_L a$ is the mass transfer coefficient (1/s), V_L is the liquid volume (m³), and C_L^* is the equilibrium-dissolved oxygen concentration corresponding to the gas phase concentration, C_G.

By definition, the time constant for the process is thus $1/k_L a$ and the dissolved oxygen response to a step change in gas concentration is given by

$$C_L = C_L^* (1 - e^{-k_L a t})$$

One has to be careful, however, in defining time constants. The first important step is to set up the correct equations appropriately. If the prime interest is not the accumulation of oxygen in the liquid as defined previously, but the depletion of oxygen from the gas bubbles, then the appropriate balance equation becomes

$$V_G \frac{dC_G}{dt} = -k_L a (C_L^* - C_L) V_L$$

Note that in this case, the gas phase concentration, C_G, relates to the total mixed gas phase volume V_G, whereas the mass transfer capacity coefficient term is more conveniently related to the liquid volume, V_L.

If we consider the case where $C_L \ll C_L^*$ and that C_L^* is related to C_G via the Henry's Law coefficient as $C_L^* = C_G/H$, we can rewrite the above equation as

$$\frac{dC_G}{dt} = \frac{-k_L a V_L}{H V_G} C_G$$

The time constant is now given by

$$\frac{H V_G}{k_L a V_L}$$

Thus for the accumulation of oxygen in the liquid phase

$$\tau_{mt} = \frac{1}{k_L a}$$

and for the depletion of oxygen from the gas phase

$$\tau_{mt} = \frac{H}{k_L a} \frac{V_L}{V_G}$$

thus representing a substantial order of difference in magnitude for the two time constants.

2.2.1.5 Heat Transfer

Heat transfer processs time constants are formulated as

$$\tau_{ht} = \frac{\text{Capacity}}{\text{Rate}} = \frac{M\,c_P\,T}{U\,A\,T} = \frac{M\,c_P}{U\,A}$$

U is the heat transfer coefficient, M the mass, c_p the heat capacity and A the heat transfer area. A knowledge and understanding of the appropriate time constants is important in interpreting many of the simulation examples.

2.2.2
Application of Time Constants

Figure 2.16 gives an example of a bubble column reactor with growing microorganisms which consume oxygen. Here the individual process time constants are plotted versus the operating variable, the superficial gas velocity. It can be seen that for high values of the superficial gas velocity (v_s) and low rates of oxygen consumption (r_{O2}), the time constants for mixing (τ_{mix}) and for the oxygen mass transfer rate (τ_{mt}) are lower in magnitude than the oxygen reaction rate time constant (τ_r). For higher reaction rates (e.g., $r_{O2} = 1 \cdot 10^{-3}$ kg/m^{3}s) and reasonable values of v_s, it is impossible to obtain a value of τ_{mix} less than τ_r, and therefore mixing and mass transfer processes can become limiting at higher reaction rates.

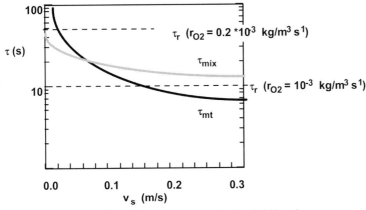

Fig. 2.16 Mixing, mass transfer and oxygen consumption in a bubble column bioreactor (Oosterhuis, 1984). τ_r reaction time constant, τ_{mt} mass transfer time constant, τ_{mix} mixing time constant. r_{O2} oxygen consumption rate, v_s superficial gas velocity.

2.3
Fundamentals of Automatic Control

Automatic process control involves the maintenance of a desired value of a measured or estimated quantity (controlled variable) within prescribed limits (deviations, errors), without the direct action of an operator. Generally, this involves three steps:
1. Measuring the present value of the controlled variable.
2. Comparing the measurement with the desired value (set point).
3. Adjusting some other variable (manipulated variable), which has influence on the controlled variable, until the set point is reached.

The most important reasons for applying process control are as follows:
- Safety for personnel and equipment.
- Uniform and high quality products.
- Increase of productivity.
- Minimisation of environmental hazards.
- Optimisation and decrease of labour costs.

Successful design of a process control system requires the following steps:
1. Selection of the control variables which are the most sensitive and easily measurable.
2. Formulation of the control objective; for example, the minimisation of some cost function.
3. Analysis of the process dynamics.
4. Selection of the optimal control strategy.

Process control is highly dynamic in nature, and its modelling leads usually to sets of differential equations which can be conveniently solved by digital simulation. A short introduction to the basic principles of process control, as employed in the simulation examples of Section 5.7, is presented.

2.3.1
Basic Feedback Control

The concept of an automatic control system is illustrated in Fig. 2.17, based on a temperature-controlled chemical reactor.

The components of the basic feedback control loop, combining the process and the controller, can be best understood using a generalised block diagram (Fig. 2.18). The information on the measured variable, temperature, taken from the system is used to manipulate the flow rate of the cooling water in order to keep the temperature at the desired constant value, or setpoint. This is illustrated by the simulation example TEMPCONT (Section 5.7.1).

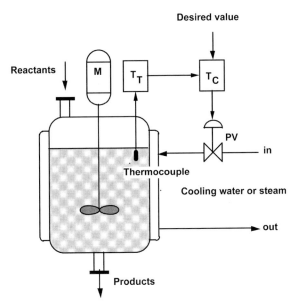

Fig. 2.17 Simple feedback temperature control system. M motor with stirrer, PV pneumatic valve, T_T temperature measurement, T_C temperature controller.

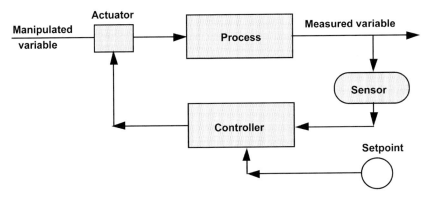

Fig. 2.18 Block diagram of a feedback control system.

2.3.2
Types of Controller Action

In the basic conventional feedback control strategy the value of the measured variable is compared with that for the desired value of that variable and if a difference exists, a controller output is generated to eliminate the error.

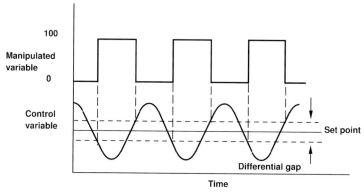

Fig. 2.19 On/off controller with differential gap or dead zone.

2.3.2.1 On/Off Control

The simplest and, despite its several drawbacks, the most widely used type of control is the on/off control system. An example is a contact thermometer, which closes or opens a heater circuit. The designation on/off means that the controller output, or the manipulated variable (electric current) is either fully on or completely off. To avoid oscillations around the setpoint, the real on/off controller has built into it a small interval on either side of the setpoint within which the controller does not respond, and which is called the differential gap or deadzone. When the controlled variable moves outside the deadzone, the manipulated variable is set either on or off. This is illustrated in Fig. 2.19. Such shifts from the set point are known as offset.

The oscillatory nature of the action and the offset make the resulting control rather imperfect, but the use of on/off control can be justified by its simplicity and low price, and the reasonable control obtained, especially for systems which respond slowly.

2.3.2.2 Proportional-Integral-Derivative (PID) Control

Three principal functional control modes are proportional (P), integral (I) and derivative (D) control. These are performed by the ideal three-mode controller (PID), described by the equation

$$P = P_0 + K_p \varepsilon(t) + \frac{K_p}{\tau_I} \int_0^t \varepsilon(t)dt + K_p \tau_D \frac{d\varepsilon(t)}{dt}$$

Controller modes: P   I   D

where:
P_0 is the controller output for zero error
K_p is the proportional gain

2.3 Fundamentals of Automatic Control

$\varepsilon(t)$ is the error or deviation of actual from desired value
τ_I is the integral time or reset time constant
τ_D is the derivative time constant

The response of a controller to an error depends on its mode. In the proportional mode (P), the output signal is proportional to the detected error, ε. Systems with proportional control often exhibit pronounced oscillations, and for sustained changes in load, the controlled variable attains a new equilibrium or steady-state position. The difference between this point and the set point is the offset. Proportional control always results in either an oscillatory behaviour or retains a constant offset error.

Integral mode controller (I) output is proportional to the sum of the error over the time. It can be seen that the corrections or adjustments are proportional to the integral of the error and not to the instantaneous value of the error. Moreover, the corrections continue until the error is brought to zero. However, the response of integral mode is slow and therefore is usually used in combination with other modes.

Derivative mode (D) output is proportional to the rate of change of the input error, as can be seen from the three-mode equation.

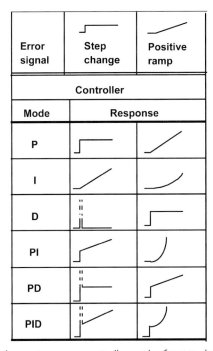

Fig. 2.20 Response of the most common controller modes for step change and ramp function of the error signal.

In industrial practice it is common to combine all three modes. The action is proportional to the error (P) and its change (D) and it continues if residual error is present (I). This combination gives the best control using conventional feedback equipment. It retains the specific advantages of all three modes: proportional correction (P), offset elimination (I) and stabilising, quick-acting character, especially suitable to overcome lag presence (D).

Simple control strategies form an integral part of many of the simulation examples, including RELUY, COOL, DEACT, REFRIG1, REFRIG2, RUN, EXTRACTCON, SULFONATION and the special control examples in Sec. 5.7, TEMPCONT, TWOTANK and CONTUN.

Figure 2.20 depicts the responses of the various control modes and their combinations to step and ramp inputs.

The performance of different feedback control modes can be seen in Fig. 2.21.

The selection of the best mode of control depends largely on the process characteristics. Further information can be found in the recommended texts listed in the reference section. Simulation methods are often used for testing control methods. The basic PI controller equations are easily programmed using a simulation language, as shown in the example programs. In the simulation examples, the general PID equation is simplified and only the P or P and I terms are used. Note that the I term can be set very low by using a high value for τ_I.

If desired, the differential term, $d\varepsilon/dt$, can be programmed as follows:

Since
$$\varepsilon = y - y_{set}$$
then
$$\frac{d\varepsilon}{dt} = \frac{dy}{dt}$$

and this derivative can be obtained directly from the model equations.

2.3.2.3 Case A: Operation of a Proportional Temperature Controller

Liquid flows continuously through a tank of volume, V, provided with an electric heater. A controller regulates the rate of heating directly in accordance with the difference between a required set point temperature, T_{set}, and the actual temperature, T_1, as shown in Fig. 2.22.

The heat balance equation for the tank is similar to that of Case A of Section 1.2.5.1, i.e.,

$$V\rho c_p \frac{dT}{dt} = F_0 \rho c_p (T_0 - T_1) + Q$$

but where Q is the rate of electric heating and is expressed by a proportional control equation (omitting the I and D terms)

$$Q = Q_0 + K_p(T_{set} - T_1)$$

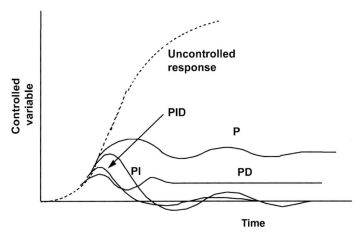

Fig. 2.21 Response of controlled variable to a step change in error using different control modes.

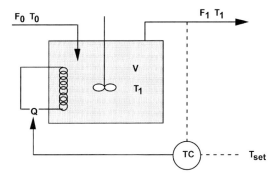

Fig. 2.22 Temperature control, TC, of a continuously operated stirred tank with an electric heater (Q).

2.3.3
Controller Tuning

The purpose of controller tuning is to choose the correct controller constants to obtain the desired performance characteristics. This usually means that the control variables should be restored in an optimal way to acceptable values, following either a change in the set point or the appearance of an input disturbance. Numerous books discussing the subject are available (e.g. Shinskey, 1996). The Internet is a good source of information, where some e-Books appeared, e.g. "The PID Controller Tuning Methods" by John Shaw. Simulation examples TEMPCONT and CONTUN provide exercises for controller tuning using the methods explained below.

2.3.3.1 Trial and Error Method

Controllers can be adjusted by changing the values of gain K_p, reset time τ_I and derivative time τ_D. The controller can be set by trial and error by experimenting, either on the real system or by simulation. Each time a disturbance is made the response is noted. The following procedure may be used to test the control with small set point or load changes:

1. Starting with a small value, K_p can be increased until the response is unstable and oscillatory. This value is called the ultimate gain K_{p0}.
2. K_p is then reduced by about one-half.
3. Integral action is brought in with high τ_I values. These are reduced by factors of 2 until the response is oscillatory, and τ_I is set at 2 times this value.
4. Include derivative action and increase τ_D until noise develops. Set τ_D at 1/2 this value.
5. Increase K_p in small steps to achieve the best results.

2.3.3.2 Ziegler–Nichols Open-Loop Method

This empirical open-loop tuning mode, known also as the "Reaction Curve" method, is implemented by uncoupling the controller. It is an empirical open-loop tuning technique, obtained by uncoupling the controller. It is based on the characteristic curve of the process response to a step change in manipulated variable of magnitude A. The response, of magnitude B, is called the process reaction curve. The two parameters important for this method are given by the slope through the inflection point normalised by A, so that S = Slope/A, and by its intersection with the time axis (lag time T_L), as determined graphically in

Table 2.2 Controller settings based on process responses.

Controller	K_p	τ_I	τ_D
Ziegler–Nichols			
P	$1/(T_L S)$		
PI	$0.9/(T_L S)$	$3.33\, T_L$	
PID	$1.2/(T_L S)$	$2\, T_L$	$T_L/2$
Cohen–Coon			
P	$\dfrac{\tau}{K\, T_L}\left(1 + \dfrac{T_L}{3\tau}\right)$		
PI	$\dfrac{\tau}{K\, T_L}\left(0.9 + \dfrac{T_L}{12\tau}\right)$	$T_L \dfrac{30 + 3T_L/\tau}{9 + 20T_L/\tau}$	
PID	$\dfrac{\tau}{K\, T_L}\left(\dfrac{4}{3} + \dfrac{T_L}{12\tau}\right)$	$T_L \dfrac{32 + 6T_L/\tau}{13 + 8T_L/\tau}$	$T_L \dfrac{4}{12 + 2T_L/\tau}$
Ultimate Gain			
P	$0.5\, K_{p0}$		
PI	$0.45\, K_{p0}$	$1/1.2\, f_0$	
PID	$0.6\, K_{p0}$	$1/2\, f_0$	$1/8\, f_0$

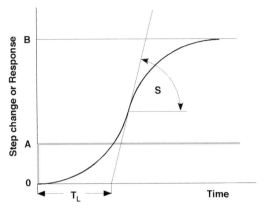

Fig. 2.23 Process reaction curve for the Ziegler–Nichols method.

Fig. 2.23. The actual tuning relations, based on empirical criteria for the "best" closed-loop response are given in Table 2.2.

2.3.3.3 Cohen–Coon Controller Settings

Cohen and Coon observed that the response of most uncontrolled (controller disconnected) processes to a step change in the manipulated variable is a sigmoidally shaped curve. This can be modelled approximately by a first-order system with time lag T_L, as given by the intersection of the tangent through the inflection point with the time axis (Fig. 2.23). The theoretical values of the controller settings obtained by the analysis of this system (e.g. Luyben and Luyben, 1997) are summarised in Table 2.2. The model parameters for a step change A to be used with this table are calculated as follows:

$$K = B/A \qquad \tau = B/S$$

where B is the extent of response, S is the slope at the inflection point, and T_L is the lag time as determined in Fig. 2.23.

2.3.3.4 Ultimate Gain Method

The previous transient-response tuning methods are sensitive to disturbances because they rely on open-loop experiments. Several closed loop methods have been developed to eliminate this drawback. One of these is the empirical tuning method, ultimate gain or continuous cycling method. The ultimate gain, K_{p0}, is the gain which brings the system with sole proportional control only to sustained oscillation (stability limits) with frequency f_{p0}, where $1/f_{p0}$ is called the ultimate period. This is determined experimentally by increasing K_p from low values in small increments until continuous cycling begins. The controller settings are then calculated from K_{p0} and f_{p0} according the tuning rules given in Table 2.2.

While this method is very simple it can be quite time consuming in terms of number of trials required and especially when the process dynamics are slow. In addition, it may be hazardous to experimentally force the system into unstable operation.

2.3.3.5 Time Integral Criteria

Several criteria may be used to estimate the quality of control (Stephanopoulos, 1984). One of these is the integral of the time-weighted absolute error (ITAE), where

$$\text{ITAE} = \int_0^t |t\varepsilon(t)|(dt)$$

Integral error criteria are ideally suited to simulation applications since only one additional program statement is required for the simulation. The optimal control parameters K_p, τ_I and τ_D can be then found at minimal ITAE. For this, it is useful to be able to apply the available optimisation tools implemented in such programs as MATLAB, ACSL-OPTIMIZE or MADONNA.

2.3.4
Advanced Control Strategies

2.3.4.1 Cascade Control

In control situations with more than one measured variable but only one manipulated variable, it is advantageous to use control loops for each measured

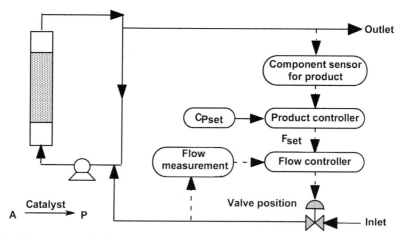

Fig. 2.24 Cascade control to maintain product concentration by manipulating the reactant concentration in the feed.

variable in a master-slave relationship. In this, the output of the primary controller is usually used as a set point for the slave or secondary loop.

An example of cascade control could be based on the simulation example DEACT and this is shown in Fig. 2.24. The problem involves a loop reactor with a deactivating catalyst, and a control strategy is needed to keep the product concentration C_P constant. This could be done by manipulating the feed rate into the system to control the product concentration at a desired level, C_{set}. In this cascade control, the first controller establishes the setpoint for flow rate. The second controller uses a measurement of flow rate to establish the valve position. This control procedure would then counteract the influence of decreasing catalyst activity.

2.3.4.2 Feedforward Control

Feedback control can never be perfect as it reacts only to disturbances in the process outlet. Feedforward control can theoretically be perfect, because the inlet disturbances are measured, and their effects on the process are anticipated via the use of a model. If the model is perfect then the calculated action to be taken will be exact.

The example simulation THERMFF illustrates this method of using a dynamic process model to develop a feedforward control strategy. At the desired setpoint the process will be at steady-state. Therefore the steady-state form of the model is used to make the feedforward calculations. This example involves a continuous tank reactor with exothermic reaction and jacket cooling. It is assumed here that variations of inlet concentration and inlet temperature will disturb the reactor operation. As shown in the example description, the steady state material balance is used to calculate the required response of flowrate and the steady state energy balance is used to calculate the required variation in jacket temperature. This feedforward strategy results in perfect control of the simulated process, but limitations required on the jacket temperature lead to imperfections in the control.

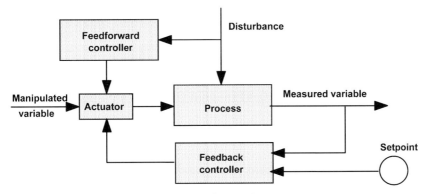

Fig. 2.25 Feed-forward control with additional feedback loop.

The success of this control strategy depends largely on the accuracy of the model prediction, which is often imperfect as models can rarely exactly predict the effects of process disturbances. For this reason, an additional feedback loop is often used as a backup or to trim the main feedforward action, as shown in Fig. 2.25. Many of the continuous process simulation examples in this book may be altered to simulate feedforward control situations.

2.3.4.3 Adaptive Control

An adaptive control system can automatically modify its behaviour according to the changes in the system dynamics and disturbances. They are applied especially to systems with non-linear and unsteady characteristics. There are a number of actual adaptive control systems. Programmed or scheduled adaptive control uses an auxiliary measured variable to identify different process phases for which the control parameters can be either programmed or scheduled. The "best" values of these parameters for each process state must be known *a priori*. Sometimes adaptive controllers are used to optimise two or more process outputs, by measuring the outputs and fitting the data with empirical functions.

2.3.4.4 Sampled Data or Discrete Control Systems

When discontinuous measurements are involved, the control system is referred to as a sampled data or discrete controller. Concentration measurements by chromatography would represent such a case.

Here a special consideration must be given to the sampling interval ΔT (Fig. 2.26). In general, the sampling time will be short enough if the sampling frequency is 2 times the highest frequency of interest or if ΔT is 0.5 times the minimum period of oscillation. If the sampling time satisfies this criterion, the system will behave as if it were continuous. Details of this and other advanced topics are given in specialised process control textbooks, some of which are listed in the references.

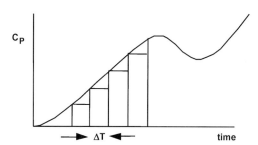

Fig. 2.26 Sampled control strategy.

2.4
Numerical Aspects of Dynamic Behaviour

2.4.1
Optimisation

Optimisation may be used, for example, to minimise the cost of reactor operation or to maximise conversion. Having set up a mathematical model of a reactor system, it is only necessary to define a cost or profit function and then to minimise or maximise this by variation of the operational parameters, such as temperature, feed flow rate or coolant flow rate. The extremum can then be found either manually by trial and error or by the use of numerical optimisation algorithms. The first method is easily applied with MADONNA, or with any other simulation software, if only one operational parameter is allowed to vary at any one time. If two or more parameters are to be optimised this method becomes extremely cumbersome. To handle such problems, MADONNA has a built-in optimisation algorithm for the minimisation of a user-defined objective function. This can be activated by the OPTIMIZE command from the Parameter menu. In MADONNA the use of parametric plots for a single variable optimisation is easy and straight-forward. It often suffices to identify optimal conditions, as shown in Case A below.

Basically two search procedures for non-linear parameter estimation applications apply (Nash and Walker-Smith, 1987). The first of these is derived from Newton's gradient method and numerous improvements on this method have been developed. The second method uses direct search techniques, one of which, the Nelder-Mead search algorithm, is derived from a simplex-like approach. Many of these methods are part of important mathematical packages, e.g., ASCL and MATLAB.

2.4.1.1 Case A: Optimal Cooling for a Reactor with an Exothermic Reversible Reaction

A reversible exothermic reaction A $\Leftrightarrow$ B is carried out in a stirred-tank reactor with cooling. The details of the model are given in the simulation example REVTEMP. Specific heats are functions of temperature. The temperature dependency of the reaction rate constants are given by the Arrhenius equation, that of the equilibrium constant by the van't Hoff equation. Adiabatic operation restricts conversion because of an unfavourable equilibrium at high temperature. Early cooling favours the equilibrium conversion but reduces reaction rates, according to the Arrhenius equation. It is assumed that the cooling water temperature is constant, and that the cooling water flow, FC, may be either on or off. At time TIMEON the cooling water flow is set to FCON. A profit function is defined as

$$\text{SPTYB} = \frac{C_B^2}{t}$$

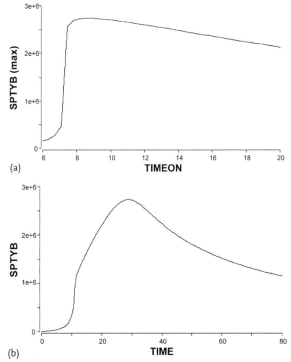

Fig. 2.27 (a) Parametric run of REVTEMP varying variable TIMEON.
(b) Run of REVTEMP with optimal value of TIMEON from (a).

This reflects the desire to have high conversion in a short time period. SPTYB always passes through a maximum during a batch run. The problem is defined as finding the optimal times to switch on the cooling water flow (TIMEON) and to harvest the tank contents (TFIN). The program listing is given on the CD.

In MADONNA this problem is easily solved in the following way. Plotting SPTYB versus TIME always gives a maximum. This would be the optimal harvesting time for a preset value of TIMEON. The optimal value for TIMEON is found by **Batch Runs**, **Initial value** = 6, **Final Value** = 20 and e.g. 40 Runs. Selecting the Mode **Parametric Plot**, choosing SPTYB as variable and selecting **Maximum value**, gives a plot as depicted in Fig. 2.27a. Increasing TIMEON from 6 to 8 increases SPTMB (max.) dramatically showing a flat maximum between 8 and 10. Inspection of the tabular output gives an optimal value of about 8.5. From making one **Run** with the optimal value of TIMEON by setting TIMEON = 8.5 in the **Parameters Window** and plotting SPTYB versus TIME the maximum value of SPTYB is directly obtained (Fig. 2.27b). From the shape of the curves it is clearly seen that the system is very robust for values of TIMEON > 8.

In non-linear systems one can usually not predict a priori whether the optimum found is global or whether the optimum obtained represents only a local condition. A good judgement on the behaviour of the model can be seen in contour and three-dimensional plots, which are easily obtained using other alternative software packages, such as ACSL-OPTIMIZE or MATLAB.

2.4.2
Parameter Estimation

Having set up a model to describe the dynamics of the system, a very important first step is to compare the numerical solution of the model with any experimental results or observations. In the first stages, this comparison might be simply a check on the qualitative behaviour of a reactor model as compared to experiment. Such questions might be answered as: Does the model confirm the experimentally found observations that product selectivity increases with temperature and that increasing flow rate decreases the reaction conversion?

Following the first preliminary comparison, a next step could be to find a set of parameters that give the best or optimal fit to the experimental data. This can be done by a manual trial-and-error procedure or by using a more sophisticated mathematical technique which is aimed at finding those values for the system parameters that minimise the difference between values given by the model and those obtained by experiment. Such techniques are general, but are illustrated here with special reference to the dynamic behaviour of chemical reactors.

Table 2.3 is used to classify the differing systems of equations, encountered in chemical reactor applications and the normal method of parameter identification. As shown, the optimal values of the system parameters can be estimated using a suitable error criterion, such as the methods of least squares, maximum likelihood or probability density function.

Table 2.3 Classification of systems of reactor equations with a set of parameters and time-dependent variables.

	Examples of linear systems	Examples of non-linear systems
Algebraic equations	Steady state of CSTR with first-order kinetics. Algebraic solution and optimisation (least squares, Draper and Smith, 1981).	Steady state of CSTR with complex kinetics. Numerical solution and optimisation (least squares or likelihood function).
Differential equations	Batch reactor with first-order kinetics. Analytical or numerical solution with analytical or numerical parameter optimisation (least squares or likelihood).	Batch reactor with complex kinetics. Numerical integration and parameter optimisation (least squares or likelihood).

2.4.2.1 Non-Linear Systems Parameter Estimation

The methods concerned with differential equation parameter estimation are, of course, the ones of most concern in this book. Generally reactor models are non-linear in their parameters, and therefore we are concerned mostly with non-linear systems.

Given a model in the form of a set of differential equations,

$$\frac{dy}{dt} = f(k_1 \ldots k_n, y)$$

A model described by this differential equation is linear in the parameters $k_1 \ldots k_n$, if

$$\frac{\partial f}{\partial k_i} \neq g(k_1 \ldots k_n)$$

but is non-linear, if for at least one of the parameters k_i

$$\frac{\partial f}{dk_i} = h(k_1 \ldots, y)$$

The application of optimisation techniques for parameter estimation requires a useful statistical criterion (e.g., least-squares). A very important criterion in non-linear parameter estimation is the likelihood or probability density function. This can be combined with an error model which allows the errors to be a function of the measured value.

If basic assumptions concerning the error structure are incorrect (e.g., non-Gaussian distribution) or cannot be specified, more robust estimation techniques may be necessary, e.g., Maria and Heinzle (1998). In addition to the above considerations, it is often important to introduce constraints on the estimated parameters (e.g., the parameters can only be positive). Such constraints are included in the simulation and parameter estimation package ACSL-OPTIMIZE and in the MATLAB Optimisation Toolbox. Because of numerical inaccuracy, scaling of parameters and data may be necessary if the numerical values are of greatly differing order. Plots of the residuals, difference between model and measurement value, are very useful in identifying systematic or model errors.

Non-linear parameter estimation is far from a trivial task, even though it is greatly simplified by the availability of user-friendly program packages such as (a) ACSL-OPTIMIZE, (b) MADONNA, (c) a set of BASIC programs (supplied with the book of Nash and Walker-Smith, 1987) or (d) by mathematical software (MATLAB). MADONNA has only limited possibilities for parameter estimation, but MADONNA programs can easily be translated into other more powerful languages.

2.4.2.2 Case B: Estimation of Rate and Equilibrium Constants in a Reversible Esterification Reaction Using MADONNA

The objective is to demonstrate the use of MADONNA in the estimation of model parameters. Here the parameters are estimated using the CURVE FIT feature of MADONNA. This allows data to be imported by clicking the IMPORT DATA in this menu and selecting an external text file. The file must have the time in the first column and the data values in the second column. Two data columns can also be used, but they must correspond to equally spaced times. The number of parameters to be estimated can be one or more.

In this example, ethanol and acetic acid react reversibly to ethyl acetate, using a catalyst, ethyl hydrogensulfate, which is prepared by reaction between sulfuric acid and ethanol.

$$\text{Acetic acid} + \text{Ethanol} \rightleftharpoons \text{Ethyl acetate} + \text{water}$$

$$A + B \underset{k_2}{\overset{k_1}{\rightleftharpoons}} C + D$$

The rate of batch reaction for reactant A (acetic acid) is modelled as

$$\frac{dC_A}{dt} = r_A = -k_1 C_A C_B + k_2 C_C C_D$$

The progress of the reaction is followed by taking samples at regular time intervals and titrating the remaining free acid with alkali (mL).

The table of measured data, time (min) versus titrated volume (mL) is imported into the program from the external file ESTERdat.txt and will be plotted after a graph window is defined. Clicking CURVE-FIT will allow the selection of the variable and the parameters. For each parameter, two preliminary guessed values and the maximum and minimum allowable values can be entered. On running under CURVE-FIT, the values of the required parameters are repeatedly updated, until the final converged values are obtained. The updated values can be found in the Parameter Window. On clicking Run, a final run is made, enabling the final simulation results to be compared with the original data values. The MADONNA program ESTERFIT and the graph of the data and the final results of the parameter estimation are shown below.

```
{ESTERFIT}
{Fitting of experimental titration data to determine the rate
constants for a reversible reaction:
ethanol + acetic Acid ↔ ethyl acetate +water}

METHOD Auto
STARTTIME = 0
STOPTIME = 1200
DT = 0.3
```

{Initial guess on rate constants, m3/kmol min}
k1 = 0.003
k2 = 0.001

INIT CA = 5.636   {initial acetic acid, kmol/m3}
INIT CB = 5.636   {initial ethanol, kmol/m3}
INIT CC = 0   {initial ethyl acetate, kmol/m3}
INIT CD = 0.634   {initial water, kmol/m3}
CA0 = 5.636   {initial acetic acid, kmol/m3}

{Batch material balances}
d/dt(CA)=-r1+r2
d/dt(CB)=-r1+r2
d/dt(CC)=+r1-r2
d/dt(CD)=+r1-r2

r1=k1*CA*CB
r2=k2*CC*CD

MLtitrated=CA/Titfact
Titfact=CA0/ML0 {Ratio (kmoles/m3)/mL 1 N NaOH}
ML0=16.25   {mL titrated for calibration}

File of data (ESTERdat) giving time (min.) versus mL titrated
0.0   16.25
2.0   15.1
4.0   14.25
6.0   13.8
8.0   12.9
10.0   11.7
15.0   12.1
20.0   11.05
25.0   10.6
30.0   10.45
35.0   10.2
40.0   9.5
45.0   9.5
55.0   8.9
70.0   8.5
103.0   7.5
143.0   7.1
1000.0   6.5
1043.0   6.5
1100.0   6.5

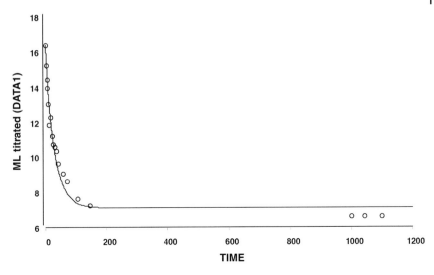

Fig. 2.28 Experimental data with fitted profile of mL titrated.
Values found: $k_1 = 0.00397$ and $k_2 = 0.00185$

From Fig. 2.28 it is obvious that a reasonable fit is easily obtained. A detailed analysis of the results, however, discloses that there seems to be a systematic deviation in the residuals with high predicted values at the equilibrium conditions (TIME > 1000) and a low prediction between TIME around 50 and 150. These differences can be caused by an inadequacy in the model or by systematic experimental errors. A more appropriate objective function may also be desirable.

2.4.3
Sensitivity Analysis

The sensitivity of a model or a real system can be determined by making changes in the parameters of interest and noting their influences on each variable. The simplest measure of sensitivity is the derivative of the variable with respect to the change in the parameter, $\partial V/\partial P$. MADONNA has an automatic means for making "Sensitivity Runs". This is done by making two runs, one at the normal value of the parameter and another run using a value which is increased by 0.1%. The sensitivity is calculated using the difference in the variable, ΔV, divided by the difference in the parameter, ΔP, to give $\partial V/\partial P = \Delta V/\Delta P$. These derivative values are plotted for all the selected variables and parameters versus time. Obviously during any simulation, or during a real experiment, the sensitivity of the process to a particular parameter will vary as the conditions vary with time. Such a sensitivity analysis provides guidance as to how accurately a parameter needs to be determined. If the process is not sensitive to a model parameter then sometimes this enables the model to be simpli-

86 | 2 Process Dynamics Fundamentals

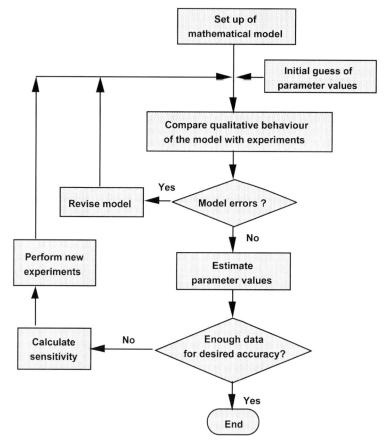

Fig. 2.29 Iterative procedure for parameter estimation, sensitivity analysis and experimentation.

fied. Figure 2.29 illustrates an iterative experimental procedure for parameter estimation, involving sensitivity analysis.

Below the results of Sensitivity Runs with MADONNA are given from the BIOREACT example that is run as a batch fermenter system. This example involves Monod growth kinetics, as explained in Section 1.4. In this example, the sensitivity of biomass concentration X, substrate concentration S and product concentration to changes in the Monod kinetic parameter, K_S, was investigated. Qualitatively, it can be deduced that the sensitivity of the concentrations to K_S should increase as the concentration of S becomes low at the end of the batch. This is verified by the results in Fig. 2.30. The results in Fig. 2.31 give the sensitivity of biomass concentration X and substrate concentration S to another biological kinetic parameter, the yield coefficient Y, as defined in Section 1.4.

Sensitivity analysis is a very important tool in analysing the relative importance of the model parameters and in the design of experiments for their optimal determination. In many cases, it is found that a model may be rather in-

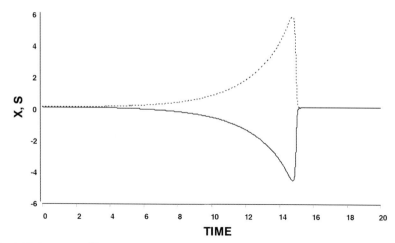

Fig. 2.30 Sensitivity of biomass concentration X and substrate concentration S ($\Delta X/\Delta K_S$ solid line and $\Delta S/\Delta K_S$ dotted line) to changes in K_S from a batch run of example BIOREACT.

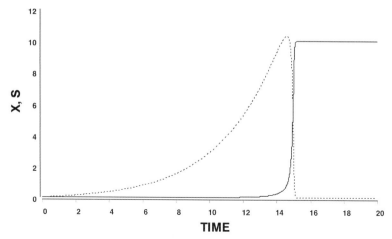

Fig. 2.31 Sensitivity of biomass X and substrate concentrations S ($\Delta X/\Delta Y$ solid line and $\Delta S/\Delta Y$ dashed line) to changes in the yield coefficient Y from the run of Fig. 2.31.

sensitive to a particular parameter value in the region of main interest, and then the parameter obviously does not need to be determined very accurately.

Model parameters are usually determined from experimental data. In doing this, sensitivity analysis is valuable in identifying the best experimental conditions for the estimation of a particular model parameter. Sensitivity analysis is easy effected with MADONNA, and sensitivity analysis is also provided in other more advanced software packages, such as ACSL-OPTIMIZE.

2.4.4
Numerical Integration

Only a very short introduction to numerical integration is given here, simply to demonstrate the basic principles and possible sources of error. In the great majority of simulation studies, the numerical integration will not be found to create problems and a detailed knowledge of the differing numerical integration methods is generally unnecessary. For more complex problems, where numerical difficulties may occur, the reader is referred to more specialist texts, e.g., Press et al. (1992), Walas (1991), Noye (1984).

In the solution of mathematical models by digital simulation, the numerical integration routine is usually required to achieve the solution of sets of simultaneous, first-order differential equations in the form

$$\frac{dy_i}{dt} = f_i(y_1, y_2, y_3, y_4, \ldots y_n) \quad \text{for } i = 1, 2, \ldots, m$$

The differential equations are often highly non-linear and the equation variables are often highly interrelated. In the above formulation, y_i represents any one of the dependent system variables and f_i is the general function relationship, relating the derivative, dy_i/dt, with the other related dependent variables. The system-independent variable, t, will usually correspond to time, but may also represent distance, for example, in the simulation of steady-state models of tubular and column devices.

In order to solve the differential equations, it is first necessary to initialise the integration routine. In the case of initial value problems, this is done by specifying the conditions of all the dependent variables, y_i, at initial time t = 0. If, however, only some of the initial values can be specified and other constant values apply at further values of the independent variable, the problem then becomes one of a split-boundary type. Split-boundary problems are inherently more difficult than the initial value problems, and although most of the examples in the book are of the initial value type, some split-boundary problems are presented.

In both types of problem, solution is usually achieved by means of a step-by-step integration method. The basic idea of this is illustrated in the information flow sheet, which was considered previously for the introductory MADONNA complex reaction model example (Fig. 1.4).

Referring to Fig. 1.4, the solution begins with the initial concentration conditions A_0, B_0, C_0 and D_0, defined at time t = 0. Knowing the magnitudes of the kinetic rate constants k_1, k_2, k_3 and k_4, thus enables the initial rates of change dC_A/dt, dC_B/dt, dC_C/dt and dC_D/dt, to be determined. Extrapolating these rates over a short period of time Δt, from the initial conditions A_0, B_0, C_0 and D_0, enables new values for A, B, C and D to be estimated at the new time, $t = t + \Delta t$. If the incremental time step Δt is sufficiently small, it is assumed that the error in the new estimated values of the concentration, A, B, C and D, will also be small. This procedure is then repeated for further small increments of time until the entire concentration versus time curves have been determined.

In this approach, the true solution is approximated as a series of discrete points along the axis of the independent variable t. The solution then proceeds step-by-step from one discrete time step to the next. In the simplest case, the time steps will be spaced at uniform intervals, but the spacing can also be varied during the course of solution. Where there are several dependent variables involved, all the variables must be updated to their new value, by projecting all the respective rates of change or concentration-time gradients over the identical time increment or integration step length, h. In order to do this, the integration method has to carry out a number of separate evaluations of the gradient terms. These evaluations are known as the "stages" of the computation.

Thus taking the single rate equation $dy/dt = f(y, t)$ and knowing the solution at any point $P_n(y_n, t_n)$, the value of the function at the next point can be predicted, knowing the local rate of change dy_n/dt_n. In the simplest case, this can be approximated by a simple difference approximation

$$\frac{dy_n}{dt} = f(y_n, t_n) = \frac{y_{n+1} - y_n}{t_{n+1} - t_n} = \frac{y_{n+1} - y_n}{h_n}$$

from which

$$y_{n+1} = y_n + h_n f(y_n, t_n)$$

This procedure is illustrated in Fig. 2.32.

Much effort has been devoted to producing fast and efficient numerical integration techniques, and there is a very wide variety of methods now available. The efficiency of an integration routine depends on the number of function evaluations, required to achieve a given degree of accuracy. The number of evaluations depends both on the complexity of the computation and on the number of integration step lengths. The number of steps depends on both the na-

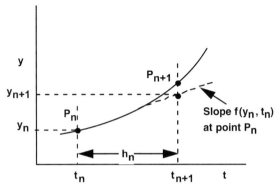

Fig. 2.32 The difference approximation for rate of change or slope.

ture and complexity of the problem and the degree of accuracy required in the solution. In practice, an over complex integration routine will require excessive computing time, owing to the many additional function evaluations that are required, and the use of an inappropriate integration algorithm can lead to an inaccurate solution, excessive computing time and, sometimes, a complete inability to solve the problem.

It is thus very important that the output of any simulation is checked, using other integration methods. Most simulation languages allow a choice of integration routine which can be made best on the basis of experience. It is important to remember that all methods generate only approximate solutions, but these must be consistent with a given error criterion. As the models themselves, however, are also approximate, errors in the numerical solution must be seen in the general context of the problem as a whole. Numerical errors occur in the approximation of the original function and also are due to limits in the numerical precision of the computation. From experience, it can be shown that most cases of "strange" behaviour, in the results of a simulation, can be attributed largely to errors in the model and inadequate model parameter selection, rather than to numerical inaccuracies. Very powerful integration routines for stiff systems are supplied by MATLAB.

Integration methods used by MADONNA

1. Fixed step Euler method (Euler).
2. Fixed step, 2nd-order, Runge-Kutta method (RK2).
3. Fixed step, 4th-order, Runge-Kutta method (RK4).
4. Variable step, 5th-order, Runge-Kutta method (AUTO).
5. Rosenbrock Method (Stiff).

Important integration parameters and default settings

METHOD	Choice of integration method as above (Euler).
DT	Calculation interval or integration step length in fixed step methods (0.25).
DTOUT	Output time interval.
STOPTIME	Value of the independent variable with which the run is terminated (12).
STARTTIME	Value of the independent variable at the start of the run (0).
TOLERANCE	Relative accuracy for the Auto and Stiff methods (0.01).

The most common numerical problem, as shown by some of the simulation examples, is that of equation stiffness. This is manifested by the need to use shorter and shorter integration step lengths, with the result that the solution proceeds more and more slowly and may come to a complete halt. Such behaviour is exhibited by systems having combinations of very fast and very slow processes. Stiff systems can also be thought of as consisting of differential equations, having large differences in the process time constants. Sometimes, the stiffness is the result of bad modelling practice and can be removed by assuming the very fast processes to be virtually instantaneous, as compared to the slower overall rate determining processes. In this way, the differential equations involving the troublesome very fast processes are replaced by steady-state algebraic equations, in which the rate of accumulation is, in effect, taken to be zero. Solving the implicit equations that result from such procedures often requires a root finder algorithm, as is supplied by MADONNA. Unfortunately, this technique is not always possible, and many systems are stiff in their own right and therefore need special integration methods.

2.4.5
System Stability

System instability can also be a problem in dynamic simulation, and this can origin either from the integration routine or from the model itself. Instability in the integration routine can arise owing to the approximation of the real functions by finite-difference approximations, which can have their own parasitical exponential solutions. When the unwanted exponentials decrease with respect to time, the numerical solution will be stable, but if the exponential is positive then this can increase very rapidly and either swamp or corrupt the solution, sometimes in a manner that may be difficult to detect. Many integration algorithms show a dependence of the stability of the solution on the integration step length. There can thus be a critical integration step, which if exceeded can lead to instability. This type of instability can be seen in many of the simulation examples, where an injudicious choice of DT can cause numerical overflow. Reducing the integration step size makes the solution run more slowly, and rounding off errors caused by the limited accuracy of the digital representation may then become important. Practically, one can try to solve such problems by changing the integration routine or by adjusting the error criteria in MADONNA.

Model instability is demonstrated by many of the simulation examples and leads to very interesting phenomena, such as multiple steady states, naturally occurring oscillations, and chaotic behaviour. In the case of a model which is inherently unstable, nothing can be done except to completely reformulate the model into a more stable form.

In general, the form of the solution to the dynamic model equations will be in the form

$$y_i(t) = \text{steady-state solution} + \text{transient solution}$$

where the transient part of the solution can be represented by a series of exponential functions

$$Y_{trans} = A_1 e^{\lambda t} + A_2 e^{\lambda_2 \cdot t} + \ldots + A_n e^{\lambda_n t}$$

In the above relationship, the coefficients A_1 to A_n depend on the initial conditions of the problem and the exponential values, λ_i, are determined by the parameters of the system and in fact represent the eigenvalues or roots of the characteristic solution of the system.

In a stable system, the above transient terms must decay to zero to give the steady-state solution. This applies when all the roots are simple negative exponentials. The system is also stable with all the roots occuring as negative real parts of complex roots, causing a decaying oscillatory approach to the eventual steady-state condition. If any of the roots are positive real numbers or complex numbers with real positive parts, the corresponding transient terms in the solution will grow in magnitude, thus directing the solution away from the unstable steady-state condition. Where the roots of the transient solution are pure imaginary numbers, the result is an oscillation of constant amplitude and frequency. The dynamic stability of such systems is often shown most conveniently on a phase-plane diagram, as shown in several of the simulation examples. Model instability is discussed further in Section 3.2.7, with regard to the stability of continuous stirred-tank reactors.

For a fuller treatment of dynamic stability problems, the reader is referred to Walas (1991), Seborg et al. (1989), Perlmutter (1972) and to the simulation examples THERM, THERMPLOT, COOL, REFRIG1, REFRIG2 and OSCIL.

3
Modelling of Stagewise Processes

3.1
Introduction

The principle of the perfectly-mixed stirred tank has been discussed previously in Section 1.2.2, and this provides an essential building block for modelling applications. In this section, the concept is applied to tank type reactor systems and stagewise mass transfer applications, such that the resulting model equations often appear in the form of linked sets of first-order difference differential equations. Solution by digital simulation works well for small problems, in which the number of equations are relatively small and where the problem is not compounded by stiffness or by the need for iterative procedures. For these reasons, the dynamic modelling of the continuous distillation columns in this section is intended only as a demonstration of method, rather than as a realistic attempt at solution. For the solution of complex distillation and extraction problems, the reader is referred to commercial dynamic simulation packages.

3.2
Stirred-Tank Reactors

3.2.1
Reactor Configurations

This section is concerned with batch, semi-batch, continuous stirred tanks and continuous stirred-tank-reactor cascades, as represented in Fig. 3.1. Tubular chemical reactor systems are discussed in Chapter 4.

Three modes of reactor operation may be distinguished: batch, semi-batch and continuous. In a batch system all reactants are added to the tank at the given starting time. During the course of reaction, the reactant concentrations decrease continuously with time, and products are formed. On completion of the reaction, the reactor is emptied, cleaned and is made ready for another batch and allows differing reactions to be carried out in the same reactor. The disadvantages are the downtime needed for loading and cleaning and possibly the changing reaction conditions. Batch operation is often ideal for small scale flex-

Chemical Engineering Dynamics: An Introduction to Modelling and Computer Simulation, Third Edition
J. Ingham, I. J. Dunn, E. Heinzle, J. E. Prenosil, J. B. Snape
Copyright © 2007 WILEY-VCH Verlag GmbH & Co. KGaA, Weinheim
ISBN: 978-3-527-31678-6

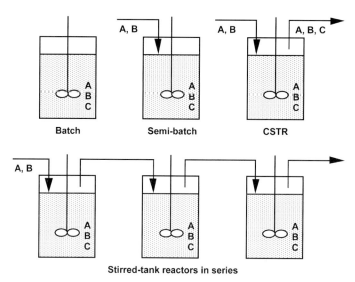

Fig. 3.1 Stirred-tank reactor configurations.

ible production and high value, low output product production, where the chemistry and reaction kinetics are not known exactly.

In many reactions pure batch operation is not possible, mainly due to safety or selectivity reasons. In semi-batch operation, one reactant may be charged to the vessel at the start of the batch, and then the others fed to the reactor at perhaps varying rates and over differing time periods. When the vessel is full, feeding is stopped and the contents allowed to discharge. Semi-batch operation allows one to vary the reactant concentration to a desired level in a very flexible way, and thus to control the reaction rates and the reactor temperature. It is, however, necessary to develop an appropriate feeding strategy. Modelling and simulation allows estimation of optimal feeding profiles. Sometimes it is necessary to adjust the feeding rates using feed-back control. The flexibility of operation is generally similar to that of a batch reactor system.

Continuous operation provides high rates of production with more constant product quality. There are no downtimes during normal operation. Reactant preparation and product treatment also have to run continuously. This requires careful flow control. Continuous operation can involve a single stirred tank, a series of stirred tanks or a tubular-type of reactor. The latter two instances give concentration profiles similar to those of batch operation, whereas in a single stirred tank, the reaction conditions are at the lowest reactant concentration, corresponding to effluent conditions.

3.2.2
Generalised Model Description

The energy and material balance equations for reacting systems follow the same principles, as described previously in Sections 1.2.3 to 1.2.5.

3.2.2.1 Total Material Balance Equation

It becomes necessary to incorporate a total material balance equation into the reactor model, whenever the total quantity of material in the reactor varies, as in the cases of semi-continuous or semi-batch operation or where volume changes occur, owing to density changes in flow systems. Otherwise the total material balance equation can generally be neglected.

The dynamic total material balance equation is represented by

$$\begin{pmatrix} \text{Rate of accumulation} \\ \text{of total mass in system} \end{pmatrix} = \begin{pmatrix} \text{Mass flow rate} \\ \text{into the system} \end{pmatrix} - \begin{pmatrix} \text{Mass flow rate} \\ \text{out of the system} \end{pmatrix}$$

3.2.2.2 Component Balance Equation

The general component balance for a well-mixed tank reactor or reaction region can be written as

$$\begin{pmatrix} \text{Rate of} \\ \text{accumulation} \\ \text{of component} \\ \text{in the} \\ \text{system} \end{pmatrix} = \begin{pmatrix} \text{Rate of} \\ \text{flow of} \\ \text{component} \\ \text{into} \\ \text{the system} \end{pmatrix} - \begin{pmatrix} \text{Rate of} \\ \text{flow of} \\ \text{component} \\ \text{from} \\ \text{the system} \end{pmatrix} + \begin{pmatrix} \text{Rate of} \\ \text{production of} \\ \text{component} \\ \text{by reaction} \end{pmatrix}$$

For batch reactors, there is no flow into or out of the system, and those terms in the component balance equation are therefore zero.

For semi-batch reactors, there is inflow but no outflow from the reactor and the outflow term in the above balance equation is therefore zero.

For steady-state operation of a continuous stirred-tank reactor or continuous stirred-tank reactor cascade, there is no change in conditions with respect to time, and therefore the accumulation term is zero. Under transient conditions, the full form of the equation, involving all four terms, must be employed.

3.2.2.3 Energy Balance Equation

For reactions involving heat effects, the total and component material balance equations must be coupled with a reactor energy balance equation. Neglecting work done by the system on the surroundings, the energy balance is expressed by where each term has units of kJ/s. For steady-state operation the accumulation

$$\begin{pmatrix} \text{Rate of} \\ \text{accumulation} \\ \text{of energy in} \\ \text{the reactor} \end{pmatrix} = \begin{pmatrix} \text{Flow of} \\ \text{energy into} \\ \text{the reactor} \\ \text{in the feed} \end{pmatrix} - \begin{pmatrix} \text{Flow of} \\ \text{energy from} \\ \text{the reactor} \\ \text{in the outlet} \end{pmatrix} + \begin{pmatrix} \text{Rate of} \\ \text{heat transfer} \\ \text{to the} \\ \text{reactor} \end{pmatrix}$$

term is zero and can be neglected. A detailed explanation of energy balancing is found in Section 1.2.5, together with a case study for a reactor.

Both flow terms are zero for the case of batch reactor operation, and the outflow term is zero for semi-continuous or semi-batch operation.

The information flow diagram for a non-isothermal, continuous-flow reactor (in Fig. 1.18, shown previously in Section 1.2.5) illustrates the close interlinking and highly interactive nature of the total material balance, component material balance, energy balance, rate equation, Arrhenius equation and flow effects F. This close interrelationship often brings about highly complex dynamic behaviour in chemical reactors.

3.2.2.4 Heat Transfer to and from Reactors

Heat transfer is usually affected by coils or jackets, but can also be achieved by the use of external loop heat exchangers and, in certain cases, heat is transported out of the reactor by the vaporization of volatile material from the reactor. The treatment here mainly concerns jackets and coils. Other examples of heat transfer are illustrated in the simulation examples of Chapter 5.

Figure 3.2 shows the case of a jacketed, stirred-tank reactor, in which either heating by steam or cooling medium can be applied to the jacket. Here V is volume, c_p is specific heat capacity, ρ is density, Q is the rate of heat transfer, U is the overall

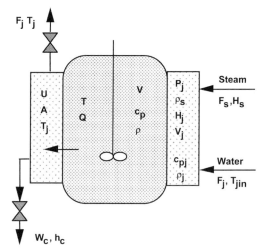

Fig. 3.2 Model representation of a stirred-tank reactor with heat transfer to the jacket.

heat transfer coefficient, A is the area for heat transfer, T is temperature, H is enthalpy of vapour, h is liquid enthalpy, F is volumetric flow rate and W is mass flow rate. The subscripts are j for the jacket, s for steam and c for condensate.

The rate of heat transfer is most conveniently expressed in terms of an overall heat transfer coefficient, the effective area for heat transfer and an overall temperature difference, or driving force, where

$$Q = UA(T - T_j) \quad kJ/s$$

Jacket or Coil Cooling

In simple cases the jacket or cooling temperature, T_j, may be assumed to be constant. In more complex dynamic problems, however, it may be necessary to allow for the dynamics of the cooling jacket, in which case T_j becomes a system variable. The model representation of this is shown in Fig. 3.3.

Under conditions, where the reactor and the jacket are well insulated and heat loss to the surroundings and mechanical work effects may be neglected

$$\begin{pmatrix} \text{Rate of} \\ \text{accumulation} \\ \text{of energy} \\ \text{in the jacket} \end{pmatrix} = \begin{pmatrix} \text{Rate of} \\ \text{energy flow} \\ \text{to the jacket} \\ \text{by convection} \end{pmatrix} - \begin{pmatrix} \text{Rate of} \\ \text{energy flow} \\ \text{from the jacket} \\ \text{by convection} \end{pmatrix} + \begin{pmatrix} \text{Rate of} \\ \text{heat transfer} \\ \text{into} \\ \text{the jacket} \end{pmatrix}$$

or

$$\begin{pmatrix} \text{Rate of} \\ \text{accumulation} \\ \text{of enthalpy} \\ \text{in the jacket} \end{pmatrix} = \begin{pmatrix} \text{Energy} \\ \text{required to} \\ \text{heat coolant} \\ \text{from } T_{jin} \text{ to } T_j \end{pmatrix} + \begin{pmatrix} \text{Rate of} \\ \text{heat transfer} \\ \text{from} \\ \text{the jacket} \end{pmatrix}$$

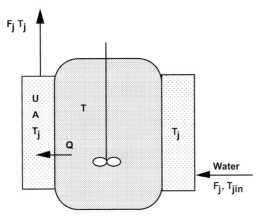

Fig. 3.3 Dynamic model representation of the cooling jacket.

3 Modelling of Stagewise Processes

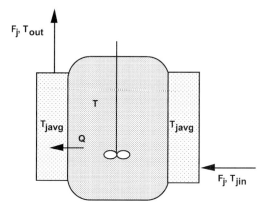

Fig. 3.4 Steady-state model representation of the cooling jacket.

Assuming the liquid in the jacket is well-mixed, the heat balance equation for the jacket becomes

$$V_j \rho_j c_{pj} \frac{dT_j}{dt} = F_j \rho_j c_{pj}(T_{jin} - T_j) + UA(T - T_j)$$

Here F_j is the volumetric flow of coolant to the jacket, T_{jin} is the inlet coolant temperature, and T_j is the jacket temperature. Under well-mixed conditions, T_j is identical to the temperature of the outlet flow.

Alternatively neglecting the jacket dynamics and assuming that the coolant in the jacket is at some mean temperature, T_{javg}, as shown in Fig. 3.4, a steady-state energy balance can be formulated as

$$\begin{pmatrix} \text{Rate of energy gain} \\ \text{by coolant flow} \end{pmatrix} = \begin{pmatrix} \text{Rate of energy transfer} \\ \text{from the reactor to the jacket} \end{pmatrix}$$

$$F_j \rho_j c_{pj}(T_{jout} - T_{jin}) = UA(T - T_{javg}) = Q$$

Assuming an arithmetic mean jacket temperature,

$$T_{javg} = \frac{T_{jin} + T_{jout}}{2}$$

Substituting for T_{jout} into the steady-state jacket energy balance, solving for T_{javg} and substituting T_{javg} into the steady-state balance, gives the result that

$$Q = UAK_1(T - T_{jin})$$

where K_1 is given by

$$K_1 = \frac{2F_j\rho_j c_{pj}}{UA + 2F_j\rho_j c_{pj}}$$

As shown in several of the simulation examples, the fact that Q is now a function of the flow rate, F_j, provides a convenient basis for the modelling of cooling effects, and control of the temperature of the reactor by regulation of the flow of coolant.

3.2.2.5 Steam Heating in Jackets

The dynamics of the jacket are more complex for the case of steam heating. The model representation of the jacket steam heating process is shown in Fig. 3.5.

A material balance on the steam in the jacket is represented by

$$\begin{pmatrix} \text{Rate of} \\ \text{accumulation} \\ \text{of steam} \\ \text{in the jacket} \end{pmatrix} = \begin{pmatrix} \text{Mass flow} \\ \text{of steam} \\ \text{to the} \\ \text{jacket} \end{pmatrix} - \begin{pmatrix} \text{Mass flow of} \\ \text{condensate} \\ \text{from the} \\ \text{jacket} \end{pmatrix}$$

$$V_j \frac{d\rho_j}{dt} = F_s \rho_s - W_c$$

The enthalpy balance on the jacket is given by

$$V_j \frac{d(H_j \rho_j)}{dt} = F_s \rho_s H_s + UA(T - T_j) - W_c H_c$$

where H_j is the enthalpy of the steam in the jacket.

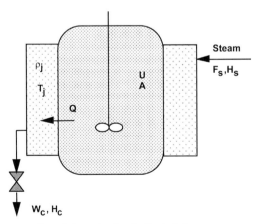

Fig. 3.5 Model representation of steam heating in the jacket.

The saturated steam density, ρ_j, depends on the jacket temperature, T_j, in the first approximation in accordance with the Ideal Gas Law and hence

$$\rho_j = \frac{18 P_j}{R T_j}$$

where R is the Ideal Gas Constant.

The jacket steam pressure, P_j, is itself a function of the jacket steam temperature, T_j, as listed in steam tables or as correlated by the Antoine equation for vapour pressure, where

$$P_j = \exp\left(\frac{A}{T_j} + B\right)$$

Here, A and B are the Antoine steam constants and T_j is the absolute steam temperature.

Combining the two equations for steam density and pressure gives an implicit equation requiring a numerical root estimation.

$$0 = P_j - \exp\left(\frac{A}{18 P_j / R \rho_j} + B\right)$$

The MADONNA root finder provides a powerful means of solution. An example of the use of root finder is employed in the simulation example RELUY for steam jacket dynamics. Other examples of use are in BUBBLE and STEAM.

3.2.2.6 Dynamics of the Metal Jacket Wall

In some cases, where the wall of the reactor has an appreciable thermal capacity, the dynamics of the wall can be of importance (Luyben, 1973). The simplest approach is to assume the whole wall material has a uniform temperature and therefore can be treated as a single lumped parameter system or, in effect, as a single well-stirred tank. The heat flow through the jacket wall is represented in Fig. 3.6.

The nomenclature is as follows: Q_m is the rate of heat transfer from the reactor to the reactor wall, Q_j is the rate of heat transfer from the reactor wall to the jacket, and

$$Q_m = U_m A_m (T - T_m)$$

$$Q_j = U_j A_j (T_m - T_j)$$

Here, U_m is the film heat transfer coefficient between the reactor and the reactor wall. U_j is the film heat transfer coefficient between the reactor wall and the jacket. A_m is the area for heat transfer between the reactor and the wall. A_j is the area for heat transfer between the wall and the jacket.

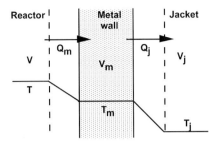

Fig. 3.6 Model representation of heat flow through a reactor wall with assumed uniform temperature.

The heat balance for the wall gives

$$V_m \rho_m c_{pm} \frac{dT_m}{dt} = Q_m - Q_j$$

and the balance for the jacket becomes

$$V_j \rho_j c_{pj} \frac{dT_j}{dt} = F_j \rho_j c_{pj} (T_{jin} - T_j) + Q_j$$

In some cases, it may be of interest to model the temperature distribution through the wall.

This might be done by considering the metal wall and perhaps also the jacket as consisting of a series of separate regions of uniform temperature, as shown in Fig. 3.7. The balances for each region of metal wall and cooling water volume then become for any region, n

$$V_{mn} \rho_m c_{pm} \frac{dT_{mn}}{dt} = Q_{mn} - Q_{jn}$$

$$V_{jn} \rho_j c_{pj} \frac{dT_{jn}}{dt} = F_j \rho_j c_{pj} (T_{jn-1} - T_{jn}) + Q_{jn}$$

where V_{mn} and V_{jn} are the respective volumes of the wall and coolant in element n. A_{mn} and A_{jn} are the heat transfer areas for transfer from the reactor to the wall and from the wall to the jacket. Hence:

$$Q_{mn} = U_m A_{mn} (T - T_{mn})$$

$$Q_{jn} = U_j A_{jn} (T_{mn} - T_{jn})$$

Note that the effects of thermal conduction along the jacket wall are assumed to be negligible in this case.

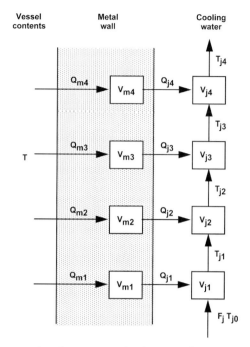

Fig. 3.7 Model representation of temperature distribution in the wall and jacket, showing wall and jacket with four lumped parameters.

3.2.3
The Batch Reactor

It is assumed that all the tank-type reactors, covered in this and the immediately following sections, are at all times perfectly mixed, such that concentration and temperature conditions are uniform throughout the tanks contents. Figure 3.8 shows a batch reactor with a cooling jacket. Since there are no flows into the reactor or from the reactor, the total material balance tells us that the total mass, within the reactor, remains constant.

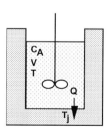

Fig. 3.8 The batch reactor with heat transfer.

Component Balance Equation

$$\begin{pmatrix} \text{Rate of accumulation} \\ \text{of reactant A} \end{pmatrix} = \begin{pmatrix} \text{Rate of production of A} \\ \text{by chemical reaction} \end{pmatrix}$$

Thus,

$$\frac{dn_A}{dt} = r_A V$$

or since

$$n_A = n_{A0}(1 - X_A)$$

$$n_{A0} \frac{dX_A}{dt} = -r_A V$$

or for constant volume

$$C_{A0} \frac{dX_A}{dt} = -r_A$$

Energy Balance Equation

$$\begin{pmatrix} \text{Rate of} \\ \text{accumulation} \\ \text{of heat in} \\ \text{the reactor} \end{pmatrix} = \begin{pmatrix} \text{Rate of heat} \\ \text{generation} \\ \text{by reaction} \end{pmatrix} - \begin{pmatrix} \text{Rate of heat} \\ \text{transfer to} \\ \text{the surroundings} \end{pmatrix}$$

$$V \rho c_p \frac{dT}{dt} = V r_Q - UA(T - T_j)$$

Here V is the volume of the reactor, ρ is the density, c_p is the mean specific heat of the reactor contents (kJ/kg K) and r_Q is the rate of generation of heat by reaction (kJ/s m^3).

3.2.3.1 Case A: Constant-Volume Batch Reactor

A constant volume batch reactor is used to convert reactant, A, to product, B, via an endothermic reaction, with simple stoichiometry, A → B. The reaction kinetics are second-order with respect to A, thus

$$r_A = -k C_A^2$$

From the reaction stoichiometry, product B is formed at exactly the same rate as that at which reactant A is decomposed.

The rate equation with respect to A is

$$-r_A = kC_A^2 = kC_{A0}^2(1 - X_A)^2$$

giving the component balance equation as

$$\frac{dX_A}{dt} = kC_{A0}(1 - X_A)^2$$

For the second-order reaction, the term representing the rate of heat production by reaction simplifies to

$$r_Q V = -r_A(-\Delta H)V = kC_{A0}^2(1 - X_A)^2(-\Delta H)V$$

This gives the resultant heat balance equation as

$$V\rho c_p \frac{dT}{dt} = kC_{A0}^2(1 - X_A)^2(-\Delta H)V + UA(T_j - T)$$

where it is assumed that heat transfer to the reactor occurs via a coil or jacket heater.

The component material balance, when coupled with the heat balance equation and temperature dependence of the kinetic rate coefficient, via the Arrhenius relation, provide the dynamic model for the system. Batch reactor simulation examples are provided by BATCHD, COMPREAC, BATCOM, CASTOR, HYDROL and RELUY.

3.2.4
The Semi-Batch Reactor

A semi-batch reactor with one feed stream and heat transfer to a cooling jacket is shown in Fig. 3.9.

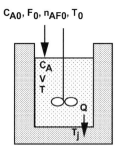

Fig. 3.9 The semi-batch reactor.

Total Material Balance

A total material balance is necessary, owing to the feed input to the reactor, where

$$\begin{pmatrix} \text{Rate of accumulation} \\ \text{of mass in the reactor} \end{pmatrix} = \begin{pmatrix} \text{Mass flow} \\ \text{rate in} \end{pmatrix}$$

$$\frac{d(\rho V)}{dt} = F_0 \rho_0$$

Here ρ_0 is the feed density.

The density in the reactor, ρ, may be a function of the concentration and temperature conditions within the reactor. Assuming constant density conditions

$$\frac{dV}{dt} = F$$

Component Balance Equation

All important components require a component balance.

For a given reactant A

$$\begin{pmatrix} \text{Rate of} \\ \text{accumulation of A} \end{pmatrix} = \begin{pmatrix} \text{Rate of flow} \\ \text{of A in} \end{pmatrix} + \begin{pmatrix} \text{Rate of production} \\ \text{of A by reaction} \end{pmatrix}$$

$$\frac{dn_A}{dt} = N_{A0} + r_A V$$

where N_{A0} is the molar feeding rate of A per unit time.

In terms of concentration, this becomes

$$\frac{d(VC_A)}{dt} = F_0 C_{A0} + r_A V$$

where F is the volumetric feed rate and C_{A0} is the feed concentration. Note that both the volumetric flow and the feed concentration can vary with time, depending on the particular reactor feeding strategy.

Energy Balance Equation

Whenever changes in temperature are to be calculated, an energy balance is needed. With the assumption of constant c_p and constant ρ, as derived in Section 1.2.5, the balance becomes

$$\rho c_p V \frac{dT}{dt} = F_0 \rho c_p (T_0 - T) + r_Q V + Q$$

Note that the available heat transfer area may also change as a function of time, and may therefore also form an additional variable in the solution. Note also that although constant ρ and c_p have been assumed here, this is not a restrictive condition and that equations showing the variations of these properties are easily included in any simulation model.

3.2.4.1 Case B: Semi-Batch Reactor

A semi-batch reactor is used to convert reactant, A, to product, B, by the reaction $A \rightarrow 2B$. The reaction is carried out adiabatically. The reaction kinetics are as before

$$r_A = -kC_A^2$$

and the stoichiometry gives

$$r_B = -2r_A = +2kC_A^2$$

The balances for the two components A and B, with flow of A, into the reactor are now

$$\frac{d(VC_A)}{dt} = FC_{A0} + r_A V$$

$$\frac{d(VC_B)}{dt} = r_B V$$

and the enthalpy balance equation is

$$V\rho c_p \frac{dT}{dt} = F\rho c_p (T_0 - T) + kC_{A0}^2 (1 - X_A)^2 (-\Delta H_A) V$$

since, for adiabatic operation, the rate of heat input into the system, Q, is zero.

With initial conditions for the initial molar quantities of A and B (VC_A, VC_B), the initial temperature, T, and the initial volume of the contents, V, specified, the resulting system of equations can be solved to obtain the time-varying quantities, $V(t)$, $VC_A(t)$, $VC_B(t)$, $T(t)$ and hence also concentrations C_A and C_B as functions of time. Examples of semi-batch operations are given in the simulation examples HMT, SEMIPAR, SEMISEQ, RUN, SULFONATION and SEMIEX.

3.2.5
The Continuous Stirred-Tank Reactor

Although continuous stirred-tank reactors (Fig. 3.10) normally operate at steady-state conditions, a derivation of the full dynamic equation for the system is nec-

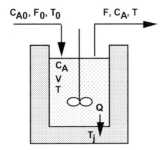

Fig. 3.10 Continuous stirred-tank reactor with heat transfer.

essary to cover the instances of plant start up, shut down and the application of reactor control.

Total Material Balance

The dynamic total material balance equation is given by

$$\begin{pmatrix} \text{Rate of accumulation} \\ \text{of mass in the reactor} \end{pmatrix} = \begin{pmatrix} \text{Mass flow} \\ \text{rate in} \end{pmatrix} - \begin{pmatrix} \text{Mass flow} \\ \text{rate out} \end{pmatrix}$$

$$\frac{dV}{dt} = F_0 \rho_0 - F\rho$$

Under constant volume and constant density conditions

$$\frac{dV}{dt} = F_0 - F = 0$$

and therefore

$$F_0 = F$$

Component Material Balance Equation

The component material balance equation is given by

$$\begin{pmatrix} \text{Rate of} \\ \text{accumulation} \\ \text{of component} \\ \text{in the reactor} \end{pmatrix} = \begin{pmatrix} \text{Rate of} \\ \text{flow in of} \\ \text{component} \end{pmatrix} - \begin{pmatrix} \text{Rate of} \\ \text{flow out of} \\ \text{component} \end{pmatrix} + \begin{pmatrix} \text{Rate of} \\ \text{production of} \\ \text{component} \\ \text{by reaction} \end{pmatrix}$$

For a given reactant A

$$\frac{dn_A}{dt} = N_{A0} - N_A - r_A V$$

where n_A is the moles of A in the reactor, N_{A0} is the molar feeding rate of A to the reactor, and N_A is the molar flow rate of A from the reactor.

Under constant density and constant volume conditions, this may be expressed as

$$V\frac{dC_A}{dt} = FC_{A0} - FC_A + r_A V$$

or

$$\frac{dC_A}{dt} = \frac{C_{A0} - C_A}{\tau} + r_A$$

where τ (= V/F) is the average holdup time or residence time of the reactor.

For steady-state conditions to be maintained, the volumetric flow rate F and inlet concentration C_{A0} must remain constant and

$$\frac{dC_A}{dt} = 0$$

hence at steady state

$$r_A V = -F(C_{A0} - C_A) = -(N_{A0} - N_A)$$

and

$$X_A = \frac{N_{A0} - N_A}{N_{A0}} = -\frac{r_A V}{N_{A0}} = -\frac{r_A V}{FC_{A0}}$$

or

$$\tau = \frac{V}{F} = \frac{C_{A0} X_A}{r_A}$$

Energy Balance Equation

This, like the other dynamic balances for the CSTR, follows the full generalised form, of Section 1.2.5, giving

$$\begin{pmatrix} \text{Rate of} \\ \text{accumulation} \\ \text{of energy in} \\ \text{the reactor} \end{pmatrix} = \begin{pmatrix} \text{Energy needed} \\ \text{to heat the} \\ \text{feed to outlet} \\ \text{temperature} \end{pmatrix} + \begin{pmatrix} \text{Rate of} \\ \text{generation} \\ \text{of heat by} \\ \text{reaction} \end{pmatrix} + \begin{pmatrix} \text{Rate of} \\ \text{heat transfer} \\ \text{from the} \\ \text{surroundings} \end{pmatrix}$$

or assuming constant c_p

$$V\rho c_p \frac{dT}{dt} = F_0 \rho c_p (T_0 - T) + r_Q V + Q$$

3.2.5.1 Case C: Constant-Volume Continuous Stirred-Tank Reactor

The chemical reaction data are the same as in the preceding example. The reaction kinetics are

$$r_A = -kC_A^2$$

with

$$r_B = -2r_A$$

and the component balances for both A and B are given by

$$V\frac{dC_A}{dt} = FC_{A0} - FC_A + r_A V$$

$$V\frac{dC_B}{dt} = FC_{B0} - FC_B + r_B V$$

Assuming both constant density and constant specific heat, the heat balance equation becomes

$$V\rho c_p \frac{dT}{dt} = F\rho c_p (T_0 - T) + kC_A^2 V(-\Delta H) - UA(T - T_j)$$

Here cooling of an exothermic chemical reaction, via a cooling coil or jacket, is included.

3.2.6 Stirred-Tank Reactor Cascade

For any continuous stirred-tank reactor, n, in a cascade of reactors (Fig. 3.11) the reactor n receives the discharge from the preceding reactor, n – 1, as its feed and discharges its effluent into reactor n + 1, as feed to that reactor.

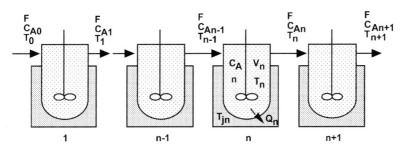

Fig. 3.11 Cascade of continuous stirred-tank reactors.

Thus the balance equations for reactor, n, simply become

$$V_n \frac{dC_{An}}{dt} = FC_{An-1} - FC_{An} + r_{An} V_n$$

$$V_n \rho c_p \frac{dT_n}{dt} = F\rho c_p (T_{n-1} - T_n) + r_{Qn} V + Q_n$$

where, for example,

$$r_{An} = -k_n C_{An}^\alpha C_{Bn}^\beta$$

$$k_n = Z e^{-E/RT_n}$$

and

$$Q_n = U_n A_n (T_{jn} - T_n)$$

Thus the respective rate expressions depend upon the particular concentration and temperature levels that exist within reactor n. The rate of production of heat by reaction r_Q was defined in Section 1.2.5 and includes all occurring reactions. Simulation examples pertaining to stirred tanks in series are CSTRPULSE, CASCSEQ and COOL.

3.2.7
Reactor Stability

Consider a simple first-order exothermic reaction, A → B, carried out in a single, constant-volume, continuous stirred-tank reactor (Fig. 3.10), with constant jacket coolant temperature, where $r_A = kC_A$.

The model equations are then given by

$$V \frac{dC_A}{dt} = F(C_{A0} - C_A) - VkC_A$$

$$V\rho c_p \frac{dT}{dt} = F\rho c_p (T_0 - T) + VkC_A(-\Delta H) - UA(T - T_j)$$

$$k = Z e^{-E/RT}$$

At steady state, the temperature and concentration in the reactor are constant with respect to time and

$$\frac{dC_A}{dt} = \frac{dT}{dt} = 0$$

Hence from the component material balance

$$F(C_{A0} - C_A) = VkC_A$$

the steady-state concentration is given by

$$C_A = \frac{C_{A0}}{1 + kV/F}$$

From the steady-state heat balance the heat losses can be equated to the heat gained by reaction, giving

$$-F\rho c_p(T_0 - T) + UA(T - T_j) = VkC_A(-\Delta H) = \frac{VkC_{A0}(-\Delta H)}{1 + kV/F}$$

i.e.,

$$\begin{pmatrix} \text{Net rate of} \\ \text{heat loss from} \\ \text{the reactor} \end{pmatrix} = \begin{pmatrix} \text{Rate of heat generation} \\ \text{by chemical reaction in} \\ \text{the reactor} \end{pmatrix}$$

or

$$H_L = H_G$$

The above equation then represents the balanced conditions for steady-state reactor operation. The rate of heat loss, H_L, and the rate of heat gain, H_G, terms may be calculated as functions of the reactor temperature. The rate of heat loss, H_L, plots as a linear function of temperature, and the rate of heat gain, H_G, owing to the exponential dependence of the rate coefficient on temperature, plots as a sigmoidal curve, as shown in Fig. 3.12. The points of intersection of the rate of heat lost and the rate of heat gain curves thus represent potential steady-

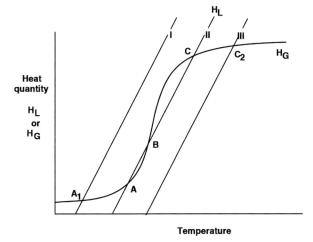

Fig. 3.12 Heat loss H_L and heat gain H_G in a steady-state continuous stirred-tank reactor.

state operating conditions that satisfy the above steady-state heat balance criterion.

Figure 3.12 shows the heat gain curve, H_G, for one particular set of system parameters, and a set of three possible heat loss, H_L, curves. Possible curve intersection points, A_1 and C_2, represent singular stable steady-state operating curves for the reactor, with cooling conditions as given by cooling curves, I and III, respectively.

The cooling conditions given by curve II, however, indicate three potential steady-state solutions at the curve intersections A, B and C. By considering the effect of small temperature variations, about the three steady-state conditions, it can be shown that points A and C represent stable, steady-state operating conditions, whereas the curve intersection point B is unstable. On start up, the reaction conditions will proceed to an eventual steady state, at either point A or at point C. Since point A represents a low temperature, and therefore a low conversion operating state, it may be desirable that the initial transient conditions in the reactor should eventually lead to C, rather than to A. However if point C is at a temperature which is too high and might possibly lead to further decomposition reactions, then A would be the desired operating point.

The basis of the argument for intersection point, B, being unstable is as follows and is illustrated in Fig. 3.13.

Consider a small positive temperature deviation, moving to the right of point B. The condition of the reactor is now such that the H_G value is greater than that for H_L. This will cause the reactor to heat up and the temperature to increase further, until the stable steady-state solution at point C is attained. For a small temperature decrease to the left of B, the situation is reversed, and the reactor will continue to cool, until the stable steady-state solution at point A is attained. Similar arguments show that points A and C are stable steady states.

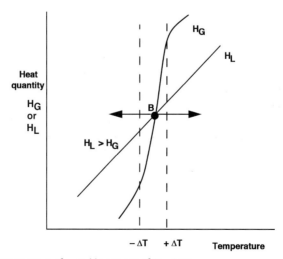

Fig. 3.13 Characteristics of unstable point B of Fig. 3.12.

System stability can also be analysed in terms of the linearised differential model equations. In this, new perturbation variables for concentration C' and temperature T' are defined. These are defined in terms of small deviations in the actual reactor conditions away from the steady-state concentration and temperature C_{ss} and T_{ss} respectively. Thus

$$C' = C - C_{ss} \quad \text{and} \quad T' = T - T_{ss}$$

The linearisation of the non-linear component and energy balance equations, based on the use of Taylor's expansion theorem, leads to two, simultaneous, first-order, linear differential equations with constant coefficients of the form

$$\frac{dC'}{dt} = a_1 C' + b_1 T'$$

$$\frac{dT'}{dt} = a_2 C' + b_2 T'$$

The coefficients of the above equations are the partial differentials of the two dynamic balance equations evaluated at C_{ss}, T_{ss} and are given by

$$a_1 = \left| \frac{\partial F(C, T)}{\partial C'} \right|_{C_{ss}, T_{ss}}$$

$$b_1 = \left| \frac{\partial F(C, T)}{\partial T'} \right|_{C_{ss}, T_{ss}}$$

$$a_2 = \left| \frac{\partial G(C, T)}{\partial C'} \right|_{C_{ss}, T_{ss}}$$

$$b_2 = \left| \frac{\partial G(C, T)}{\partial T'} \right|_{C_{ss}, T_{ss}}$$

where F(C, T) represents the dynamic component balance equation and G(C, T) represents the dynamic heat balance equation.

The two linearised model equations have the general solution of the form

$$C' = a e^{\lambda_1 t} + b e^{\lambda_2 t}$$

and

$$T' = c e^{\lambda_1 t} + d e^{\lambda_2 t}$$

The dynamic behaviour of the system is thus determined by the values of the exponential coefficients, λ_1 and λ_2, which are the roots of the characteristic equation or eigenvalues of the system and which are also functions of the system parameters.

If λ_1 and λ_2 are real numbers and both have negative values, the values of the exponential terms and hence the magnitudes of the perturbations away from the steady-state conditions, c' and T', will reduce to zero with increasing time. The system response will therefore decay back to its original steady-state value, which is therefore a stable steady-state solution or stable node.

If λ_1 and λ_2 are real numbers and both or one of the roots are positive, the system response will diverge with time and the steady-state solution will therefore be unstable, corresponding to an unstable node.

If the roots are, however, complex numbers, with one or two positive real parts, the system response will diverge with time in an oscillatory manner, since the analytical solution is then one involving sine and cosine terms. If both roots, however, have negative real parts, the sine and cosine terms still cause an oscillatory response, but the oscillation will decay with time, back to the original steady-state value, which, therefore remains a stable steady state.

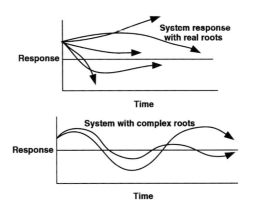

Fig. 3.14 Systems response with real and complex roots.

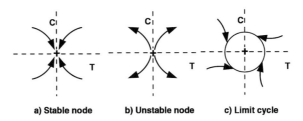

Fig. 3.15 Phase-plane representations of reactor stability. In the above diagrams the point + represents a possible steady-state solution, which (a) may be stable, (b) may be unstable or (c) about which the reactor produces sustained oscillations in temperature and concentration.

If the roots are pure imaginary numbers, the form of the response is purely oscillatory, and the magnitude will neither increase nor decay. The response, thus, remains in the neighbourhood of the steady-state solution and forms stable oscillations or limit cycles.

The types of system behaviour predicted by the above analysis are depicted in Figs. 3.14 and 3.15. The phase-plane plots of Fig. 3.15 give the relation of the dependent variables C and T. Detailed explanation of phase-plane plots is given in control textbooks (e.g., Stephanopoulos, 1984). Linearisation of the reactor model equations is used in the simulation example, HOMPOLY.

Thus it is possible for continuous stirred-tank reactor systems to be stable, or unstable, and also to form continuous oscillations in output, depending upon the system, constant and parameter, values.

This analysis is limited, since it is based on a steady-state criterion. The linearisation approach, outlined above, also fails in that its analysis is restricted to variations, which are very close to the steady state. While this provides excellent information on the dynamic stability, it cannot predict the actual trajectory of the reaction, once this departs from the near steady state. A full dynamic analysis is, therefore, best considered in terms of the full dynamic model equations and this is easily effected, using digital simulation. The above case of the single CSTR, with a single exothermic reaction, is covered by the simulation examples THERMPLOT and THERM. Other simulation examples, covering aspects of stirred-tank reactor stability are COOL, OSCIL, REFRIG1 and REFRIG2. Phase-plane plots are very useful for the analysis of such systems.

3.2.8
Reactor Control

Two simple forms of a batch reactor temperature control are possible, in which the reactor is either heated by a controlled supply of steam to the heating jacket, or cooled by a controlled flow of coolant (Fig. 3.16). Other control schemes

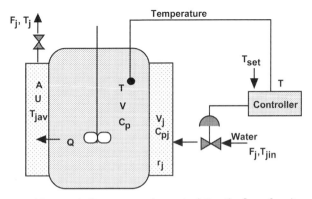

Fig. 3.16 Reactor with control of temperature by manipulating the flow of cooling water.

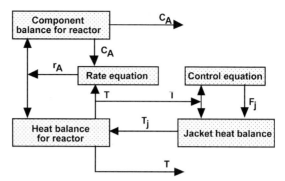

Fig. 3.17 Information flow diagram for the above reactor control implementation.

would be to regulate the reactor flow rate or feed concentration, in order to maintain a given reaction rate (see simulation example SEMIEX).

Figure 3.17 represents an information flow diagram for the above control scheme.

Assuming the jacket is well-mixed, a heat balance on the jacket gives

$$V_j \rho_j c_{pj} \frac{dT_j}{dt} = F_j \rho_j c_{pj}(T_{jin} - T_j) + UA(T - T_j)$$

Consider the case of a proportional controller, which is required to maintain a desired reactor temperature, by regulating the flow of coolant. Neglecting dynamic jacket effects, the reactor heat balance can then be modified to include the effect of the varying coolant flow rate, F_j, in the model equation as:

$$V \rho c_p \frac{dT}{dt} = -r_Q V + UAK_1(T - T_{jin})$$

where the mean temperature of the jacket is accounted for by the term K_1, as shown in Section 3.2.2.4, and given by

$$K_1 = \frac{2F_j \rho_j c_{pj}}{UA + 2F_j \rho_j c_{pj}}$$

For a proportional controller

$$F_j = F_{jss} + K_c(T - T_{set})$$

where K_c is the proportional gain of the controller and the temperature difference term $(T - T_{set})$, represents the error between the reactor temperature T and controller set point T_{set}. Note that in this conventional negative feedback system, when the reactor temperature, T, is below the setpoint temperature, T_{set}, the coolant flow is decreased, in order to reduce the rate of heat loss from the reactor to the jacket.

Note also that the incorporation of the controller equation and parameter value, K_c, into the dynamic model also alters the stability parameters for the system and thus can also change the resultant system stability characteristics. A stable system may be made to oscillate by the use of high values of K_c or by the use of a positive feedback action, obtained with the use of the controller equation with K_c negative. Thus the reactor acts to provide greater degrees of cooling when $T<T_{set}$, and conversely reduced degrees of cooling when $T>T_{set}$. This phenomenon is shown particularly in simulation example OSCIL.

This example shows that the reactor may oscillate either naturally according to the system parameters or by applied controller action. Owing to the highly non-linear behaviour of the system, it is sometimes found that the net yield from the reactor may be higher under oscillatory conditions than at steady state (see simulation examples OSCIL and COOL). It should be noted also that, under controlled conditions, T_{set} need not necessarily be set equal to the steady-state value T and T_{set}, and that the control action may be used to force the reactor to a more favourable yield condition than that simply determined by steady-state balance considerations.

The proportional and integral controller equation

$$F_j = F_{jss} + K_c \varepsilon + \frac{K_c}{\tau_I} \int_0^t \varepsilon \, dt$$

where τ_I is the integral time constant and ε is the error = $(T - T_{set})$ or $(C_A - C_{Aset})$ can similarly be incorporated into the reactor simulation model.

Luyben (1973) (see simulation example RELUY) also demonstrates a reactor simulation including the separate effects of the measuring element, measurement transmitter, pneumatic controller and valve characteristics which may in some circumstances be preferable to the use of an overall controller gain term.

3.2.9
Chemical Reactor Safety

A necessary condition for chemical production is the safe operation of chemical reactors and other unit processes (Grewer, 1994). This is best achieved by the design of inherently safe processes, i.e. processes with negligible potential for the release of harmful chemicals. This is often impossible owing to the requirement for active substrates or because of the inherently high reactivity of certain products. The next possible improvement may be made by minimizing the amount of material held under critical conditions or by maintaining process conditions, e.g. pressure and temperature, in the ranges of non-critical conditions. Modeling and simulation provide modern tools for the support of risk analysis and are of major importance for the design of safe chemical reactors and safe operation of chemical reactors (Stoessel, 1995). In the following, safety aspects of batch reactor design and operation are discussed.

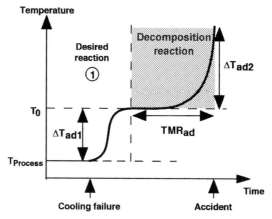

Fig. 3.18 Scenario of cooling failure with thermal runaway. ΔT_{ad1} is the adiabatic temperature rise by desired reaction. ΔT_{ad2} is the adiabatic temperature increase by the decomposition reaction. The time required for this increase is TMR_{ad}.

3.2.9.1 The Runaway Scenario

A most critical event for an exothermic chemical reaction is that of a cooling failure. Depending on the reactive potential of the process the temperature of the reactor will rise and may thus possibly trigger the formation of unwanted decomposition reactions with often very large potential for possible reactor failure (Fig. 3.18). The thermal runaway of a polymerisation reaction is described in the simulation example RUN. The reactor operation, i.e. feeding of reactants, the selection of process temperature etc., has to be such that a cooling failure will not cause a reaction runaway or that the time-to-maximum rate criterion, TMR_{ad}, is sufficiently long, e.g. 24 hours, to permit the appropriate safety actions to be taken. See example SULFONATION where substrate feed is adjusted such as to keep TMR_{ad} always above 24 hours. A proper safety assessment should start with calorimetric measurements to allow the estimation of $\Delta T_{ad,2}$, the adiabatic temperature rise by decomposition reaction, and also of the time required for this, TMR_{ad}. This study is preferably achieved via the use of differential scanning calorimetry, DSC.

Other simulation examples involving various safety aspects are HMT, THERM, REFRIG1, REFRIG2 and DSC.

3.2.9.2 Reaction Calorimetry

In differential scanning calorimetry, the selected chemical reaction is carried out in a crucible and the temperature difference ΔT compared to that of an empty crucible is measured. The temperature is increased by heating and from the measured ΔT the heat production rate, q, can be calculated (Fig. 3.19). Integration of the value of q with respect to time yields measures of the total heats

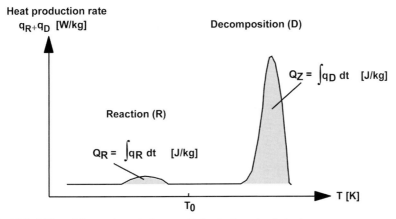

Fig. 3.19 Differential scanning calorimetry with heat release by desired reaction, R and by decomposition reaction D.

released, Q_R, and Q_D, from which ΔH_R and ΔH_D as well as ΔT_{ad1} and ΔT_{ad2} can be calculated from Fig. 3.18. The estimation of the activation energy, E_a, is possible by the use of isothermal DSC based on the following equation

$$q = k_0 f(c)(-\Delta H_R) \exp\left(-\frac{E_a}{RT}\right)$$

or in its logarithmic form

$$\ln q = \ln[k_0 f(c)(-\Delta H_R)] - \frac{E_a}{RT}$$

by plotting ln q versus 1/T at constant conversion. The preexponential factor, k_0, can be determined knowing f(c), that means conversion and kinetics have to be known. These data permit then a computer simulation of a reactor runaway scenario. Keller et al. (1997) employed a series of chemical kinetic models to simulate the DSC experimental technique suitable for preliminary safety assessment and as an aid to preliminary screening (see simulation example DSC).

3.2.10
Process Development in the Fine Chemical Industry

As many other industries, the fine chemical industry is characterized by strong pressures to decrease the time-to-market. New methods for the early screening of chemical reaction kinetics are needed (Heinzle and Hungerbühler, 1997). Based on the data elaborated, the digital simulation of the chemical reactors is possible. The design of optimal feeding profiles to maximize predefined profit functions and the related assessment of critical reactor behavior is thus possible, as seen in the simulation examples RUN and SELCONT.

3.2.11
Chemical Reactor Waste Minimisation

The optimal design and operation of chemical reactors lies at the highest point of the waste minimisation hierarchy, in facilitating a reduction in the production of waste directly at source though eventually the behaviour of a production process as a whole is decisive (Heinzle et al., 2006). Ullmann's Encyclopedia (1995) describes many significant industrial process improvements that have resulted from direct waste reduction at source initiatives. These include the development of new process routes, equilibrium shifts to improve productivities, improvements in selectivity, new catalyst developments, process optimisation changes, reaction material changes, raw material purity changes, the employment of less harmful process materials and improved recycle and reuse of waste residues. In carrying out such new process initiatives, the combination of chemical reactor design theory and process chemistry is one that provides considerable opportunities for the exercise of flair and imagination. The chemical reactor is now recognised as the single most important key item in the chemical process and especially in regard to waste minimisation, where the emphasis is being applied increasingly to the maximum utilisation of reactants, maximum production of useful products and minimisation of wasteful by-products. In this, it is important to choose the optimum process chemistry, the best chemical reaction conditions and the most appropriate form and mode of operation for the actual chemical reactor. The importance of the reactor, in this respect, lies in the dependence of chemical reaction rates on reactant concentration and temperature. Thus the manipulation of concentration and temperature levels within the reactor provides an opportunity to directly manipulate relative reaction rates and to force reactions towards higher degrees of conversion and towards higher selectivities for the desired products as compared to byproduct wastes.

The above principles have been recognised since the very beginnings of chemical reactor design theory and are well treated in conventional reactor design textbooks. The topic is also given special emphasis with respect to the minimisation of waste by Smith (1995). The importance of the very early consideration of waste minimisation in an integrated process development strategy has also been highlighted by Heinzle and Hungerbühler (1997). In this latter respect, digital simulation forms a very important adjunct in enabling a detailed study of many prospective alternatives at the earliest design stage. An essential part of this process is the application of modelling and simulation techniques, particularly using models that are formulated as simply as possible with minimum complexity; a topic stressed extensively throughout this book. Important aspects of waste minimisation, as allied to chemical reactor design and operation, are illustrated in many of the accompanying digital simulation examples as follows: BATSEQ involving optimum reaction time; BATCOM, HYDROL, RELUY, BENZHYD, ANHYD, REVREACT and REVTEMP having optimal temperature or optimum temperature profiling strategies; CASCSEQ, SEMIPAR and SEMISEQ with optimal feed distribution policies; REXT featuring increased conversion via in situ product removal and BATSEQ, SEMISEQ and COMPSEQ describing feed segregation effects.

In all these examples, the relevant waste minimisation aspects can be usefully studied simply in terms of the relative distribution of reaction compounds produced or by an extension of the programs to include possible quantification. A useful method of comparison is the "environmental index" as defined by Sheldon (1994), as the mass ratio of the total waste to that of useful product He showed that this index is very dependent both on the scale and chemical complexity of the process, with values ranging from about 0.1 for oil refining applications, up to 50 for fine chemical production and up to 100 plus for pharmaceutical operations. It thus reflects both the smaller scale of production, the greater usage of batch and semi batch operations and the greater complexity of the chemical processes. Sheldon also suggested the use of an environmental quotient, defined as $E_Q = E\ Q$, where Q is an arbitrary quotient related to the degree of environmental damage, with Q varying from 1 for simple compounds and up to 1000 for very destructive compounds.

Heinzle et al. (1998), Koller et al. (1998), Biwer and Heinzle (2004) and Heinzle et al. (2007) have described a simple but slightly more elaborate methodology to calculate such an environmental quotient based on more detailed material balances. Alternatively in terms of the present simulation examples, the analysis may perhaps be extended simply by means of incorporating a simple weighting or cost factor for each material to account for positive sales value or negative cost values representing the potential damage to the environment, as suggested in simulation example BATSEQ. A full analysis of any reactor or process simulation must of course take full account of all the relevent aspects, and Heinzle et al. (1998) and Koller et al. (1998) describe detailed studies as to how this might best be achieved.

Simulation Considerations

The waste minimisation reaction related examples in this text are represented mainly by combinations of consecutive and parallel type reactions. Although the major details of such problems are dealt with in conventional textbooks, it may be useful to consider the main aspects of such problems from the viewpoint of solution by digital simulation.

Consider the following first order consecutive reaction sequence

$$A \xrightarrow{k_1} B \xrightarrow{k_2} C$$

Solving the kinetic equations clearly demonstrates that the concentration of B passes through a maximum in respect to reaction time. If B is the desired product and C is waste, an optimal time t_{opt} can be defined for the maximum concentration of B, and where both the optimal yield and optimum reaction time are functions of the kinetic rate constants k_1 and k_2. In waste minimisation terms, however, the quantity of B obtained both in relation to the unreacted A and waste product C is important, since these may represent quite distinct separation problems and may also have quite distinct associated environmental loadings. In general, one wants the rate of decomposition of A to B to be high relative to the rate of decomposition of B to C. Since these rates are also temperature dependent, a favourable product distribution can also be effected by varying the reaction temperature.

Now consider the following parallel reaction where A and B are reactants, P is useful product and Q is by-product waste.

$$A + B \xrightarrow{k_1} P$$

$$A + B \xrightarrow{k_2} Q$$

It is obviously important to achieve complete reaction for A and B and high selectivity for the formation of P with respect to Q.

The individual rates of reaction may be given by

$$r_P = k_1 C_A^{nA1} C_B^{nB1}$$

and

$$r_Q = k_2 C_A^{nA2} C_B^{nB2}$$

where n_{A1} and n_{A2} and n_{B1} and n_{B2} are the respective reaction for reactions 1 and 2.

The relative selectivity for the reaction is given by

$$\frac{r_P}{r_Q} = \frac{k_1}{k_2} C_A^{(nA1-nA2)} C_B^{(nB1-nB2)}$$

Thus if $n_{A1} > n_{A2}$, high selectivity for P is favoured by maintaining the concentration of A high. Conversely if $n_{A1} < n_{A2}$, high selectivity is favoured by low concentration of B.

With the orders of reaction being equal in both reactions

$$\frac{r_P}{r_Q} = \frac{k_1}{k_2}$$

where k_1 and k_2 are both functions of temperature, again showing that a favourable selectivity can be obtained by appropriate adjustment of the reactor temperature.

3.2 Stirred-Tank Reactors

These and other waste minimisation considerations can be explored more fully both by reference to conventional texts and by simulation.

3.2.12
Non-Ideal Flow

The previous analysis of stirred-tank reactors has all been expressed in terms of the idealised concept of perfect mixing. In actual reactors, the mixing may be far from perfect and for continuous flow reactors may lie somewhere between the two idealised instances of perfect mixing and perfect plug flow, and may even include dead zones or short-cut flow. The concept of ideal plug flow is usually considered in terms of continuous tubular or continuous column type devices and the application of this to continuous flow reactors is discussed in Section 4.3.1. In ideal plug-flow reactors, all elements of fluid spend an identical period of time within the reactor, thus giving a zero distribution of residence times. All elements of fluid thus undergo an equal extent of reaction, and temperatures, concentrations and fluid velocities are completely uniform across any flow cross section. For a well-mixed, continuous stirred-tank reactor, however, there will be elements of fluid with, theoretically, a whole range of residence times varying from zero to infinity. Practical reactors will normally exhibit a distribution of residence times, which lie somewhere between these two extreme conditions and which effectively determine the performance of the reactor. The form of the residence time distribution curve can therefore be used to characterise the nature of the flow in the reactor, and models of the mixing behaviour can be important in simulations of reactor performance. Often different combinations of tanks in series or in parallel can be used to represent combinations of differing mixed flow regions and provide a powerful tool in the analysis of actual reactor behaviour.

Using Tracer Information in Reactor Design

Residence time distributions can be determined in practice by injecting a non-reactive tracer material into the input flow to the reactor and measuring the output response characteristics in a similar manner to that described previously in Section 2.1.1.

Unfortunately RTD studies cannot distinguish between early mixing and late mixing sequences of different types. Whereas the mixing history does not influence a first-order reaction, other reaction types are affected; the more complex the reaction kinetics the more the reaction selectivity or product distribution generally will be influenced. Thus for certain kinetic cases a detailed knowledge of the mixing can be important to the reactor performance. In practice this information may be difficult to obtain. In principle, tracer injection and sampling at different points in the reactor can supply the needed information. In practice, the usual procedure is to develop a model based on RTD experiments and the modeller's intuition. A comparison of the actual reactor performance with the

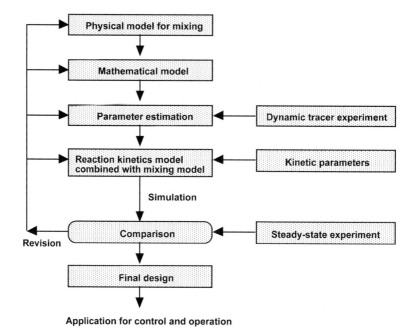

Fig. 3.20 Reactor design procedure with reactors having residence time distributions deviating from those of ideal reactors.

model predictions for known reaction kinetics reveals then whether the model assumptions were correct. The procedure is outlined above in Fig. 3.20.

Simulation examples demonstrating non-ideal mixing phenomenon in tank reactors are CSTRPULSE, NOCSTR and TUBEMIX. Other more general examples demonstrating rank-based residence time distributions are MIXFLO1, MIXFLO2, GASLIQ1, GASLIQ2 and SPBEDRTD.

3.2.13
Tank-Type Biological Reactors

Fermentation systems obey the same fundamental mass and energy balance relationships as do chemical reaction systems, but special difficulties arise in biological reactor modelling, owing to uncertainties in the kinetic rate expression and the reaction stoichiometry. In what follows, material balance equations are derived for the total mass, the mass of substrate and the cell mass for the case of the stirred tank bioreactor system (Dunn et al., 2003).

As indicated below in Fig. 3.21, feed enters the reactor at a volumetric flow rate F_0, with cell concentration X_0 and substrate concentration S_0. The vessel contents, which are well-mixed, are defined by volume V, substrate concentration S_1 and cell concentration X_1. These concentrations are identical to those of the outlet stream, which has a volumetric flow rate F_1.

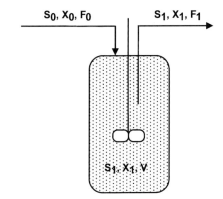

Fig. 3.21 Tank fermenter variables.

As shown previously, the general balance form can be derived by setting:

(Rate of accumulation) = (Input rate) − (Output rate) + (Production rate)

and can be applied to the whole volume of the tank contents.
Expressing the balance in equation form gives:

Total mass balance:
$$\frac{d(V\rho)}{dt} = \rho(F_0 - F_1)$$

Substrate balance:
$$\frac{d(VS_1)}{dt} = F_0 S_0 - F_1 S_1 + r_S V$$

Organism balance:
$$\frac{d(VX_1)}{dt} = F_0 X_0 - F_1 X_1 + r_X V$$

where the units are: V (m³), ρ (kg/m³), F (m³/s), S (kg/m³), X (kg/m³) with r_S and r_X (kg/m³ s).

The rate expressions can be simply given by the Monod Equation:

$$r_X = \mu X_1$$

and

$$\mu = \frac{\mu_m S_1}{K_S + S_1}$$

using a constant yield coefficient

$$r_S = \frac{r_X}{Y_{X/S}}$$

but other forms of rate equation may equally apply.

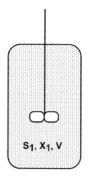

Fig. 3.22 The batch fermenter and variables.

The above generalised forms of equations can be simplified to fit particular cases of bioreactor operation.

3.2.13.1 The Batch Fermenter

Starting from an inoculum (X at t=0) and an initial quantity of limiting substrate, S at t=0, the biomass will grow, perhaps after a short lag phase, and will consume substrate. As the substrate becomes exhausted, the growth rate will slow and become zero when substrate is completely depleted. The above general balances can be applied to describe the particular case of a batch fermentation (constant volume and zero feed). Thus,

Total balance:
$$\frac{dV}{dt} = 0$$

Substrate balance:
$$V\frac{dS_1}{dt} = r_S V$$

Organism balance:
$$V\frac{dX_1}{dt} = r_X V$$

Suitable rate expressions for r_S and r_X and the specification of the initial conditions would complete the batch fermenter model, which describes the exponential and limiting growth phases but not the lag phase.

3.2.13.2 The Chemostat

The term chemostat refers to a tank fermentation which is operated continuously. This bioreactor mode of operation normally involves sterile feed ($X_0=0$), constant volume and steady state conditions, meaning that $dV/dt=0$, $d(VS_1)/dt=0$, $d(VX_1)/dt=0$.

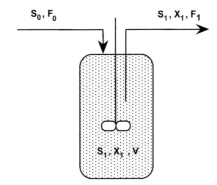

Fig. 3.23 The chemostat and its variables.

For constant density the total mass balance simplifies to

$$0 = F_0 - F_1$$

which means that the flow rates in and out of the bioreactor must be equal. The dynamic component balance equations are then

Substrate balance:
$$V\frac{dS_1}{dt} = F(S_0 - S_1) + r_S V$$

Cell balance:
$$V\frac{dX_1}{dt} = -FX_1 + r_X V$$

where F is the volumetric flow through the system. At steady state, $dS_1/dt = 0$ and $dX_1/dt = 0$.
Hence for the substrate balance:

$$0 = F(S_0 - S_1) + r_S V$$

and for the cell balance:
$$0 = -FX_1 + r_X V$$

Inserting the Monod-type rate expressions gives:

For the cell balance

$$\frac{FX_1}{V} = r_X = \mu X_1$$

or simply

$$\mu = \frac{F}{V} = D$$

Here D is the dilution rate and is equal to $1/\tau$, where $\tau = V/F$ and is equal to the mean residence time of the tank.

For the substrate balance:

$$F(S_0 - S_1) = \frac{r_X}{Y_{X/S}} V$$

from which

$$X_1 = Y_{X/S}(S_0 - S_1)$$

Thus, the specific growth rate in a chemostat is controlled by the feed flow rate, since μ is equal to D at steady state conditions. Since μ, the specific growth rate, is a function of the substrate concentration, and since μ is also determined by dilution rate, then the flow rate F also determines the outlet substrate concentration S_1. The last equation is, of course, simply a statement that the quantity of cells produced is proportional to the quantity of substrate consumed, as related by the yield factor $Y_{X/S}$.

If the flow rate F is increased, D will also increase, which causes the steady state value of S_1 to increase and the corresponding value of X_1 to decrease. It can be seen by simulation that when D nears μ_m, X_1 will become zero and S_1 will rise to the inlet feed value S_0. This corresponds to a complete removal of the cells by flow out of the tank, and this phenomenon is known as "washout".

3.2.13.3 The Feed Batch Fermenter

This bioreactor mode refers to a tank fermenter operated semi-continuously. The rate of the feed flow, F_0, may be variable, and there is no outlet flow rate from the fermentor. As a consequence of feeding the reactor volume will change with respect to time.

The balance equations then become for constant density

$$\frac{dV}{dt} = F_0$$

$$\frac{d(VS_1)}{dt} = F_0 S_0 + r_S V$$

$$\frac{d(VX_1)}{dt} = r_X V$$

Here the quantities VS_1 and VX_1 represent the masses of substrate and biomass, respectively, in the reactor. In a simulation, dividing these masses by the volume V gives the concentrations S_1 and X_1 as a function of time and which are needed in the appropriate kinetic relationships to calculate r_S and r_X.

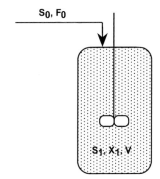

Fig. 3.24 Fed batch fermenter variables.

It can be shown by simulation that a quasi-steady state can be reached for a fed-batch fermenter, where $dX_1/dt=0$ and $\mu=F/V$ (Dunn and Mor, 1975). Since V increases, μ must therefore decrease, and thus the reactor moves through a series of changing steady states for which $\mu=D$, during which S_1 and μ decrease, and X_1 remains constant. A detailed analysis of fed batch operation has been made by Keller and Dunn (1978).

All three bioreactor modes described above can be simulated using the example BIOREACT.

3.3
Stagewise Mass Transfer

3.3.1
Liquid-Liquid Extraction

Liquid-liquid extraction is an important chemical engineering separation process, and a knowledge of the process dynamics is important since many solvent extraction operations are still carried out batchwise. In addition, although most continuous solvent extraction plants are still designed on a steady-state basis, there is an increasing awareness of the need to assess possible safety and environmental risks at the earliest possible design stage. For this, a knowledge of the probable dynamic behaviour of the process becomes increasingly important. This applies, especially, in the fields of nuclear reprocessing and heavy metal extraction.

The modelling of solvent extraction is also of interest, since the dynamic behaviour of both liquid phases can be important, and because of the wide range of equipment types that can be employed and the wide range of dynamic behaviour that results. The equipment is typified by mixer–settlers at one extreme, often representing high capacity, stagewise contacting devices, in which near equilibrium conditions are achieved with slow, stable, but long-lasting dynamic characteristics. At the other extreme are differential column devices, representing high throughput, low volume, non-equilibrium, differential contacting devices, with fast-acting dynamic behaviour. These columns have a limited range of permissible operating conditions and often an inherent lack of stability, especially when plant conditions are changing rapidly. Truly differential column devices are considered in Chapter 4, but some types of extraction columns can be regarded basically as stagewise in character, since the modelling of the dynamic characteristics of this type of device leads quite naturally from the equilibrium stagewise approach.

The treatment is confined to the use of two completely immiscible liquid phases, the feed or aqueous phase and the solvent or organic phase. No attempt is made to apply the modelling methodology to the case of partially miscible systems. Although one of the phases, the dispersed phase, will be in the form of droplets, dispersed in a continuum of the other, this is simplified by assuming each liquid phase to consist of separate well-mixed stage volumes. The modelling approach, shown in this chapter, follows the general modelling methodology in that it starts with the simplest case of a single component, batch extraction and then builds in further complexities. Finally a complex model of a non-ideal flow in a multistage, multicomponent, extraction cascade, which includes a consideration of both hydrodynamic effects and control, is achieved.

3.3.1.1 Single Batch Extraction

Volumes V_L and V_G of the two immiscible liquid phases are added to the extraction vessel and a single solute distributes itself between the phases as concentrations X and Y, respectively, at a rate, Q, as shown in Fig. 3.25.

For batch extraction, with no feed into the system, the component balances on each phase are given by:

$$\begin{pmatrix} \text{Rate of accumulation} \\ \text{of solute in} \\ \text{the given phase} \end{pmatrix} = \pm \begin{pmatrix} \text{Effective rate of} \\ \text{mass transfer to} \\ \text{or from the phase} \end{pmatrix}$$

Neglecting the effects of concentration changes on solvent density, the phase volumes will remain constant. Thus for the liquid phase with volume V_L

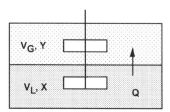

Fig. 3.25 Single-solute batch extraction between immiscible liquid phases.

$$V_L \frac{dX}{dt} = -Q$$

and for liquid phase with volume V_G

$$V_G \frac{dY}{dt} = +Q$$

where Q is the rate of solute transfer with units (mol/s) or (kg/s)

$$Q = K_L a (X - X^*) V$$

K_L is the mass transfer coefficient for the L phase (m/s), a is the interfacial area per unit volume (m²/m³), referred to the total liquid volume of the extractor, V is the total holdup volume of the tank, and is equal to $(V_L + V_G)$. X^* is the equilibrium concentration, corresponding to concentration Y, given by

$$X^* = f_{eq}(Y)$$

as illustrated in Fig. 3.26.

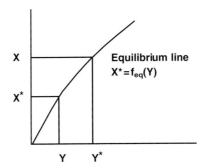

Fig. 3.26 Equilibrium relationship between two liquid phases.

The information flow diagram (Fig. 3.27) for this system shows the two component material balance relations to be linked by the equilibrium and transfer rate relationships.

Note that the transfer rate equation is based on an overall concentration driving force $(X - X^*)$ and overall mass transfer coefficient K_L. The two-film theory for interfacial mass transfer shows that the overall mass transfer coefficient, K_L, based on the L-phase is related to the individual film coefficients for the L and G-phase films, k_L and k_G, by the relationship

$$\frac{1}{K_L} = \frac{1}{k_L} + \frac{m}{k_G}$$

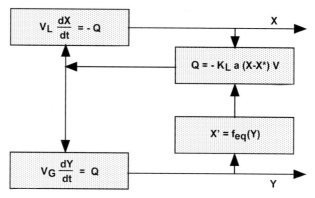

Fig. 3.27 Information flow diagram for simple batch extraction.

where m is the slope of the equilibrium curve and

$$\frac{dY^*}{dX} = m$$

For a linear equilibrium curve with constant film coefficients, k_L and k_G, the overall coefficient, K_L, will also be constant, but for the case of a non-linear equilibrium relationship, the value of m, which is the local slope of the equilibrium curve, will vary with solute concentration. The result is that the overall coefficient, K_L, will also vary with concentration, and therefore in modelling the case of a non-linear equilibrium extraction, further functional relationships relating the mass transfer coefficient to concentration will be required, such that

$$K_L = f(X)$$

3.3.1.2 Multisolute Batch Extraction

Two solutes distribute themselves between the two phases as concentrations X_A and Y_A, and X_B and Y_B and with rates Q_A and Q_B, respectively, as shown in Fig. 3.28. The corresponding equilibrium concentrations X_A^* and X_B^* are functions of both the interacting solute concentrations, Y_A and Y_B, and can be expressed by functional relationships of the form

$$X_A^* = f_{Aeq}(Y_A, Y_B)$$

$$X_B^* = f_{Beq}(Y_A, Y_B)$$

Typical representations of the way that the two differing equilibrium relationships can interact are shown in Fig. 3.29, and it is assumed that the equilibria can be correlated by appropriate, explicit equation forms.

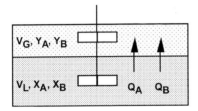

Fig. 3.28 Two-solute batch extraction.

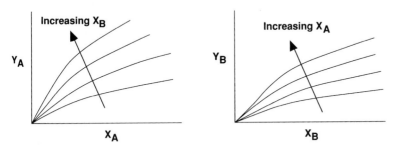

Fig. 3.29 Interacting solute equilibria for the two solutes A and B.

For multi-component systems, it is necessary to write the dynamic equation for each phase and for each solute, in turn. Thus, for phase volume V_L, the balances for solute A and for solute B are

$$V_G \frac{dY_A}{dt} = K_{LA} a (X_A - X_A^*) V_L$$

$$V_G \frac{dY_B}{dt} = K_{LB} a (X_B - X_B^*) V_L$$

The overall mass transfer coefficients are also likely to vary with concentration, owing to the complex multisolute equilibria, such that

$$K_{LA} = f_A(X_A, X_B)$$

$$K_{LB} = f_B(X_A, X_B)$$

Again, these functional relationships should ideally be available in an explicit form in order to ease the numerical method of solution. Two-solute batch extraction is covered in the simulation example TWOEX.

3.3.1.3 Continuous Equilibrium Stage Extraction

Here the extraction is carried out continuously in a single, perfectly mixed, extraction stage as shown in Fig. 3.30. It is assumed that the outlet flow concen-

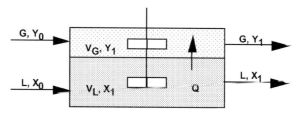

Fig. 3.30 Continuous equilibrium stage extraction.

trations, X_1 and Y_1, achieve equilibrium and that density variations are negligible.

Following an initial transient, the extractor will achieve a steady state operating condition, in which the outlet concentrations remain constant with respect to time.

At steady state, the quantity of solute entering the extractor is equal to the quantity of solute leaving. A steady-state balance for the combined two-phase system gives

$$LX_0 + GY_0 = LX_1 + GY_1$$

where for an equilibrium stage extraction

$$X_1 = f_{eq}(Y_1)$$

Here, L and G are the volumetric flow rates of the heavy and the light phases, respectively, X_0 and Y_0 are the respective inlet solute concentrations of the two phases, X_1 and Y_1 are the respective outlet solute concentrations.

For a linear equilibrium relationship

$$Y_1 = mX_1$$

A simple substitution of the value Y_1 in the balance equation enables the steady-state concentration X_1 to be determined, where

$$X_1 = \frac{(LX_0 + GY_0)}{(L + mG)}$$

The steady-state approach, however, provides no information on the initial transient conditions, whereby the extractor achieves eventual steady state or on its dynamic response to disturbances.

The eventual steady state solution may often be also very difficult to calculate for cases in which the equilibrium is non-linear or where complex interacting equilibria for multicomponent mixtures are involved. In such instances, we have found a dynamic solution to provide a very simple means of solution.

3.3 Stagewise Mass Transfer

The dynamic component balance equations for each of the two phases in turn.

$$\begin{pmatrix} \text{Rate of} \\ \text{accumulation} \\ \text{of solute} \end{pmatrix} = \begin{pmatrix} \text{Rate of} \\ \text{flow in} \\ \text{of solute} \end{pmatrix} - \begin{pmatrix} \text{Rate of} \\ \text{flow out} \\ \text{of solute} \end{pmatrix} \pm \begin{pmatrix} \text{Rate} \\ \text{of solute} \\ \text{transfer} \end{pmatrix}$$

The sign of the transfer term will depend on the direction of mass transfer. Assuming solute transfer again to proceed in the direction from volume V_L to volume V_G, the component material balance equations become for volume V_L

$$V_L \frac{dX_1}{dt} = LX_0 - LX_1 - Q$$

and for volume V_G

$$V_G \frac{dY_1}{dt} = GY_0 - GY_1 + Q$$

where

$$Q = K_L a (X_1 - X_1^*) V_L$$

For an equilibrium stage, the outlet concentrations leaving the stage are in equilibrium, i.e.

$$X_1^* = f_{eq}(Y_1)$$

Here an arbitrarily high value for the mass transfer coefficient K_L can be used to force a close approach to equilibrium. Thus for a finite value of the rate of transfer Q, the driving force will be very small, and hence the value of X_n^* is forced to be very close to X_n. The final near equilibrium condition is thus achieved as a result of the natural cause and effect in which equilibrium is favoured by a high mass transfer coefficient, which is used here simply as a high gain forcing factor, in the manner originally suggested by Franks (1972). Using this technique, some additional problems in solution may be experienced due to stiffness caused by a too high value for K_L, but using the fast numerical integration routines of MADONNA, such difficulties become rather minimal.

This dynamic approach to equilibrium method is used in later examples to illustrate its further application to the solution of complex steady state problems.

Continuous single-stage extraction is treated in the simulation example EQEX. Chemical reaction with integrated single stage extraction is demonstrated in the simulation example REXT.

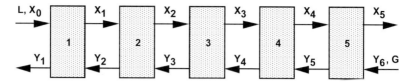

Fig. 3.31 Multistage countercurrent extraction cascade.

3.3.1.4 Multistage Countercurrent Extraction Cascade

For a high degree of extraction efficiency, it is usual to connect several continuous flow stages to form a countercurrent flow extraction cascade, as indicated in Fig. 3.31.

In each stage, it is assumed that the two phases occupy well-mixed, constant volumes V_L and V_G. The phase volumes V_L and V_G can, however, vary from stage to stage along the cascade. This effect is easily included into any simulation program. Additional complexity, in the formulation of the model, is now provided by the requirement of having to write balance equations for each of the stages of the cascade. The total number of equations to be solved is thus increased, but the modelling procedure remains straightforward.

For any given stage, n, the component material balance equations for each phase are thus defined by

$$V_{Ln}\frac{dX_n}{dt} = L(X_{n-1} - X_n) - Q_n$$

$$V_{Gn}\frac{dY_n}{dt} = G(Y_{n+1} - Y_n) + Q_n$$

where

$$Q_n = K_{Ln}A_n(X_n - X_n^*)V_n$$

and

$$X_n^* = f_{eq}(Y_n)$$

as shown in Fig. 3.32.

Note that the rate of transfer is defined by the local concentrations X_n and X_n^* appropriate to the particular stage, n. It is straightforward in the formulation of the model to allow for variations of the parameter values V_L, V_G, K_L and a from stage to stage and for both K_L and a to vary with respect to the local concentration. In order to do this, it is necessary to define new constant values for V_L and V_G for each stage and to have functional relationships, relating the mass transfer capacity coefficient to stage concentration. In order to model equilibrium stage behaviour, actual values of the mass transfer capacity product term $(K_{Ln}\,a_n)$ would be again replaced by an arbitrary high value of the gain coefficient, K_{Ln}, to force actual stage concentrations close to the equilibrium. Multi-

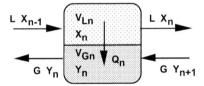

Fig. 3.32 Flow and composition inputs to stage n of the cascade.

stage extraction is treated in the simulation example EQMULTI, which permits calculation of the dynamics and steady state for situations in which the numbers of stages, flow rate and mass transfer conditions can all become variables in the simulation solution.

3.3.1.5 Countercurrent Extraction Cascade with Backmixing

The extension of the modelling approach to allow for backmixing between stages, cascades with side streams, or multiple feeds is also accomplished, relatively easily, by an appropriate modification of the inflow and outflow terms in the component balance equations (Ingham and Dunn, 1974). Backmixing reduces the efficiency of countercurrent mass transfer cascades, owing to its effect on the concentration profiles within the cascade and in decreasing the effective concentration driving forces. The effects of backmixing are especially severe in the case of solvent extraction columns. The stagewise model with backmixing is a well-known model representation, but the analytical solution is normally mathematically very complex and analytical solutions, both for steady-state and unsteady-state operating conditions, only apply for single-solute extraction where parameter values remain constant and furthermore where a linear equilibrium relationship applies. Compared to this, the solution of the dynamic model equations by digital simulation using MADONNA is far more general, since this has the ability to encompass varying parameter values, non-linear equilibria-multisolute systems plus a variable number of stages.

A multistage extraction cascade with backmixing is shown in Fig. 3.33. Here the backmixing flow rates L_B and G_B act in the reverse direction to the main phase flows, between the stages and along the cascade. One important factor in the modelling process is to realise that, as a consequence of the backmixing flows, since phase volumes remain constant, then the interstage flow rates along the cascade, in the forward direction, must also be increased by the magnitude of the appropriate backmixing flow contribution. With a backmixing flow L_B in the aqueous phase, the resultant forward flow along the cascade must now be $(L+L_B)$, since the backmixing does not appear exterior to the column. Similarly with a backmixed flow G_B, the forward flow for the organic phase is also increased to $(G+G_B)$. Taking into account the changed flow rates, however, the derivation of the component balance equations follows normal procedures.

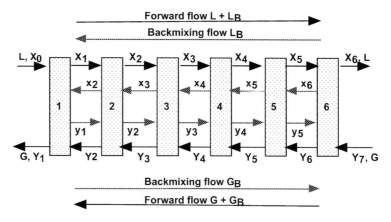

Fig. 3.33 Multistage extraction cascade with backmixing of both phases.

The relative inflow and outflow contributions for each phase of any stage n of the cascade is shown in Fig. 3.34.

Allowing for the additional backmixing flow contributions, the component balance equation for the two phases in stage n of the cascade are now

$$V_{Ln}\frac{dX_n}{dt} = (L+L_B)X_{n-1} - (L+L_B)X_n + L_B X_{n+1} - L_B X_n - Q_n$$

or

$$V_{Ln}\frac{dX_n}{dt} = (L+L_B)(X_{n-1} - X_n) + L_B(X_{n+1} - X_n) - Q_n$$

and

$$V_{Gn}\frac{dY_n}{dt} = (G+G_B)(Y_{n+1} - Y_n) + G_B(Y_{n-1} - Y_n) + Q_n$$

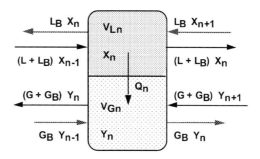

Fig. 3.34 Stage n of a multistage extraction cascade with backmixing.

3.3.1.6 Countercurrent Extraction Cascade with Slow Chemical Reaction

A countercurrent extraction cascade with reaction $A+B\rightarrow C$ is shown in Fig. 3.35. The reaction takes place between a solute A in the L-phase, which is transferred to the G-phase by the process of mass transfer, where it then reacts with a second component, B, to form an inert product, C, such that A, B and C are all present in the G-phase.

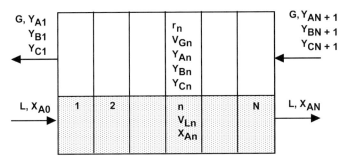

Fig. 3.35 Solute A is transferred from the L-phase to the G-phase, where it reacts with a component B, to form C.

The general component balance form of equation gives

$$\begin{pmatrix} \text{Rate of} \\ \text{accumulation} \\ \text{of solute} \\ \text{in phase i} \end{pmatrix} = \begin{pmatrix} \text{Inflow} \\ \text{rate of} \\ \text{solute} \\ \text{in phase i} \end{pmatrix} - \begin{pmatrix} \text{Outflow} \\ \text{rate of} \\ \text{solute} \\ \text{in phase i} \end{pmatrix} \pm \begin{pmatrix} \text{Rate of} \\ \text{solute} \\ \text{mass} \\ \text{transfer} \end{pmatrix} \pm \begin{pmatrix} \text{Rate of} \\ \text{solute} \\ \text{reaction} \end{pmatrix}$$

where for phase L

$$V_{Ln}\frac{dX_{An}}{dt} = L(X_{An-1} - X_{An}) - Q_{An}$$

and for phase G

$$V_{Gn}\frac{dY_{An}}{dt} = G(Y_{An+1} - Y_{An}) + Q_{An} - r_n V_{Gn}$$

Components B and C are both immiscible in phase L and remain in phase G. Therefore

$$V_{Gn}\frac{dY_{Bn}}{dt} = G(Y_{Bn+1} - Y_{Bn}) - r_n V_{Gn}$$

$$V_{Gn} \frac{dY_{Cn}}{dt} = G(Y_{Cn+1} - Y_{Cn}) + r_n V_{Gn}$$

where the transfer rate is

$$Q_{An} = K_{ALn} a_n (X_{An} - X_{An}^*) V_n$$

and the reaction rate is

$$r_n = k Y_{An} Y_{Bn}$$

Figure 3.36 shows the graphical output in the G-phase concentrations of component A with respect to time, starting the cascade at time t=0 with initially zero concentrations throughout. The maximum in the Y_{A1} profile for stage 1 is due to a delay of reactant B in reaching A. This is because A and B are fed from the opposite ends of the cascade, and thus a certain time of passage through the extractor is required before B is able to react with A in stage 1.

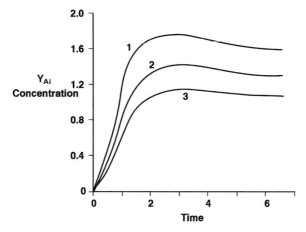

Fig. 3.36 Concentrations in the solvent phases of stages 1, 2 and 3 in a countercurrent extraction column with slow chemical reaction.

3.3.1.7 Multicomponent Systems

Assuming the liquid phases remain immiscible, the modelling approach for multicomponent systems remains the same, except that it is now necessary to write additional component balance equations for each of the solutes present, as for the multistage extraction cascade with backmixing in Section 3.2.2. Thus for component j, the component balance equations become

$$V_{Ln} \frac{dX_j}{dt} = (L + L_B)(X_{jn-1} - X_{jn}) + L_B (X_{jn+1} - X_{jn}) - Q_{jn}$$

$$V_{Gn}\frac{dY_j}{dt} = (G + G_B)(Y_{jn+1} - Y_{jn}) + G_B(Y_{jn-1} - Y_{jn}) + Q_{jn}$$

where

$$Q_{jn} = K_{Ljn}a_n(X_{jn} - X_{jn}^*)V_n$$

The additional number of differential equations and increased complexities of the equilibrium relationships may also be compounded by computational problems caused by widely differing magnitudes in the equilibrium constants for the various components. As discussed in Section 3.3.2, it is shown that this can lead to widely differing values in the equation time constants and hence to stiffness problems for the numerical solution.

3.3.1.8 Control of Extraction Cascades

A typical control problem might be the maintenance of a required raffinate outlet concentration, Y_N, with the controller action required to compensate the effect of variations in the feed concentration Y_0, as indicated in Fig. 3.37.

The proportional-integral control equation, as given in Section 2.3.2.2, is:

$$L = L_0 + K_p\varepsilon(t) + \frac{K_p}{\tau_I}\int_0^t \varepsilon(t)dt$$

where L is the manipulated variable, i.e., the solvent flow rate, L_0 is the base value for L, K_p is the proportional gain, τ_I is the integral action time and ε is the error between the actual concentration Y_N and the desired value Y_{set}. These relationships can be directly incorporated into a digital simulation program as shown in example EXTRACTCON. If required, it is also possible to include the

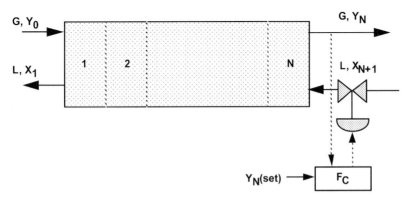

Fig. 3.37 Maintenance of raffinate outlet concentration, by regulation of solvent flow rate.

dynamic effects of the measuring elements, measurement transmitters, and the control valve characteristics into the simulation program as shown by Franks (1972) and Luyben (1990).

3.3.1.9 Mixer–Settler Extraction Cascades

The archetypal stagewise extraction device is the mixer–settler. This consists essentially of a well-mixed agitated vessel, in which the two liquid phases are mixed and brought into intimate contact to form a two-phase dispersion, which then flows into the settler for the mechanical separation of the two liquid phases by continuous decantation. The settler, in its most basic form, consists of a large empty tank, provided with weirs to allow the separated phases to discharge. The dispersion entering the settler from the mixer forms an emulsion band, from which the dispersed phase droplets coalesce into the two separate liquid phases. The mixer must adequately disperse the two phases, and the hydrodynamic conditions within the mixer are usually such that a close approach to equilibrium is obtained within the mixer. The settler therefore contributes little mass transfer function to the overall extraction device.

Ignoring the quite distinct functions and hydrodynamic conditions which exist in the actual mixer and settler items of the combined mixer–settler unit, it is possible, in principle, to treat the combined unit simply as a well-mixed equilibrium stage. This is done in exactly the way as considered previously in Sections 3.2.1 to 3.2.6. A schematic representation of an actual mixer–settler device is shown in Fig. 3.38 and an even more simplified representation of the equivalent simple well-mixed stage is given in Fig. 3.39.

A realistic description of the dynamic behaviour of an actual mixer–settler plant item should however also involve some consideration of the hydrodynamic characteristics of the separate mixer and settler compartments and the possible flow interactions between mixer and settler along the cascade.

The notation for separate mixer–settler units is shown in Fig. 3.40, for stage n of the cascade.

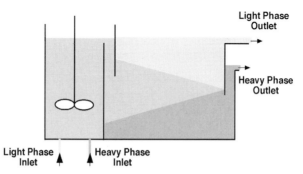

Fig. 3.38 Schematic representation of a mixer–settler unit.

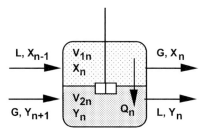

Fig. 3.39 The well-mixed stage representation of a mixer–settler unit.

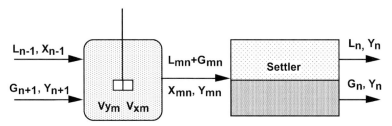

Fig. 3.40 The separate mixer and settler compartments for stage n of a mixer–settler cascade.

In this representation, the heavy phase with a flow rate, L_{n-1}, enters the mixer from the preceding stage n–1, together with solvent flow, G_{n+1}, from stage n+1. The corresponding phase flow rates, in the dispersion, leaving the mixer and entering the settler are shown as L_{mn} and G_{mn} and with concentrations X_{mn} and Y_{mn} respectively. This is to allow for possible changes in the volumetric holdup of the mixer following changes in flow rate. The modelling of the separate mixer and settler compartments follows that of Wilkinson and Ingham (1983).

Mixer Dynamics

Owing to the intensive agitation conditions and intimate phase dispersion, obtained within the mixing compartment, the mixer can usually be modelled as a single perfectly mixed stage in which the rate of mass transfer is sufficient to attain equilibrium. As derived previously in Section 3.3.1.3, the component balance equations for the mixer, based on the two combined liquid phases, is thus given by

$$\frac{d(V_{Lm}X_{mn} + V_{Gm}Y_{mn})}{dt} = L_{n-1}X_{n-1} + G_{n+1}Y_{n+1} - G_{mn}Y_{mn} - L_{mn}X_{mn}$$

where subscript m refers specifically to the conditions within the mixer and hence to the effluent flow, leaving the mixer and entering the settler.

The total material balance for the mixer is expressed by

$$\begin{pmatrix} \text{Rate of change of} \\ \text{mass in mixer} \end{pmatrix} = \begin{pmatrix} \text{Mass} \\ \text{flow in} \end{pmatrix} - \begin{pmatrix} \text{Mass} \\ \text{flow out} \end{pmatrix}$$

Neglecting the effects of any density changes, the total material balance then provides the relationship for the change of total volume in the mixer with respect to time.

$$\frac{d(V_{Lmn} + V_{Gmn})}{dt} = L_{n-1} + G_{n+1} - L_{mn} - G_{mn}$$

Under well-mixed flow conditions it is reasonable to assume that the mixer holdup volumes, V_{Lmn} and V_{Gmn}, will vary in direct proportion to the appropriate phase flow rate, and that the total liquid holdup in the mixer will vary as a function of the total flow rate to the mixer.

The total flow rate ($L_{mn} + G_{mn}$), leaving the mixer, will be related to the total phase volumes V_{Lm} and V_{Gm} by a hydrostatic equation, which will depend on the net difference in the head of liquid between the levels in the mixer and in the settler. The actual form of this relationship might need to be determined experimentally, but could, for example, follow a simple square-root relationship of the form in which flow rate is proportional to the square root of the difference in liquid head, or indeed to the total volume of liquid in the mixer, e.g.,

$$L_{mn} + G_{mn} \propto (V_{Lm} + V_{Gm})^{1/2}$$

The further assumptions are that the respective phase volumes are in direct proportion to the phase flow rate, i.e.,

$$\frac{V_{Lmn}}{V_{Gmn}} = \frac{L_{mn}}{G_{mn}}$$

and the concentrations leaving the mixer are in equilibrium according to

$$Y_{mn} = f_{eq}(X_{mn})$$

These equations complete a preliminary model for the mixer. Note that it is also possible, in principle, to incorporate changing density effects into the total material balance equation, provided additional data, relating liquid density to concentration, are available.

Settler Dynamics

The simplest settler model is that in which it is assumed that each phase flows through the settler in uniform plug flow, with no mixing and constant velocity. This has the effect that the concentrations leaving the settler, X_n and Y_n, are simply the time-delayed values of the exit mixer concentrations X_{mn} and Y_{mn}.

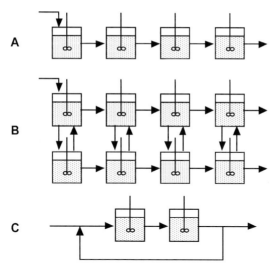

Fig. 3.41 Alternative settler-flow representations for the separate phases.

In this the magnitude of the time delay is thus simply the time required for the phase to pass through the appropriate settler volume.

The outlet phase flow rates, L_n and G_n, may again be related to the inlet settler phase flow rates, L_{mn} and G_{mn}, and settler phase volumes by hydraulic considerations, using similar formulations to those proposed for the mixer, as required.

In practice, some mixing will, however, occur in each phase of the settler, and various models involving either an arbitrary number of perfect mixing stages or various flow combinations, with and without recycle effects, can be postulated. Some of these are indicated in Fig. 3.41, where Fig. 3.41 A represents settler mixing, given by a series of stirred tanks, Fig. 3.41 B a series of well-mixed tanks interconnected to stagnant regions and Fig. 3.41 C a series of two tanks with recycle. The actual representation adopted for a given situation, would, of course, have to depend very much on the actual mechanical arrangement and flow characteristics of the particular settler design, together with actual observations of the flow behaviour.

Figure 3.42 shows one possible representation in which a proportion of each phase passes through the settler in plug flow, while the remaining proportion is well mixed.

The resultant outlet concentration from the settler is then given by the combined plug-flow and well-mixed flow streams.

The notation for the above flow model, in respect of the aqueous-phase settler volume, is shown in Fig. 3.43.

If L_{mn} is the volumetric flow rate of the heavy phase entering the settler from the mixer, and f is the fraction of flow passing through the plug-flow region with time delay t_{Dn}, then $L_{mn}f$ is the volumetric flow passing through the plug-

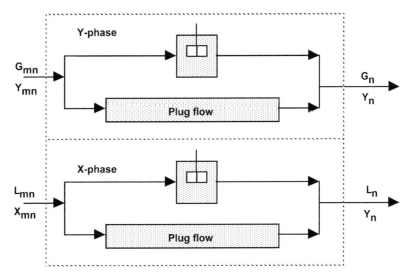

Fig. 3.42 Combined plug-flow and well-mixed settler-flow representation.

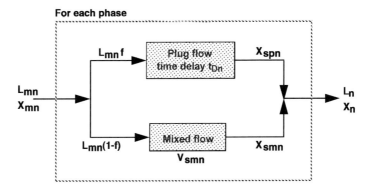

Fig. 3.43 Combined plug-flow and well-mixed flow representation for the heavy phase settler flow.

flow region. The concentration at the plug-flow region outlet, X_{spn}, is the inlet concentration at time $t-t_{Dn}$ and is given by

$$X_{spn} = X_{mn(t-t_{Dn})}$$

The fractional flow rate $L_{mn}(1-f)$ is then also the volumetric flow passing through the well-mixed region of settler phase volume, V_{mix}. The flows leaving the plug-flow and well-mixed regions X_{spn} and X_{smn}, respectively, then combine to give the actual exit concentration from the settler X_n.

The model equations for the heavy phase settler region then become for the well-mixed region

$$V_{smn} \frac{dX_{smn}}{dt} = L_{mn}(1-f)(X_{mn} - X_{smn})$$

and for the combined outlet phase flow

$$L_n X_n = L_{mn}(1-f)X_{smn} + L_{mn}f X_{spn}$$

Mixer–Settler Cascade

The individual mixer and settler model representations can then be combined into an actual countercurrent-low, multistage, extraction scheme representation as shown in Fig. 3.44. This includes an allowance for backmixing, between the stages of the cascade caused by inefficient phase disengagement in the settlers, such that a fraction f_L or f_G of the appropriate phase flow leaving the settler is entrained. This means it is actually carried back in the reverse direction along the cascade by entrainment in the other phase. It is assumed, for simplicity, that the total flows of each phase, L and G, remain constant throughout all the stages of the cascade. Entrainment fractions f_L and f_G are also assumed constant for all settlers. The above conditions, however, are not restrictive in terms of the capacity of the solution by digital simulation.

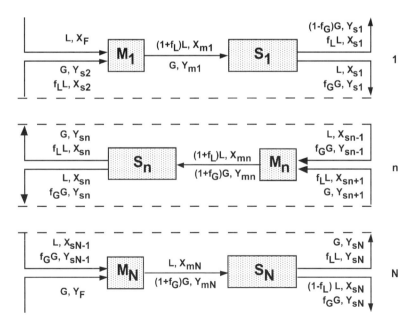

Fig. 3.44 Multistage mixer–settler cascade with entrainment backmixing. M and m refers to mixer and S and s to settler; stages are 1 to N.

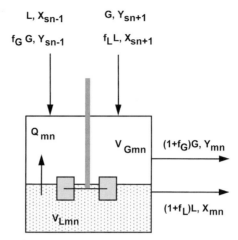

Fig. 3.45 One mixer–settler stage, n.

The conditions in the mixer of any stage n are represented in Fig. 3.45.

Allowing for the additional flow contributions due to the entrainment backmixing, the component balance equations, for any mixer n along the cascade, are now expressed by

$$V_{Lmn}\frac{dX_{mn}}{dt} = LX_{sn-1} + f_L LX_{sn+1} - (1+f_L)LX_{mn} - Q_{mn}$$

and

$$V_{Gmn}\frac{dY_{mn}}{dt} = GY_{sn+1} + f_G GY_{sn-1} - (1+f_G)GY_{mn} + Q_{mn}$$

where

$$Q_{mn} = K_L a_m (X_{mn} - X_{mn}^*) V_{Lmn}$$

$$V_{mn} = V_{Lmn} + V_{Gmn}$$

The settler equations are as shown previously, but must, of course, be applied to both phases.

Figure 3.46 shows the output obtained from a full solution of the mixer–settler model. The effect of the time delay in the settlers, as the disturbance, as propagated through the system from stage to stage, is very evident.

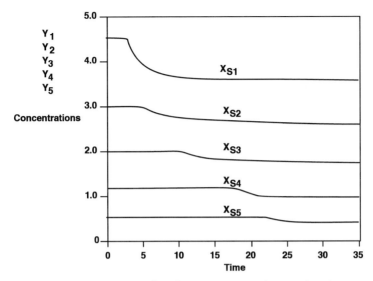

Fig. 3.46 Computer simulation output for a five-stage mixer–settler cascade with entrainment.

3.3.1.10 Staged Extraction Columns

A wide variety of extraction column forms are used in solvent extraction applications. Many of these, such as rotary-disc contactors (RDC), Oldshue–Rushton columns, and sieve-plate column extractors, have rather distinct compartments and a geometry which lends itself to an analysis of column performance in terms of a stagewise model. As the compositions of the phases do not come to equilibrium at any stage, however, the behaviour of the column is therefore basically differential in nature.

At the prevailing high levels of dispersion normally encountered in such types of extraction columns, the behaviour of these essentially differential type contactors, however, can be represented by the use of a non-equilibrium stagewise model.

The modelling approach to multistage countercurrent equilibrium extraction cascades, based on a mass transfer rate term as shown in Section 1.4, can therefore usefully be applied to such types of extractor column. The magnitude of the mass transfer capacity coefficient term, now used in the model equations, must however be a realistic value corresponding to the hydrodynamic conditions, actually existing within the column and, of course, will be substantially less than that leading to an equilibrium condition.

In Fig. 3.47 the column contactor is represented by a series of N non-equilibrium stages, each of which is of height H and volume V. The effective column height, Z, is thus given by $Z = N\,H$.

The stagewise model with backmixing is an essential component of any model representation of a stagewise extraction column. As shown in Section 3.3.1.5 the non-ideal flow behaviour is represented by the presence of the N stages

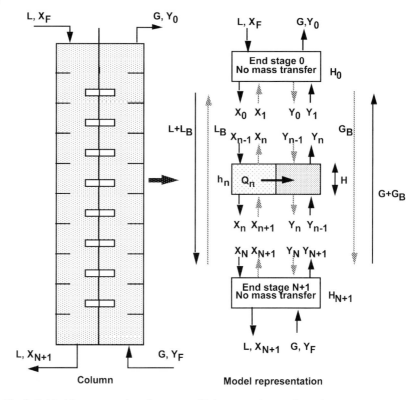

Fig. 3.47 Model representation of a non-equilibrium-staged extraction column.

in series and the constant backflow contributions, L_B and G_B, as indicated in Fig. 3.47 appropriate for each phase.

Special attention has to be given to the end compartments of an extraction column, since the phase inlet and outlet points are usually located at different points of the column. These are complicated by the presence of phase distributors and at one end by the coalescence zone for the dispersed phase droplets.

In Fig. 3.47 the end sections are represented quite simply as well-mixed zones, in which some limited degree of mass transfer may be present, but at which the mass transfer rate is much lower than in the main body of the column.

The standard equations for a stagewise extraction cascade with backmixing as developed in Section 3.3.1.5 are

$$V_{Ln}\frac{dX_n}{dt} = (L + L_B)(X_{n-1} - X_n) + L_B(X_{n+1} - X_n) - Q_n$$

$$V_{Gn}\frac{dY_n}{dt} = (G + G_B)(Y_{n+1} - Y_n) + G_B(Y_{n-1} - Y_n) + Q_n$$

where
$$Q_n = K_{Ln}a_n(X_n - X_n^*)V$$
and
$$X_n^* = f_{eq}(Y_n)$$

Here a_n is the interfacial area per unit volume.

In extraction column design the model equations are normally expressed in terms of superficial phase velocities, L′ and G′, based on unit cross-sectional area. The volume of any stage in the column is then A H, where A is the cross-sectional area of the column. Thus the volume occupied by the total dispersed phase is h A H, where h is the fractional holdup of dispersed phase, i.e., the droplet volume in the stage divided by the total volume of the stage. The volume occupied by the continuous phase in the stage is (1−h) A H.

Taking the phase flow rate G′ to represent the dispersed phase, the component balance equations now become for any stage n

$$A H_n(1 - h_n)\frac{dX_n}{dt} = (L' + L'_B)(X_{n-1} - X_n) + L'_B(X_{n+1} - X_n) - Q_n$$

$$A H_n h_n \frac{dY_n}{dt} = (G' + G'_B)(Y_{n+1} - Y_n) + G'_B(Y_{n-1} - Y_n) + Q_n$$

In the above equations K_L is the overall mass transfer coefficient (based on phase L), a is the specific interfacial area for mass transfer related to unit column volume, X and Y are the phase solute concentrations, X^* is the equilibrium concentration corresponding to concentration Y and subscript n refers to stage n of the extractor.

Normally the backmixing flow rates L_B and G_B are defined in terms of constant backmixing factors $a_L = L_B/L$ and $a_G = G_B/G$. The material balance equations then appear in the form

$$H_n(1 - h_n)\frac{dX_n}{dt} = L'(1 + a_L)X_{n-1} - L'(1 + 2a_L)X_n + a_L L' X_{n+1} - Q_n$$

$$H_n h_n \frac{dY_n}{dt} = G'(1 + a_G)Y_{n+1} - G'(1 + 2a_G)Y_n + a_G G' Y_{n-1} - Q_n$$

Considering the end regions of the column as well-mixed stages with small but finite rates of mass transfer, component balance equations can be derived for end stage 0

$$V_0(1 - h_0)\frac{dX_0}{dt} = LX_F + L_B X_1 - (L + L_B)X_0 - Q_0$$

$$V_0 h_0 \frac{dY_0}{dt} = (G + G_B)(Y_1 - Y_0) + Q_0$$

and for end stage N

$$V_{N+1}(1 - h_{N+1})\frac{dX_{N+1}}{dt} = (L + L_B)(X_N - X_{N+1}) - Q_{N+1}$$

$$V_{N+1}h_{N+1}\frac{dY_{N+1}}{dt} = GY_F + G_BY_N - (G + G_B)Y_{N+1} + Q_{N+1}$$

The correct modelling of the end sections is obviously of great importance and, depending on the geometrical arrangement, it is possible to consider the column end sections as combinations of well-mixed tanks, exterior to the actual column.

3.3.1.11 Column Hydrodynamics

Under changing flow conditions it can be important to include some consideration of the hydrodynamic changes within the column (Fig. 3.48), as manifested by changes in the fractional dispersed phase holdup h_n and the phase flow rates L_n and G_n which, under dynamic conditions, can vary from stage to stage. Such variations can have a considerable effect on the overall dynamic characteristics of an extraction column, since variations in h_n also affect the solute transfer rate terms Q_n by virtue of the corresponding variation in the specific interfacial area a_n.

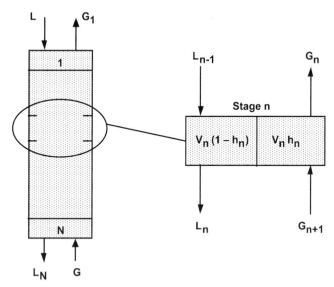

Fig. 3.48 Hydrodynamic model representation of an extraction column.

A dynamic balance for the dispersed phase holdup in stage n gives

$$V_n \frac{dh_n}{dt} = G_{n+1} - G_n$$

Since liquid phases are incompressible

$$L_{n-1} + G_{n+1} = L_n + G_n$$

and for the overall column

$$L + G = L_N + G_1$$

The fractional dispersed phase holdup h is normally correlated on the basis of a characteristic velocity equation, which is based on the concept of a slip velocity for the drops v_{slip}, which then can be related to the free rise velocity of single drops, using some correctional functional dependence on holdup f(h). The normal method of correlating dispersed phase holdup is normally of the form

$$v_{slip} = \frac{L'}{(1-h)} - \frac{G'}{h} = v_{char} f(h)$$

where v_{char} is the characteristic velocity for the dispersed phase droplets. Knowing the value of v_{char}, the value of h can be determined for any values of L' and G', using an iterative procedure.

In some cases, the characteristic velocity can cause difficulties in solution, owing to the presence of an implicit equation. In this the appropriate value of L_n or G_n satisfying the value of h_n generated by the differential material balance equation must be found by root finding algorithms increasing computation time required.

If necessary, the implicit nature of the calculation may, however, be avoided by a reformulation of the holdup relationship into an explicit form. The resulting calculation procedure then becomes much more straightforward and the variation of holdup in the column may be combined into a fuller extraction column model in which the inclusion of the hydrodynamics now provides additional flexibility. The above modelling approach to the column hydrodynamics, using an explicit form of holdup relationship, is illustrated by the simulation example HOLDUP.

3.3.2
Stagewise Absorption

Gas–liquid contacting systems can be modelled in a manner similar to liquid–liquid contactors. There are however some modelling features which are peculiar to gas–liquid systems. The single well-mixed contacting stage is shown in Fig. 3.49.

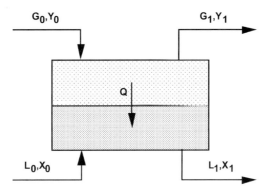

Fig. 3.49 Well-mixed gas–liquid contacting stage.

In this, G and L are the volumetric flow rates of the two phases, and X and Y are the concentrations of any component in each phase. Q is the transfer rate of the component.

For single-solute gas–liquid mass transfer the component balances are as before

$$V_L \frac{dX_1}{dt} = L_0 X_0 - L_1 X_1 + Q$$

$$V_G \frac{dY_1}{dt} = G_0 Y_0 - G_1 Y_1 - Q$$

where V_G is the volume of the well-mixed gas phase, and V_L is the volume of the well-mixed liquid phase.

In the preceding solvent extraction models, it was assumed that the phase flow rates L and G remained constant, which is consistent with a low degree of solute transfer relative to the total phase flow rate. For the case of gas absorption, normally the liquid flow is fairly constant and L_0 is approximately equal to L_1, but often the gas flow can change quite substantially, such that G_0 no longer equals G_1. For highly concentrated gas phase systems, it is therefore often preferable to define flow rates, L and G, on a solute-free mass basis and to express concentrations X and Y as mass ratio concentrations. This system of concentration units is used in the simulation example AMMONAB.

The transfer term Q is written as

$$Q = K_L a (X_1^* - X_1) V_L$$

where a is the transfer surface per volume of liquid, K_L is the overall mass transfer coefficient for phase L, V_L is the liquid phase volume and X_1^* is given by the equilibrium relation

$$X_i^* = f_{eq}(Y_1)$$

As shown for the case of extraction, a high value of $K_L a$ will result in X_1 approaching very close to the value X_i^*, and therefore the outlet concentrations of the two phases will be close to equilibrium.

Owing to the substantially greater density of liquids, as compared to gases, the volumetric flow rate of the gas is usually much greater than that of the liquid or $G \gg L$ as a general consequence

$$\frac{V_L}{L} \gg \frac{V_G}{G}$$

meaning that

$$\tau_L \gg \tau_G$$

The significance of the large difference in the relative magnitudes of the time constants, for the two phases, is that the gas concentrations will reach steady state much faster than the liquid phase.

In the component balance equations dY_1/dt will therefore be zero, whereas dX_1/dt may still be quite large. This can obviously cause considerable difficulties in the integration procedure, owing to equation stiffness.

For gas absorption this problem can often be circumvented by the assumption of a quasi-steady-state condition for the gas phase. In this, the dynamics of the gas phase are effectively neglected and the steady state, rather than the dynamic form of component balance, is used to describe the variation in gas phase concentration.

The gas phase balance then becomes for the above situation

$$0 = G_0 Y_0 - G_1 Y_1 - Q$$

Hence

$$Y_1 = \frac{G_0 Y_0 - K_L a (X_1^* - X_1) V_L}{G_1}$$

Thus Y_1 is obtained not as the result of the numerical integration of a differential equation, but as the solution of an algebraic equation, which now requires an iterative procedure to determine the equilibrium value X_1^*. The solution of algebraic balance equations in combination with an equilibrium relation has again resulted in an implicit algebraic loop. Simplification of such problems, however, is always possible, when X_1^* is simply related to Y_1, as for example

$$X_1^* = m Y_1$$

Combining the two equations then gives an explicit solution for concentration Y_1 and hence also X_1

$$Y_1 = \frac{G_0 Y_0 + K_L a X_1 V_L}{G_1 + K_L a m V_L}$$

and the implicit algebraic loop is eliminated from the solution procedure.

Assuming equilibrium conditions and a linear equilibrium relationship, where $Y_1 = m X_1$ and a quasi-steady-state conditions in the gas with $dY_1/dt = 0$ to be achieved, a component balance for the entire two phase system of Fig. 3.49 gives

$$V_L \frac{dX_1}{dt} = GY_0 + LX_0 - LX_1 - GmX_1$$

which can be expressed as

$$\frac{dX_1}{dt} = \frac{\frac{GY_0 + LX_0}{L + Gm} - X_1}{\frac{V_L}{L + Gm}}$$

This equation has the form

$$\frac{dX_1}{dt} = \frac{A - X_1}{\tau}$$

where the time constant for the system, τ, is thus shown to be dependent on the value of the equilibrium constant m. Since the value of m depends on the nature of the particular solute concerned, this has the consequence that in multicomponent applications the value of the time constant will vary according to the system component. This can cause problems of equation stiffness in the solution of the often quite large sets of simultaneous multicomponent balance equations. The importance of eliminating unnecessary stiffness, by careful consideration of the relative magnitudes of the various system time constants, thus becomes very apparent.

3.3.3
Stagewise Distillation

3.3.3.1 Simple Overhead Distillation
A simple overhead topping distillation process, without fractionation, is illustrated in Fig. 3.50.

The total material balance is given by

$$\begin{pmatrix} \text{Rate of accumulation} \\ \text{of mass in the still} \end{pmatrix} = \begin{pmatrix} \text{Rate of mass} \\ \text{input to the still} \end{pmatrix}$$

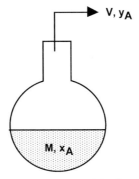

Fig. 3.50 Model representation of a simple overhead distillation.

giving

$$\frac{dM}{dt} = -V$$

where M is the total moles of liquid in the still and V is the vapour removal rate in moles/time.

For a simple binary distillation, the component balance equation becomes

$$\frac{d(Mx_A)}{dt} = -Vy_A$$

where x_A and y_A are the liquid and vapour mole fraction of component A of the liquid and vapour phases, respectively, where A is the more volatile component.

The relative volatility α is usually related to the compound having the higher boiling point, which in this case is B and hence

$$\alpha_{A/B} = \frac{y_A/x_A}{y_B/x_B}$$

Assuming that the liquid and vapour compositions in the still are in equilibrium, i.e., that the still acts as a theoretical stage

$$y_A = f_{eq}(x_A)$$

or in terms of relative volatility α

$$y_A = \frac{\alpha_A x_A}{1 + (\alpha_A - 1)x_A}$$

The combination of the two material balance equations, together with an explicit form of equilibrium relationship gives a system that is very easily solvable by direct numerical integration, as demonstrated in the simulation example BSTILL.

Extending the method to a multicomponent mixture, the total material balance remains the same, but separate component balance equations must now be written for each individual component i, giving

$$\frac{dMx_i}{dt} = -Vy_i$$

and where now the equilibrium condition is given by

$$y_i = \frac{a_i x_i}{\Sigma a_i x_i}$$

Again solution is straightforward, as illustrated in the simulation examples DIFDIST and MCSTILL.

3.3.3.2 Binary Batch Distillation

A batch distillation represents a complete dynamic process since everything, apart from the geometry of the column and the nature of the equilibrium relationship, varies with time. Owing to the removal of a distillate containing more of the volatile component, the compositions of the vapour and the liquid on all plates of the column vary with time. The total quantity of liquid in the still decreases with time, and its composition becomes successively depleted in the more volatile component. This makes the separation more and more difficult, requiring the use of higher reflux ratios to maintain a high distillate composition. The increased reflux increases the liquid flow down the column, and hence the liquid holdup on each plate. As a result of the increasing concentration of less volatile component in the still, the still temperature increases during distillation, thus reducing the rate of heat transfer to the still by reducing the temperature driving force in the reboiler and hence reducing the vapour boil-up rate. Despite this, conventional textbooks persist in analysing batch distillation in terms of quasi-steady-state graphical techniques applied at different concentration levels during the distillation process. These are also based on rather idealised and unrealistic conditions of operating a batch distillation process, i.e.:

1. Distillation at constant reflux ratio but varying top product composition.
2. Distillation at constant top product composition but varying reflux ratio.

Compared to this a solution approach based on digital simulation is much more realistic.

Consider the binary batch distillation column, represented in Fig. 3.51, and based on that of Luyben (1973, 1990). The still contains M_B moles with liquid mole fraction composition x_B. The liquid holdup on each plate n of the column is M_n with liquid composition x_n and a corresponding vapour phase composition y_n. The liquid flow from plate to plate L_n varies along the column with consequent variations in M_n. Overhead vapours are condensed in a total condenser and the condensate collected in a reflux drum with a liquid holdup volume M_D

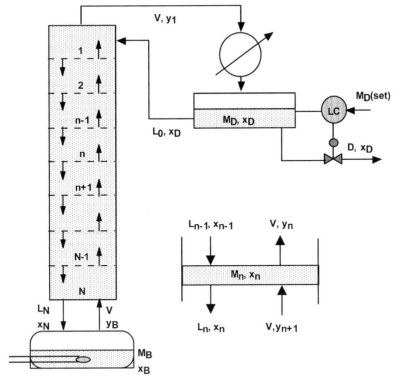

Fig. 3.51 Model representation of a batch distillation column and typical plate n as per Luyben (1973).

and liquid composition x_D. From here part of the condensate is returned to the top plate of the column as reflux at the rate L_0 and composition x_D. Product is removed from the reflux drum at a composition x_D and rate D, which is controlled by a simple proportional controller acting on the reflux drum level and is proportional to M_D.

For simplicity the following assumptions are made, although a more general model could easily be derived, in which these assumptions could be relaxed.

1. The system is ideal, with equilibrium described by a constant relative volatility, the liquid components have equal molar latent heats of evaporation and there are no heat losses or heat of mixing effects on the plates. Hence the concept of constant molar overflow (excluding dynamic effects) and the use of mole fraction compositions are allowable.
2. The liquid volumes in the still, reflux drum and on the column plates are well-mixed regions of uniform composition.
3. The dynamics of the overhead pipework and condenser are negligible.
4. The dynamics of the vapour phase in the column are much faster than that of the liquid phase and are neglected.

5. The still provides a constant vapour boil-up rate, which remains constant with respect to time.
6. The column plates have 100% plate efficiency and act as theoretical plates.

Since the vapour phase dynamics are negligible, the vapour flow rate through the column is constant from plate to plate, at the rate of V (kmol/s). The liquid flow rates L_n and the liquid holdup on the plate, however, will vary under changing hydrodynamic conditions in the column. The corresponding notation, for any plate n in the column, is as indicated in Fig. 3.51.

Total mass and component material balance equations are written for all the plates of the column, for the still and for the top reflux drum.

Here L, G and D are the molar flow rates and x and y are mol fraction compositions.

For plate n the total material balance is given by

$$\frac{dM_n}{dt} = L_{n-1} + V_{n+1} - L_n - V_n$$

Since vapour phase dynamics are neglected and "Constant Molal Overflow" conditions also apply, $V_{n+1} = V_n = V$, and

$$\frac{dM_n}{dt} = L_{n-1} - L_n$$

The component balance equation is given by

$$\frac{d(M_n x_n)}{dt} = L_{n-1} x_{n-1} - L_n x_n + V(y_{n+1} - y_n)$$

The corresponding equations for the boiler are

$$\frac{dM_B}{dt} = L_N - V$$

and

$$\frac{d(M_B x_B)}{dt} = L_N x_N - V y_B$$

where N refers to conditions on the bottom plate of the column.

For the reflux drum

$$\frac{dM_D}{dt} = V - D - L_0$$

and
$$\frac{d(M_D x_D)}{dt} = Y y_1 - (L_0 + D) x_D$$

where $L_0/D = R$ is the reflux ratio for the column.

Assuming theoretical plate behaviour, i.e., equilibrium between the gas and liquid phases, for plate n

$$y_n = \frac{a x_n}{1 + (a-1) x_n}$$

where a is the relative volatility.

The above equation also applies to the liquid and vapour compositions of the still, where equilibrium plate behaviour is again assumed.

The maintenance of constant liquid level in the reflux drum can be expressed by the following proportional control equation

$$D = K_p (M_D - M_{D(set)})$$

where K_p is the controller gain and $M_{D(set)}$ is the level controller set point.

Automatic control of distillate composition (x_D) may also be affected by control of the reflux ratio, for example to maintain the distillate composition at constant set point (x_{Dset}).

$$R = K_{pR} (x_D - x_{Dset})$$

Batch distillation with continuous control of distillate composition via the regulation of reflux ratio is illustrated in the simulation example BSTILL. In this an initial total reflux condition, required to establish the initial concentration profile with the column, is represented in the simulation by a high initial value of R, which then changes to the controller equation for conditions of distillate removal.

Changes in the hydraulic hold-up of liquid on the column plates is known to have a significant effect on the separating efficiency of batch distillation columns, and may be relatively easily incorporated into the batch simulation model. The hydraulic condition of the plates is represented in Fig. 3.52.

A material balance for the liquid on plate n is given by

$$\frac{dM_n}{dt} = L_{n-1} - L_n$$

In this simplified model, it is assumed that liquid may leave the plate, either by flow over the weir $L_{n(weir)}$ or by weepage $L_{n(weep)}$. Both these effects can be described by simple hydraulic relations, in which the flow is proportional to the square root of the available hydrostatic liquid head. The weir flow depends on the liquid head above the weir, and hence

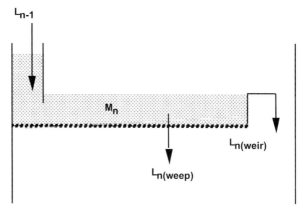

Fig. 3.52 Model representation of the plate hydraulics.

$$L_{n(weir)} \propto (M_n - M_{ns})^{0.5}$$

$L_{n(weir)}$ is zero for the condition $M_n < M_{ns}$. M_n is the mass of liquid on plate n, and M_{ns} is the mass of liquid on the plate corresponding to the weir height or static liquid holdup on the plate. The rate of loss of liquid from the plate by weepage, however, will depend on the total mass of liquid on the plate

$$L_{n(weep)} \propto M_n^{0.5}$$

The total flow of liquid from the plate is therefore given by

$$L_n = K_1(M_n - M_{ns})^{0.5} + K_2 M_n^{0.5}$$

where K_1 is an effective weir discharge constant for the plate, K_2 is the weepage discharge constant and M_{ns} is the static holdup on the plate.

3.3.3.3 Continuous Binary Distillation

The continuous binary distillation column of Fig. 3.53 follows the same general representation as that used previously in Fig. 3.51. The modelling approach again follows closely that of Luyben (1990).

The relationships for the section of column above the feed plate, i.e., the enriching section of the column, are exactly the same as those derived previously for the case of the batch distillation column.

The material balance relationships for the feed plate, the plates in the stripping section of the column and for the reboiler must, however, be modified, owing to the continuous feed to the column and the continuous withdrawal of bottom product from the reboiler. The feed is defined by its mass flow rate, F, its composition x_F and the thermal quality or q-factor, q. The column bottom prod-

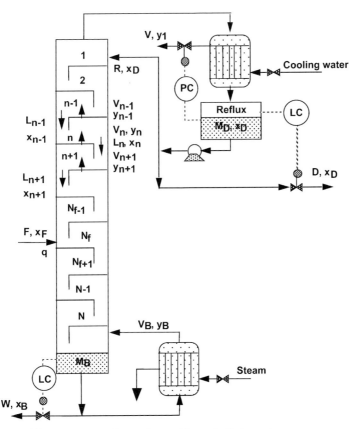

Fig. 3.53 Model representation of a continuous binary distillation column. PC is the cooling water controller, LC the reflux controller.

uct is defined by its mass flow rate W and composition x_W and is controlled to maintain constant liquid level in the reboiler.

The liquid and vapour molar flow rates in the enriching section are denoted by L and V, as previously and in the stripping section as L' and V'. The relationship between L, V, L' and V' is determined by the feed rate F and the thermal quality of the feed "q".

The thermal quality of the feed is defined as the heat required to raise 1 mole of feed from the feed condition to vapour at the feed plate condition divided by the molar latent heat, and the following values apply: q=0 for saturated liquid feed; q=1 for saturated vapour feed, and q>1 for cold feed. The value of q affects the relative liquid and vapour flow rates (L and V) above and (L' and V') below the feed plate, as indicated in Fig. 3.54.

Thus an energy balance around the feed plate can be employed to show that under certain conditions

3 Modelling of Stagewise Processes

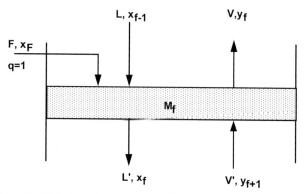

Fig. 3.54 Column feedplate flow rates and compositions.

$$L' = L + qF$$

$$V = V' - (1-q)F$$

and where for a saturated liquid feed with q = 1

$$L' = L + F$$

and

$$V = V'$$

For any plate n above the feed as shown previously for constant liquid holdup conditions

$$M_n \frac{dx_n}{dt} = L(x_{n-1} - x_n) + V(y_{n+1} - y_n)$$

The component balance for the feed plate is given by

$$M_n \frac{dx_F}{dt} = Lx_{f-1} - L'x_f + V'y_{F+1} - Vy_f + Fx_F$$

and for any plate m, in the stripping section, below the feed

$$M_m \frac{dx_n}{dt} = L'(x_{m-1} - x_m) + V'(y_{m+1} - y_m)$$

For the reboiler

$$M_B \frac{dx_B}{dt} = L'x_N - Wx_B - V'y_B$$

The controller equations

$$W = f(M_B)$$

$$V' = f(x_B)$$

are also required to complete the model. The relationships around the top part of the column and control of reflux drum level remain the same as those for the batch situation, described in Section 3.3.3.2.

If necessary the hydraulic relationships, previously derived for batch distillation, are also easily implemented into a continuous distillation model.

Continuous binary distillation is illustrated by the simulation example CONSTILL. Here the dynamic simulation example is seen as a valuable adjunct to steady state design calculations, since with MADONNA the most important column design parameters (total column plate number, feed plate location and reflux ratio) come under the direct control of the simulator as facilitated by the use of sliders. Provided that sufficient simulation time is allowed for the column conditions to reach steady state, the resultant steady state profiles of composition versus plate number are easily obtained. In this way, the effects of changes in reflux ratio or choice of the optimum plate location on the resultant steady state profiles become almost immediately apparent.

3.3.3.4 Multicomponent Separations

As discussed previously in Section 3.3.1.7, each additional component of the feed mixture must be expressed by a separate component material balance equation and by its own equilibrium relationship.

Thus for component i of a system of j components, the component balance equation, on the nth plate, becomes

$$M_n \frac{dx_{in}}{dt} = L(x_{in-1} - x_{in}) + V(y_{in+1} - y_{in})$$

where i=1 to j.

Assuming the equilibrium to be expressed in terms of relative volatilities a_i and theoretical plate behaviour, the relation between the vapour and liquid mole fraction compositions leaving the plate is given by

$$y_{in} = \frac{a_{in} x_{in}}{\sum_{1}^{j} a_{in} x_{in}}$$

or where the equilibrium can be based in terms of K values, the relationship becomes

$$y_{in} = K_{in} x_{in}$$

Using the above form of equilibrium relationship, the component balance equations now become

$$M_n \frac{dx_{in}}{dt} = Lx_{in-1} - Lx_{in} + VK_{in}(x_{in+1} - x_{in})$$

This again shows that the component balance may thus have different time constants, which depend on the relative magnitudes of the equilibrium constants K_i, which again can lead to problems of numerical solution due to equation stiffness.

One way of dealing with this is to replace those component balance differential equations, having low time constants (i.e., high K values) and fast rates of response, by quasi-steady-state algebraic equations, obtained by setting

$$M \frac{dx_i}{dt} = 0$$

and effectively neglecting the dynamics in the case of those components, having very fast rates of response.

Continuous multicomponent distillation simulation is illustrated by the simulation example MCSTILL, where the parametric runs facility of MADONNA provides a valuable means of assessing the effect of each parameter on the final steady state. It is thus possible to rapidly obtain the optimum steady state settings for total plate number, feed plate number and column reflux ratio via a simple use of sliders.

3.3.3.5 Plate Efficiency

The use of a plate efficiency correction enables the simulation of columns with a real number of plates to be simulated. This may be important in the study of real columns, when incorporating an allowance for plate hydrodynamic behaviour.

The situation for any plate n, with liquid composition x_n corresponding to an equilibrium vapour composition y_n^*, but with actual vapour composition y_n, is represented on a small section of the McCabe–Thiele diagram in Fig. 3.55.

The actual plate efficiency can be defined as

$$\eta = \frac{\text{Actual change of composition}}{\text{Maximum possible change of composition}}$$

where

$$\eta = \frac{(y_n - y_{n1})}{(y_n^* - y_{n1})}$$

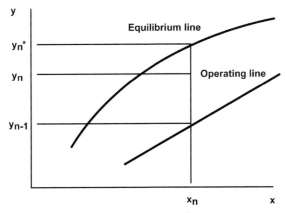

Fig. 3.55 Actual and theoretical plate compositions for plate n.

Hence by simple algebra

$$y_n = \eta(y_n^* - y_{n-1}) + y_{n-1}$$

Additional equations, as above, can thus be used to correct the values of y_n^*, obtained from the equilibrium data to give actual plate values y_n.

3.3.3.6 Complex Column Simulations

More complex situations where ideal behaviour can no longer be assumed require the incorporation of activity coefficient terms in the calculation of the equilibrium vapour compositions. Assuming ideal behaviour in the gas phase, the equilibrium relation for component i is

$$y_i = \frac{\gamma_i x_i P_i}{P}$$

where P is the total pressure in the column. Since the saturated vapour pressure of the pure compound i, P_i, is a function of temperature, the calculation of the equilibrium vapour composition requires that a plate temperature must be determined such that the condition $\Sigma y_i = 1$ is obtained. Examples of this technique are illustrated in the corresponding simulation examples STEAM and BUBBLE.

Furthermore heat effects on the plates may also have to be accounted for, by means of a dynamic heat balance for each plate, including allowances for the enthalpies of the liquid and vapour streams, entering and leaving the plate, heat of mixing, etc. This thus represents a much more complicated and time-consuming computational procedure than has been considered so far. In such cases, it obviously becomes much more meaningful to employ larger simulation packages, with their sophisticated physical property data-bases and estimation procedures. The general principles of the modelling procedure, however, remain very much the same.

Multicomponent equilibria combined with distillation heat effects are discussed in more detail in Section 3.3.4 below.

3.3.4
Multicomponent Steam Distillation

Steam distillation is a process whereby organic liquids may be separated at temperatures sufficiently low to prevent their thermal decomposition or whereby azeotropes may be broken. Fats or perfume production are examples of applications of this technique. The vapour–liquid equilibria of the three-phase system is simplified by the usual assumption of complete immiscibility of the liquid phases and by the validity of the Raoult and Dalton laws. Systems containing more than one volatile component are characterised by complex dynamics (e.g., boiling point is not constant).

Steam distillation is normally carried out as a semi-batch process whereby the organic mixture is charged into the still and steam is bubbled through continuously, as depicted in Fig. 3.56.

As discussed, modelled and simulated by Prenosil (1976), the dynamics of the process bring in the question of steam consumption, steam flow rate, starting time of the distillation, and shut-down time when the desired degree of separation has been reached. The modelling of steam distillation often involves the following assumptions.

1. Ideal behaviour of all components in pure state or mixture.
2. Complete immiscibility of the water and the organic phases.
3. Zero temperature gradients in the bulk phases (ideal mixing in the boiler).
4. Equilibrium between the organic vapour and its liquid at all times.

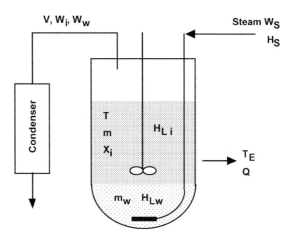

Fig. 3.56 Schematic drawing of the apparatus for steam distillation.

3.3 Stagewise Mass Transfer

The mathematical model is divided into two time periods:

(a) The heating period until boiling point is reached.
(b) The distillation period after boiling has started.

Heating Period

To describe the dynamic behaviour of this semi-batch process, unsteady-state mass and energy balances are needed. Their interrelationships are depicted in Fig. 3.57.

For the water phase,

$$\frac{dm_w}{dt} = W_S$$

Here, it is assumed that all the steam condenses in the distillation vessel. In this period, the organic phase component masses remain constant.

The rate of heat accumulation is balanced by the heat of condensation and the heat losses. An energy balance therefore gives

$$\frac{d(m_w H_{Lw} + m\Sigma x_i H_{Li})}{dt} = W_S H_S + Q$$

The enthalpy changes are calculated from molar heat capacities given by the usual functions of temperature, according to

$$(m_w c_{pLw} + m\Sigma x_i c_{pLi})\frac{dT}{dt} = W_S(H_S - H_{Lw}) + Q$$

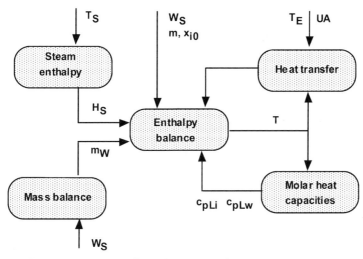

Fig. 3.57 Information flow diagram for the heating period.

from which it follows that

$$\frac{dT}{dt} = \frac{W_S(H_S - H_{Lw}) + Q}{m_w c_{pLw} + \Sigma x_i c_{pLi}}$$

The heat transfer Q to the surroundings is calculated from the simple relation

$$Q = UA(T_E - T)$$

The solution of the above model gives the temperature of the mixture at any time during the heating period.

Distillation Period

The distillation starts when the boiling point is reached. Then a vapour stream at flow rate V is obtained, which condenses as a distillate. The material balances can be written as follows:

For water

$$\frac{dm_w}{dt} = W_S - V y_v$$

For the organic compound i

$$\frac{d(m x_i)}{dt} = -V y_i$$

and for the total organic phase

$$\frac{dm}{dt} = -V \Sigma y_i$$

The energy balance is now

$$\frac{d(m_w H_{Lw} + m \Sigma x_i H_{Li})}{dt} = W_S H_S - V H_V + Q$$

where

$$H_V = y_w H_{Vw} + \Sigma y_i H_{Vi}$$

The vapour enthalpies are calculated from the molar heat capacity functions for the vapour components and the latent heats of vaporisation at standard temperature. The vapour overflow, V, is then obtained from the energy balance as

$$V = W_S H_S + Q - \frac{d(m_w H_{Lw} + m \Sigma x_i H_{Li})}{dt} \frac{1}{H_V}$$

Phase Equilibria

Assuming ideal liquid behavior, the total partial pressure of the organic phase is given by the sum of the partial pressures of its components according to Raoult's Law.

$$\frac{y_i}{x_i} = \frac{P_i}{P}$$

where P_i is the vapour pressure of pure component i. For non-idealities, this must be modified with appropriate activity expressions (Prenosil, 1976). For water, the vapour pressure varies only with temperature so that

$$y_w = \frac{P_w}{P}$$

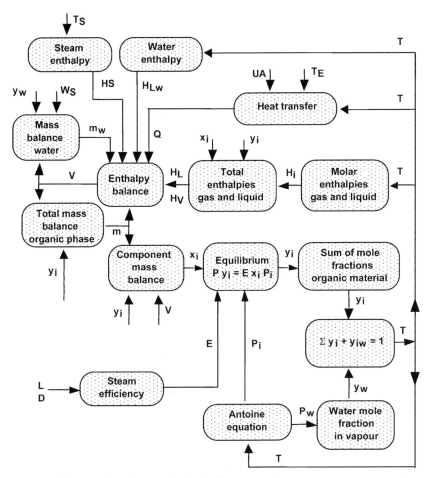

Fig. 3.58 Information flow diagram for the distillation period.

Boiling will commence when the sum of the organic partial pressures and the water vapour pressure is equal to the total pressure or in terms of the mole fractions

$$\Sigma y_i + y_w = 1$$

As boiling proceeds the loss of the lightest organic vapours will cause the boiling point to increase with time. The vapour pressures P_i and P_w of the pure components can be calculated using the Antoine equation

$$\log P = A - \frac{B}{C+T}$$

The highly interactive nature of the balance and equilibria equations for the distillation period are depicted in Fig. 3.58. An implicit iterative algebraic loop is involved in the calculation of the boiling point temperature at each time interval. This involves guessing the temperature and calculating the sum of the partial pressures or mole fractions. The condition required is that $\Sigma y_i + y_w = 1$. The model of Prenosil (1976) also included an efficiency term E for the steam heating, dependent on liquid depth L and bubble diameter D.

Multicomponent steam distillation is illustrated in simulation example STEAM.

4
Differential Flow and Reaction Applications

4.1
Introduction

4.1.1
Dynamic Simulation

The main process variables in differential contacting devices vary continuously with respect to distance. Dynamic simulations therefore involve variations with respect to both time and position. Thus two independent variables, time and position, are now involved. Although the basic principles remain the same, the mathematical formulation for the dynamic system now results in the form of partial differential equations. As most digital simulation languages permit the use of only one independent variable, the second independent variable, either time or distance, is normally eliminated by the use of a finite-differencing procedure. In this chapter, the approach is based very largely on that of Franks (1967), and the distance coordinate is treated by finite differencing.

In this procedure, the length coordinate of the system is divided into N finite-difference elements or segments, each of length ΔZ, where N times ΔZ is equal to the total length or distance. It is assumed that within each element any variation with respect to distance is relatively small. The conditions at the mid-point of the element can therefore be taken to represent the conditions of the element as a whole. This is shown in Fig. 4.1, where the average concentration of any element n is identified by the midpoint concentration C_n. The actual continuous variation in concentration with respect to length is therefore approximated by a series of discontinuous variations.

The dynamic behaviour of element n is affected by the conditions in its neighbouring elements $n-1$ and $n+1$, and each original partial differential equation is approximated by a system of N simultaneous difference differential equations. In practice, the length of each element ΔZ may be kept constant or may be varied from segment to segment. A greater number of elements usually improves the approximation of the profile, but the computational effort required is also greater. The approach is demonstrated in the simulation examples DISRET, DISRE, AXDISP, CHROMDIFF, MEMSEP, HEATEX, DRY, ENZDYN, BEAD, SOIL and LEACH.

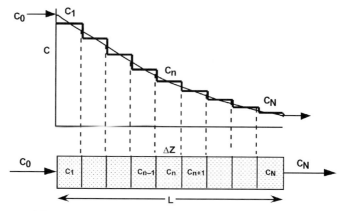

Fig. 4.1 Finite-differencing a tubular reactor with the stepwise approximation of the continuous concentration profile.

4.1.2
Steady-State Simulation

Under steady-state conditions, variations with respect to time are eliminated and the steady-state model can now be formulated in terms of the one remaining independent variable, length or distance. In many cases, the model equations now become simultaneous first-order differential equations, for which solution is straightforward. Simulation examples of this type are the steady-state tubular reactor models TUBE and TUBEDIM, TUBTANK, ANHYD, BENZHYD and NITRO.

Some models, however, take the form of second-order differential equations, which often give rise to problems of the split boundary type. In order to solve this type of problem, an iterative method of solution is required, in which an unknown condition at the starting point is guessed, the differential equation integrated. After comparison with the second boundary condition a new starting point is estimated, followed by re-integration. This procedure is then repeated until convergence is achieved. MADONNA provides such a method. Examples of the steady-state split-boundary type of solution are shown by the simulation examples ROD and ENZSPLIT.

In order to overcome the problem of split boundaries, it is sometimes preferable to formulate the model dynamically, and to obtain the steady-state solution, as a consequence of the dynamic solution, leading to the eventual steady state. This procedure is demonstrated in simulation example ENZDYN.

4.2
Diffusion and Heat Conduction

This section deals with problems involving diffusion and heat conduction. Both diffusion and heat conduction are described by similar forms of equation. Fick's Law for diffusion has already been met in Section 1.2.2 and the similarity of this to Fourier's Law for heat conduction is apparent.

With Fick's Law

$$j_A = -D\frac{dC_A}{dZ}$$

and Fourier's Law

$$q = -k\frac{dT}{dZ}$$

Here j_A is the diffusional flux of component A (kmol/m² s), D is the diffusion coefficient (m²/s), C_A is the concentration of component A (kmol/m³), q is the heat transfer flux (kJ/m² s), k is the thermal conductivity (kJ/m s K), T is the temperature (K) and Z is the distance (m).

In diffusional mass transfer, the transfer is always in the direction of decreasing concentration and is proportional to the magnitude of the concentration gradient, the constant of proportionality being the diffusion coefficient for the system.

In conductive heat transfer, the transfer is always in the direction of decreasing temperature and is proportional to the magnitude of the temperature gradient, the constant of proportionality being the thermal conductivity of the system.

The analogy also extends to Newton's equation for momentum transport, where

$$\tau = -\mu\frac{dv}{dZ}$$

where, for Newtonian liquids, μ is the viscosity, τ is the shear stress, v is velocity and Z is again distance.

4.2.1
Unsteady-State Diffusion through a Porous Solid

This problem illustrates the solution approach to a one-dimensional, non-steady-state, diffusional problem, as demonstrated in the simulation examples, DRY and ENZDYN. The system is represented in Fig. 4.2. Water diffuses through a porous solid, to the surface, where it evaporates into the atmosphere.

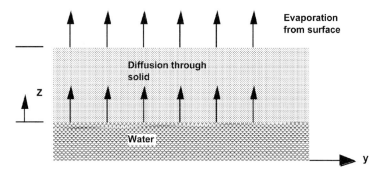

Fig. 4.2 Unsteady-state diffusion through a porous solid.

It is required to determine the water concentration profile in the solid, under drying conditions. The quantity of water is limited and, therefore, the solid will eventually dry out and the drying rate will reduce to zero.

The movement of water through a solid, such as wood, in the absence of chemical reaction, is described by the following time-dependent diffusional equation.

$$\frac{\partial C}{\partial t} = -D \frac{\partial^2 C}{\partial Z^2}$$

where, at steady state, $\partial C/\partial t = 0$, and

$$0 = D \frac{d^2 C}{dZ^2}$$

integrating

$$\frac{dC}{dZ} = \text{constant}$$

Thus at steady state the concentration gradient is constant.

Note that since there are two independent variables of both length and time, the defining equation is written in terms of the partial differentials, $\partial C/\partial t$ and $\partial C/\partial Z$, whereas at steady state only one independent variable, length, is involved and the ordinary derivative function is used. In reality the above diffusion equation results from a combination of an unsteady-state material balance, based on a small differential element of solid length dZ, combined with Fick's Law of diffusion.

To set up the problem for simulation involves discretising one of the independent variables, in this case length, and solving the time-dependent equations, obtained for each element, by means of a simulation language. By finite-differencing the length coordinate of the solid, as shown in Fig. 4.3, the drying process is approximated to that of a series of finite-differenced solid segments.

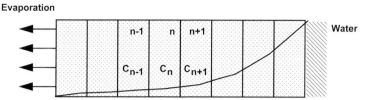

Fig. 4.3 Finite-differenced equivalent of the depth of solid.

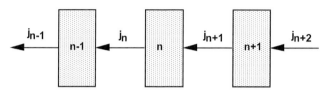

Fig. 4.4 Diffusional fluxes from segment to segment.

The diffusional fluxes from segment to segment are indicated in Fig. 4.4.

Note that the segments are assumed to be so small, that any variation in concentration within the segment, with respect to length, can effectively be ignored. The effective concentration of the segment can therefore be taken as that at the midpoint.

A component material balance is written for each segment, where

$$\begin{pmatrix} \text{Rate of accumulation} \\ \text{in the segment} \end{pmatrix} = \begin{pmatrix} \text{Diffusional} \\ \text{flow in} \end{pmatrix} - \begin{pmatrix} \text{Diffusional} \\ \text{flow out} \end{pmatrix}$$

or

$$\Delta V_n \frac{dC_n}{dt} = (j_{n+1} - j_n)A$$

Here j_n is the mass flux leaving segment n (kg/m² s), C_n is the concentration of segment n (kg/m³), A is the cross-sectional area (m²), t is time (t) and ΔV_n is the volume of segment n (m³).

By Fick's Law

$$j = -D \frac{dC}{dZ}$$

with dimensions

$$\frac{M}{L^2 T} = \frac{L^2}{T} \frac{M}{L^3 L}$$

The concentration gradient terms, dC/dZ, both in and out of segment n, can be approximated by means of their finite-differenced equivalents. Substituting these into the component balance equation gives

$$\Delta V_n \frac{dC_n}{dt} = D\frac{(C_{n+1} - C_n)}{\Delta Z}A - D\frac{(C_n - C_{n-1})}{\Delta Z}A$$

where ΔZ is the length of the segment and $\Delta V_n = A\,\Delta Z$. Thus

$$\frac{dC_n}{dt} = D\frac{(C_{n+1} - 2C_n + C_{n-1})}{\Delta Z^2}$$

The above procedure is applied to all the finite-difference segments in turn. The end segments ($n = 1$ and $n = N$), however, often require special attention according to particular boundary conditions: For example, at $Z = L$ the solid is in contact with pure water and $C_{N+1} = C_{eq}$, where the equilibrium concentration C_{eq} would be determined by prior experiment.

At the air–solid surface, $Z = 0$, the drying rate is determined by the convective heat and mass transfer drying conditions and the surrounding atmosphere of the drier. Assuming that the drying rate is known, the component balance equation for segment 1 becomes

$$\begin{pmatrix}\text{Rate of accumulation}\\ \text{in segment 1}\end{pmatrix} = \begin{pmatrix}\text{Rate of input}\\ \text{by diffusion}\\ \text{from segment 2}\end{pmatrix} - \begin{pmatrix}\text{Rate of}\\ \text{drying}\\ \text{from surface}\end{pmatrix}$$

Alternatively, the concentration in segment 1, C_1, may be taken simply as that in equilibrium with the surrounding air.

The unsteady model, originally formulated in terms of a partial differential equation, is thus transformed into N difference differential equations. As a result of the finite-differencing, a solution can be obtained for the variation with respect to time of the water concentration, for every segment throughout the bed.

The simulation example DRY is based directly on the above treatment, whereas ENZDYN models the case of unsteady-state diffusion combined with chemical reaction. Unsteady-state heat conduction can be treated in an exactly analogous manner, though for cases of complex geometry, with multiple heat sources and sinks, the reader is referred to specialist texts, such as Carslaw and Jaeger (1959).

4.2.2
Unsteady-State Heat Conduction and Diffusion in Spherical and Cylindrical Coordinates

Although the foregoing example in Section 4.2.1 is based on a linear coordinate system, the methods apply equally to other systems, represented by cylindrical and spherical coordinates. An example of diffusion in a spherical coordinate system is provided by simulation example BEAD. Here the only additional complication in the basic modelling approach is the need to describe the geometry of the system, in terms of the changing area for diffusional flow through the bead.

4.2.3
Steady-State Diffusion with Homogeneous Chemical Reaction

The following example, taken from Welty et al. (1976), illustrates the solution approach to a steady-state, one-dimensional, diffusion or heat conduction problem.

As shown in Fig. 4.5, an inert gas containing a soluble component S stands above the quiescent surface of a liquid, in which the component S is both soluble and in which it reacts chemically to form an inert product. Assuming the concentration of S at the gas–liquid surface to be constant, it is desired to determine the rate of solution of component S and the subsequent steady-state concentration profile within the liquid.

Under quiescent conditions, the rate of solution of S within the liquid is determined by molecular diffusion and is described by Fick's Law, where

$$j_S = -D \frac{dC_S}{dZ} \quad \text{kmol/m}^2\text{s}$$

At steady-state conditions, the rate of supply of S by diffusion is balanced by the rate of consumption by chemical reaction, where assuming a first-order chemical reaction

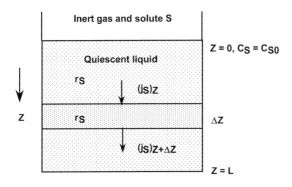

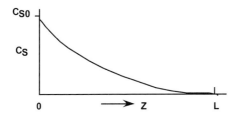

Fig. 4.5 Steady-state diffusion with chemical reaction.

$$r_S = -kC_S$$

with typical units of kmol/m^{3}s.

Thus considering a small differential element of liquid volume, dV, and depth, dZ, the balance equation becomes

$$j_S A_S = r_S dV$$

where A_S is the cross-sectional area of the element and $dV = A_S dZ$.

Hence

$$-D\frac{dC_S}{dZ} A_S = -kC_S A_S dZ$$

or

$$-D\frac{d^2C_S}{dZ^2} + kC_S = 0$$

where each term has the dimensions mass/time or units (kmol/s).

The above second-order differential equation can be solved by integration. At the liquid surface, where Z=0, the bulk gas concentration, C_{S0}, is known, but the concentration gradient dC_S/dZ is unknown. Conversely at the full liquid depth, the concentration C_{S0} is not known, but the concentration gradient is known and is equal to zero. Since there can be no diffusion of component S from the bottom surface of the liquid, i.e., j_S at Z=L is 0 and hence from Fick's Law dC_S/dZ at Z=L must also be zero.

The problem is thus one of a split boundary value type, and it can be solved by an iterative procedure based on an assumed value for one of the unknown boundary conditions. Assuming a value for dC_S/dZ at the initial condition Z=0, the equation can be integrated twice to produce values of dC_S/dZ and C_S at the terminal condition, Z=L. If the correct initial value has been chosen, the integration will lead to the correct final boundary condition, i.e., that $dC_S/dZ=0$ at Z=L and hence give the correct values of C_S. The value of the concentration gradient dC_S/dZ is also obtained for all values of Z throughout the depth of liquid.

Further applications of this method are given in the simulation examples ENZSPLIT and ROD.

4.3
Tubular Chemical Reactors

Mathematical models of tubular chemical reactor behaviour can be used to predict the dynamic variations in concentration, temperature and flow rate at various locations within the reactor. A complete tubular reactor model would however be extremely complex, involving variations in both radial and axial positions, as well as perhaps spatial variations within individual catalyst pellets. Models of such complexity are beyond the scope of this text, and variations only with respect to both time and axial position are treated here. Allowance for axial dispersion is however included, owing to its very large influence on reactor performance, and the fact that the modelling procedure using digital simulation is relatively straightforward.

4.3.1
The Plug-Flow Tubular Reactor

Consider a small element of volume, ΔV, of an ideal plug-flow tubular reactor, as shown in Fig. 4.6.

Component Balance Equation
A component balance equation can be derived for the element ΔV, based on the generalised component balance expression, where, for any reactant A

$$\begin{pmatrix} \text{Rate of} \\ \text{accumulation} \\ \text{of A} \end{pmatrix} = \begin{pmatrix} \text{Mass} \\ \text{flow} \\ \text{of A in} \end{pmatrix} - \begin{pmatrix} \text{Mass} \\ \text{flow} \\ \text{of A out} \end{pmatrix} + \begin{pmatrix} \text{Rate of} \\ \text{formation} \\ \text{of A by reaction} \end{pmatrix}$$

The rate of accumulation of component A in element ΔV is $(\Delta V \, dC_A/dt)$, where dC_A/dt is the rate of change of concentration.

The mass rate of flow of A into element ΔV is $F \, C_A$, and the rate of flow of A from element ΔV is $F \, C_A + \Delta(F \, C_A)$, where F is the volumetric flow rate. The rate of formation of A by reaction is $r_A \Delta V$, where r_A is the rate per unit volume.

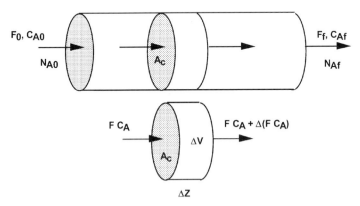

Fig. 4.6 Component balancing for a tubular plug-flow reactor.

Substituting these quantities gives the resulting component balance equation as

$$\Delta V \frac{dC_A}{dt} = F\,C_A - [F\,C_A + \Delta(F\,C_A)] + r_A \Delta V$$

or

$$\frac{dC_A}{dt} = -\frac{\Delta(FC_A)}{\Delta V} + r_A$$

The above equation may also be expressed in terms of length, since

$$\Delta V = A_c \Delta Z$$

where A_c is the cross-sectional area of the reactor.

Allowing ΔV to become very small, the above balance equation is transformed into the following partial differential equation, where

$$\frac{\partial C_A}{\partial t} = -\frac{1}{A_c}\frac{\partial(FC_A)}{\partial Z} + r_A$$

For constant volumetric flow rate, F, throughout the reactor

$$\frac{\partial C_A}{\partial t} = -\frac{F}{A_c}\frac{\partial C_A}{\partial Z} + r_A$$

where F/A_c is the superficial linear fluid velocity v, through the reactor.

Under steady-state conditions

$$\frac{\partial C_A}{\partial t} = 0$$

and hence, at steady state

$$\frac{dC_A}{dZ} = \frac{1}{v} r_A$$

This equation can be integrated to determine the resulting steady-state variation of C_A with respect to Z, knowing the reaction kinetics $r_A = f(C_A)$ and the initial conditions C_A at $Z=0$.

Cases with more complex multicomponent kinetics will require similar balance equations for all the components of interest.

The component balance equation can also be written in terms of fractional conversion, X_A, where for constant volumetric flow conditions

$$C_A = C_{A0}(1 - X_A)$$

and C_{A0} is the inlet reactor feed concentration. Thus

$$dC_A = -C_{A0} dX_A$$

4.3 Tubular Chemical Reactors

The material balance, in terms of X_A, is thus given by

$$C_{A0}\frac{dX_A}{dZ} = -\frac{1}{v}r_A$$

Reactant A is consumed, so r_A is negative, and the fractional conversion will increase with Z.

In terms of molar flow rates

$$N_{A0}\frac{dX_A}{dZ} = -A_c\, r_A$$

and

$$N_{A0}\frac{dX_A}{dV} = -r_A$$

where N_{A0} is the molar flow of reactant A entering to the reactor.

Energy Balance Equation

The energy balance for element ΔV of the reactor again follows the generalised form, derived in Section 1.2.5. Thus

$$\begin{pmatrix}\text{Rate of}\\\text{accumulation}\\\text{of energy}\end{pmatrix} = \begin{pmatrix}\text{Rate of energy}\\\text{required to}\\\text{heat the incoming}\\\text{stream from}\\\text{T to T}+\Delta T\end{pmatrix} + \begin{pmatrix}\text{Rate of}\\\text{energy}\\\text{generated}\\\text{by reaction}\\\text{at T}+\Delta T\end{pmatrix} - \begin{pmatrix}\text{Rate of}\\\text{energy}\\\text{out by}\\\text{transfer}\end{pmatrix}$$

Referring to Fig. 4.7, the general energy balance for segment n with S components and R reactions is

$$\sum_{i=1}^{S}(n_{in}c_{pin})\frac{dT_n}{dt} = -\sum_{i=1}^{S}N_{in-1}\int_{T_{n-1}}^{T_{n-1}+\Delta T}c_{pin-1}dT + \sum_{j=1}^{R}\frac{R_{ijn}}{v_{ij}}(-\Delta H_{jn}) - Q_n$$

With more complex cases, c_p and ΔH are functions of temperature, then the substitutions would be as shown in Case C of Section 1.2.5.3. This general form, which may give rise to very complex expressions, may be important for gas phase reactions (see also Section 4.3.3). Simplifying by assuming only one reaction (R = 1) and constant c_{pi} gives

$$\sum_{i=1}^{S}(n_{in}c_{pin})\frac{dT_n}{dt} = -\sum_{i=1}^{S}(N_{in-1}c_{pin-1}\Delta T_n) + \frac{R_{in}}{v_i}(-\Delta H_n) - Q_n$$

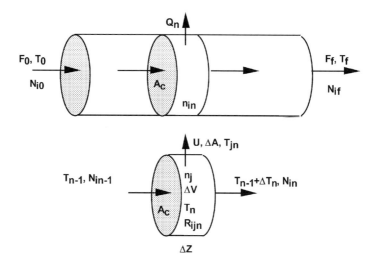

Fig. 4.7 Energy balancing for the tubular plug-flow reactor.

Assuming the total heat capacity to be constant, $\sum_{i=1}^{S}(N_i c_{pi}) = F \rho c_p$, and replacing $\sum_{i=1}^{S}(n_{in} c_{pin})$ by $\Delta V \rho c_p$, the total heat capacity of one element gives

$$\Delta V \rho c_p \frac{dT_n}{dt} = -F \rho c_p \Delta T_n + \Delta V \frac{r_{in}}{v_i}(-\Delta H_n) - Q_n$$

The heat loss through the wall to the jacket is

$$Q_n = U \Delta A (T_n - T_j)$$

The term ΔT can be approximated by $(dT/dZ)\Delta Z$ (see also Section 4.3.5). For a tube

$$\frac{\Delta A}{\Delta V} = \frac{d \pi \Delta Z}{\left(\frac{d}{2}\right)^2 \pi \Delta Z} = \frac{4}{d}$$

where d is the tube diameter. Noting further that $\Delta V = A_c \Delta Z$, that the linear velocity $v = F/A_c$ and letting ΔZ approach zero, gives the defining partial differential equation

$$\frac{\partial T}{\partial t} = -v \frac{\partial T}{\partial Z} + \frac{1}{\rho c_p} \frac{r_i}{v_i}(-\Delta H) - \frac{4U}{d \rho c_p}(T - T_j)$$

For steady-state conditions, the above equation reduces to

$$\frac{dT}{dZ} = \frac{1}{v\rho c_p} \frac{r_i}{v_i}(-\Delta H) - \frac{4U}{v\rho c_p d}(T - T_j)$$

This equation can be integrated together with the component balances and the reaction kinetic expressions where the kinetics could, for example, be of the form

$$r_i = k_i C_i^n = k_0 e^{-E/RT} C_i^n$$

thus including variation of the rate constant k with respect to temperature.

The component material balance equation, combined with the reactor energy balance equation and the kinetic rate equation, provide the basic model for the ideal plug-flow tubular reactor.

4.3.2
Liquid-Phase Tubular Reactors

Assuming the case of a first-order chemical reaction ($r_A = -kC_A$) and a non-compressible liquid system, the generalised mass and energy balance equations reduce to

$$\frac{dC_A}{dZ} = -\frac{k}{v} C_A$$

and

$$\frac{dT}{dZ} = \frac{(-\Delta H)}{v\rho c_p} k C_A - \frac{4U}{v\rho c_p d}(T - T_j)$$

where v, ρ and c_p are assumed constants and k is given by the Arrhenius equation $k = k_0 e^{-E/RT}$.

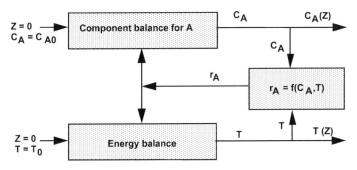

Fig. 4.8 Flow diagram for the simultaneous integration of the balances.

The general solution approach, to this type of problem, is illustrated by the information flow diagram, shown in Fig. 4.8. The integration thus starts with the initial values at Z=0, and proceeds with the calculation of r_A, along the length of the reactor, using the computer updated values of T and C_A, which are also produced as outputs.

The simultaneous integration of the two continuity equations, combined with the chemical kinetic relationships, thus gives the steady-state values of both, C_A and T, as functions of reactor length. The simulation examples BENZHYD, ANHYD and NITRO illustrate the above method of solution.

4.3.3
Gas-Phase Tubular Reactors

In gas-phase reactors, the volume and volumetric flow rate frequently vary, owing to the molar changes caused by reaction and the effects of temperature and pressure on gas phase volume. These influences must be taken into account when formulating the mass and energy balance equations.

The Ideal Gas Law can be applied both to the total moles of gas, n, or to the moles of a given component of the gas mixture n_i, where

$$PV = nRT$$

and

$$p_i V = n_i RT$$

Here P is the total pressure of the system, p_i is the partial pressure of component i, V is the volume of the system, T is temperature and R is the Ideal Gas Constant.

Using Dalton's Law

$$p_i = y_i P$$

the relationship for the concentration C_i, in terms of mole fraction y_i, and total pressure is obtained as

$$C_i = \frac{n_i}{V} = \frac{y_i P}{RT}$$

This can also be expressed in terms of the molar flow rate N_i, and the volumetric flow rate G, where

$$N_i = \frac{y_i P G}{RT}$$

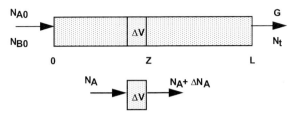

Fig. 4.9 Material balancing for a gas-phase tubular reactor.

Assuming first-order kinetics for the reaction $A \rightarrow mB$

$$r_A = -kC_A = -\frac{ky_A P}{RT}$$

and by stoichiometry

$$r_B = mkC_A = \frac{mky_A P}{RT}$$

The steady-state material balance – for a volume element, ΔV, as shown in Fig. 4.9, for reactant A – is given by

$$0 = N_A - (N_A + \Delta N_A) + r_A \Delta V$$

which, since $\Delta V = A_c \Delta Z$, then gives

$$\frac{dN_A}{dZ} = r_A A_c$$

Similarly for component B

$$\frac{dN_B}{dZ} = r_B A_c$$

where A_c is again the cross-sectional flow area of the reactor.

The variation in molar flow can be written as

$$\frac{dN_A}{dZ} = \frac{d\left(\frac{y_A PG}{RT}\right)}{dZ}$$

which for constant temperature and pressure conditions becomes

$$\frac{dN_A}{dZ} = \frac{P}{RT}\frac{d(y_A G)}{dZ}$$

Substituting this and the reaction kinetics r_A into the component balance equation for reactant A gives

$$\frac{d(y_A G)}{dZ} = -k y_A A_c \quad \text{(I)}$$

Similarly for B

$$\frac{d(y_B G)}{dZ} = m k y_A A_c \quad \text{(II)}$$

The volumetric flow rate depends on the total molar flow of the gas and the temperature and pressure of reaction, where

$$G = \frac{(N_A + N_B + N_{\text{inerts}})RT}{P} \quad \text{(III)}$$

where the molar flow rates of A and B are given by

$$N_A = \frac{y_A G P}{RT} \quad \text{(IV)}$$

and

$$N_B = \frac{y_B G P}{RT} \quad \text{(V)}$$

The model equations, I to V above, provide the basis for solution, for this case of constant temperature and pressure with a molar change owing to chemical reaction. This is illustrated by the information flow diagram (Fig. 4.10). The step-by-step calculation procedure is as follows:

1. The initial molar flow rates of each component at the reactor inlet, $(y_A G)_0$ and $(y_B G)_0$, are known.
2. The component balance equations (I) and (II) are integrated with respect to distance to give the volumetric flow rate of each component, $(y_A G)_Z$ and $(y_B G)_Z$, at any position Z, along the length of the reactor.
3. The total molar flow rate of each component, $(N_A)_Z$ and $(N_B)_Z$, can then be calculated at position Z, from equations (IV) and (V).
4. The total volumetric flow rate G is calculated at each position, using equation (III) and hence:
5. The composition of the gas mixture at position Z is obtained by dividing the individual molar flow rate by the volumetric flow rate.

The results of the calculation are thus the mole fraction compositions y_A and y_B, together with the total volumetric flow rate G, as steady-state functions of reactor length.

The step-by-step evaluation is, of course, effected automatically by the computer, as shown in the simulation example VARMOL.

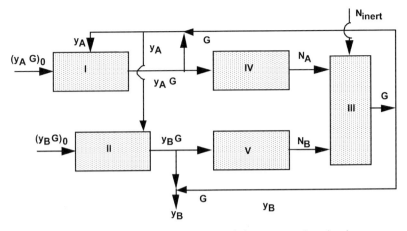

Fig. 4.10 Information flow diagram for a gas-phase tubular reactor with molar change.

To obtain the fractional conversion at any position along the reactor the appropriate equation is

$$N_A = N_{A0}(1 - X_A)$$

where

$$y_A N_{Tot} = y_{A0} N_{Tot0}(1 - X_A)$$

In the case of non-isothermal situations with significant pressure drop through the reactor, the term

$$\frac{d\left(\dfrac{y_i G P}{RT}\right)}{dZ}$$

must be retained in the model equations. The variation of reactor pressure with reactor length, $(P)_Z$, can be obtained by the use of available pressure drop-flow correlations, appropriate to the reactor geometry and flow conditions. The variation in temperature, with respect to length, $(T)_Z$ must be obtained via a steady-state energy balance equation, as described in Section 4.3.1.

4.3.4
Batch Reactor Analogy

The ideal plug-flow reactor is characterised by the concept that the flow of liquid or gas moves with uniform velocity similar to that of a plug moving through the tube. This means that radial variations of concentration, temperature and flow velocity are neglected and that axial mixing is negligible. Each element of fluid flows through the reactor with the same velocity and therefore remains in the reactor for the same length of time, which is given by the flow volume of the reactor divided by the volumetric flow rate. The residence time of fluid in the ideal tubular reactor is thus analogous to the reaction time in a batch reactor.

With respect to reaction rates, an element of fluid will behave in the ideal tubular reactor, in the same way, as it does in a well-mixed batch reactor. The similarity between the ideal tubular and batch reactors can be understood by comparing the model equations.

For a batch reactor, under constant volume conditions, the component material balance equation can be represented by

$$\frac{dC_A}{dt} = r_A$$

For a plug-flow tubular reactor, the flow velocity v through the reactor can be related to the distance travelled along the reactor or tube Z, and to the time of passage t, where

$$dt = \frac{dZ}{v}$$

Equating the time of passage through the tubular reactor to that of the time required for the batch reaction, gives the equivalent ideal-flow tubular reactor design equation as

$$\frac{dC_A}{dZ} = \frac{r_A}{v}$$

4.3.5
Dynamic Simulation of the Plug-Flow Tubular Reactor

The coupling of the component and energy balance equations in the modelling of non-isothermal tubular reactors can often lead to numerical difficulties, especially in solutions of steady-state behaviour. In these cases, a dynamic digital simulation approach can often be advantageous as a method of determining the steady-state variations in concentration and temperature, with respect to reactor length. The full form of the dynamic model equations are used in this approach, and these are solved up to the final steady-state condition, at which condition

$$\frac{dT}{dt} = \frac{dC_A}{dt} = 0$$

The procedure is to transform the defining model partial differential equation system into sets of difference-differential equations, by dividing the length or volume of the reactor into disc-shaped segments. The concentrations and temperatures at the boundaries of each segment are approximated by a midpoint average. Each segment can therefore be thought of as behaving in a similar manner to that of a well-stirred tank.

4.3 Tubular Chemical Reactors

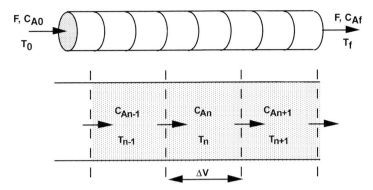

Fig. 4.11 Finite-differencing for a dynamic tubular reactor model.

This is shown above in Fig. 4.11, where segment n, with volume ΔV, is identified by its midpoint concentration C_{An} and midpoint temperature T_n.

The concentration of reactant entering segment n, from segment n−1, is approximated by the average of the concentrations in the two segments and is given by

$$\frac{C_{An-1} + C_{An}}{2}$$

Similarly, the concentration of reactant leaving segment n, and entering segment n+1, is approximated by

$$\frac{C_{An} + C_{An+1}}{2}$$

This averaging procedure has the effect of improving the approximation. Alternatively, each segment may be treated as a well-mixed tank with the outflow variables equal to the tank values. This simpler approach would require a greater number of segments for the same accuracy.

Applying the generalised component balance equation for component A to each segment n gives

$$\begin{pmatrix} \text{Rate of} \\ \text{Accumulation} \\ \text{of reactant} \end{pmatrix} = \begin{pmatrix} \text{Flow of} \\ \text{reactant} \\ \text{in} \end{pmatrix} - \begin{pmatrix} \text{Flow of} \\ \text{reactant} \\ \text{out} \end{pmatrix} + \begin{pmatrix} \text{Rate of} \\ \text{Production} \\ \text{of reactant} \end{pmatrix}$$

and results in

$$\Delta V \frac{dC_{An}}{dt} = F\left(\frac{C_{An-1} + C_{An}}{2} - \frac{C_{An} + C_{An+1}}{2}\right) + r_{An}\Delta V$$

where $\Delta V = A\, \Delta Z$.

The application of the energy balance equation to segment n results similarly in the relationship

$$\Delta V \rho c_p \frac{dT_n}{dt} = F \rho c_p \left(\frac{T_{n-1} + T_n}{2} - \frac{T_n + T_{n+1}}{2} \right) + \frac{r_{An}}{v_A}(-\Delta H)\Delta V - U\Delta A_t(T_n - T_j)$$

where the outer surface area for heat transfer in segment n is given by

$$\Delta A_t = \frac{A_t}{N}$$

A_t is the total surface for heat transfer and for a single tube is given by

$$A_t = \pi D \Delta Z$$

where D is the tube diameter and ΔZ is the length of segment.

The resulting forms of the component and energy balance equations are thus

$$\frac{dC_{An}}{dt} = \frac{C_{An-1} - C_{An+1}}{2\tau_n} + r_{An}$$

$$\frac{dT_n}{dt} = \frac{T_{n-1} - T_{n+1}}{2\tau_n} + \frac{r_{An}}{v_A}\frac{(-\Delta H)}{\rho c_p} - \frac{U\Delta A_t(T_n - T_j)}{\Delta V \rho c_p}$$

The above equations are linked by the reaction rate term r_A, which depends on concentration and temperature.

In the above equation, τ_n is the mean residence time in segment n and is equal to the volume of the segment divided by the volumetric flow rate

$$\tau_n = \frac{\Delta V}{F}$$

The modelling of the end sections, however, needs to be handled separately, according to the appropriate boundary conditions. The concentration and temperature conditions at the inlet to the first segment, n=1, are, of course with no axial dispersion, identical to those of the feed, C_{A0} and T_0, which gives rise to a slightly different form of the balance equations for segment 1. The outlet stream conditions may be taken as C_{AN} and T_N or, more accurately, as suggested by Franks (1967), one may assume an extrapolated value of C_{AN+1} and T_{N+1} through segments N−1, N and N+1. This is treated in greater detail in the following section.

4.3.6
Dynamics of an Isothermal Tubular Reactor with Axial Dispersion

Axial and radial dispersion or non-ideal flow in tubular reactors is usually characterised by analogy to molecular diffusion, in which the molecular diffusivity is replaced by eddy dispersion coefficients, characterising both radial and longitudinal dispersion effects. In this text, however, the discussion will be limited to that of tubular reactors with axial dispersion only. Otherwise the model equations become too complicated and beyond the capability of a simple digital simulation language.

Longitudinal diffusion can be analysed using the unsteady-state diffusion equation

$$\frac{\partial C}{\partial t} = -\frac{\partial}{\partial Z}\left(D\frac{\partial C}{\partial Z}\right)$$

based on Fick's Law.

In the above case, D is an eddy dispersion coefficient and Z is the axial distance along the reactor length. When combined with an axial convective flow contribution and considering D as constant the equation takes the form

$$\frac{\partial C}{\partial t} = vC - D\frac{\partial^2 C}{\partial Z^2}$$

where v is the linear flow velocity.

Written in dimensionless form, this equation is seen to depend on a dimensionless group vL/D, which is known as the Peclet number. The inverse of the Peclet number is called the Dispersion Number. Both terms represent a measure of the degree of dispersion or axial mixing in the reactor. Thus low values of the Peclet number correspond to high dispersion coefficients, low velocities and short lengths of tube and thus characterise conditions approximating to those of perfect mixing. For high values of the Peclet number the converse conditions apply and thus characterise conditions approximating to perfect plug flow.

In the extreme, the Peclet number corresponds to the following conditions:

Pe → 0   perfect mixing prevails
Pe → ∞   plug flow prevails

4.3.6.1 Dynamic Difference Equation for the Component Balance Dispersion Model

The development of the equations for the dynamic dispersion model starts by considering an element of tube length ΔZ, with a cross-sectional area of A_c, a superficial flow velocity of v and an axial dispersion coefficient or diffusivity D. Convective and diffusive flows of component A enter and leave the element, as shown by the solid and dashed arrows, respectively, in Fig. 4.12.

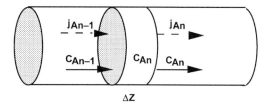

Fig. 4.12 Fluxes for the axial dispersion model.

For each element, the material balance is

$$\begin{pmatrix}\text{Rate of}\\\text{accumulation}\\\text{of A}\end{pmatrix} = \begin{pmatrix}\text{Convective}\\\text{flow of}\\\text{A in}\end{pmatrix} - \begin{pmatrix}\text{Convective}\\\text{flow of}\\\text{A out}\end{pmatrix} + \begin{pmatrix}\text{Diffusive}\\\text{flow of}\\\text{A in}\end{pmatrix} -$$

$$- \begin{pmatrix}\text{Diffusive}\\\text{flow of}\\\text{A out}\end{pmatrix} - \begin{pmatrix}\text{Rate of}\\\text{loss of A}\\\text{due to}\\\text{reaction}\end{pmatrix}$$

As before, the concentrations are taken as the average in each segment and the diffusion fluxes are related to the concentration gradients at the segment boundaries.

The concentrations of reactant entering and leaving section n are

$$C_{Ain} = \frac{C_{An-1} + C_{An}}{2}$$

and

$$C_{Aout} = \frac{C_{An} + C_{An+1}}{2}$$

The concentration gradients at the inlet and outlet of the section are

$$\left(\frac{dC}{dZ}\right)_{in} = \frac{C_{An-1} - C_{An}}{\Delta Z}$$

and

$$\left(\frac{dC}{dZ}\right)_{out} = \frac{C_{An} - C_{An+1}}{\Delta Z}$$

The convective mass flows in and out are obtained by multiplying the respective concentrations by the volumetric flow rate, which is equal to $A_c v$. The diffusive

4.3 Tubular Chemical Reactors

mass flows are calculated from the inlet and outlet concentration gradients using the multiplying factor of $A_c D$.

Dropping the A subscript for concentration, the component balance for reactant A, in section n, becomes

$$A_c \Delta Z \frac{dC_n}{dt} = A_c v \left(\frac{C_{n-1} + C_n}{2} - \frac{C_n + C_{n+1}}{2} \right) +$$
$$+ A_c D \left(\frac{C_{n-1} - C_n}{\Delta Z} - \frac{C_n - C_{n+1}}{\Delta Z} \right) - k_n C_n A_c \Delta Z$$

where here the reaction rate is taken as first order, $r_A = k C_A$.

Dividing by $A_c \Delta Z$ gives the defining component material balance equation for segment n, as

$$\frac{dC_n}{dt} = \frac{v}{\Delta Z} \left(\frac{C_{n-1} - C_{n+1}}{2} \right) + D \frac{(C_{n-1} - 2C_n + C_{n+1})}{\Delta Z^2} - k_n C_n$$

A dimensionless form of the balance equation can be obtained by substituting the following dimensionless variables

$$\overline{C}_n = \frac{C_n}{C_0}, \quad \Delta \overline{Z} = \frac{\Delta Z}{L}, \quad \overline{t} = \frac{tv}{L}$$

The model for section n, now in dimensionless form, yields

$$\frac{d\overline{C}_n}{d\overline{t}} = \frac{\overline{C}_{n-1} - \overline{C}_{n+1}}{2 \Delta \overline{Z}} + \frac{D}{vL} \left(\frac{\overline{C}_{n-1} - 2\overline{C}_n + \overline{C}_{n+1}}{\Delta \overline{Z}^2} \right) - \frac{k_n L}{v \overline{C}_n}$$

where $D/Lv = 1/Pe$, is the inverse Peclet number.

The boundary conditions determine the form of balance equation for the inlet and outlet sections. These require special consideration as to whether diffusion fluxes can cross the boundaries in any particular physical situation. The physical situation of closed ends is considered here. This would be the case if a smaller pipe were used to transport the fluid in and out of the reactor, as shown in Figs. 4.13 and 4.14.

Since no diffusive flux enters the closed entrance of the tube, the component balance for the first section becomes

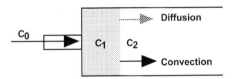

Fig. 4.13 Inlet section for the tubular reactor.

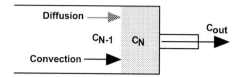

Fig. 4.14 Outlet section of the tubular reactor.

$$A_c \Delta Z \frac{dC_1}{dt} = A_c v \left(C_0 - \frac{C_1 + C_2}{2} \right) - A_c D \frac{(C_1 - C_2)}{\Delta Z} - k_1 C_1 A_c \Delta Z$$

Dividing by $A_c \Delta Z$ gives

$$\frac{dC_1}{dt} = \frac{v(2C_0 - C_1 - C_2)}{2\Delta Z} - D\frac{C_1 - C_2}{\Delta Z^2} - k_1 C_1$$

Similarly, the outlet of the reactor is closed for diffusion as shown in Fig. 4.14.

An extrapolation of the concentration profile over the last half of element N is used to calculate the outlet concentration C_{out}, giving

$$C_{out} = C_N - \frac{C_{N-1} - C_N}{2}$$

with the balance for section N becoming

$$A_c \Delta Z \frac{dC_N}{dt} = A_c v(C_{N-1} - C_N) + \frac{A_c D}{\Delta Z}(C_{N-1} - C_N) - k_N C_N A_c \Delta Z$$

$$\frac{dC_N}{dt} = v\frac{C_{N-1} - C_N}{\Delta Z} + D\frac{C_{N-1} - C_N}{\Delta Z^2} - k_N C_N$$

A similar finite-differenced equivalent for the energy balance equation (including axial dispersion effects) may be derived. The simulation example DISRET involves the axial dispersion of both mass and energy and is based on the work of Ramirez (1976). A related model without reaction is used in the simulation example FILTWASH.

4.3.7
Steady-State Tubular Reactor Dispersion Model

Letting the element distance ΔZ approach zero in the finite-difference form of the dispersion model, gives

$$\frac{\partial C_A}{\partial t} = -v\frac{\partial C_A}{\partial Z} + D\frac{\partial^2 C_A}{\partial Z^2} - kC_A$$

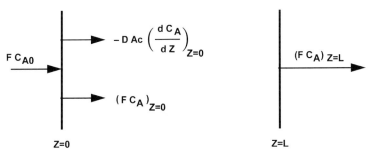

Fig. 4.15 Convective and diffusive fluxes at the entrance (Z=0) and exit (Z=L) of the tubular reactor.

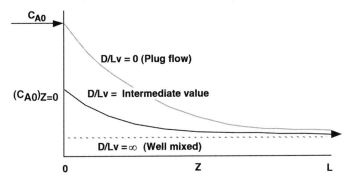

Fig. 4.16 Concentration profiles in the tubular reactor for extreme and intermediate values of the dispersion number.

Defining the following dimensionless variables

$$\overline{C}_A = \frac{C_A}{C_{A0}}, \quad \overline{Z} = \frac{Z}{L}, \quad \overline{t} = \frac{t}{\tau}$$

where $\tau = L/v$.

The dimensionless form for an nth-order reaction is

$$\frac{\partial \overline{C}_A}{\partial \overline{t}} = -\frac{\partial \overline{C}_A}{\partial \overline{Z}} - \left(\frac{D}{Lv}\right) \frac{\partial^2 \overline{C}_A}{\partial \overline{Z}^2} - k\tau C_{A0}^{n-1} (\overline{C}_A)^n$$

At steady state $\partial \overline{C}_A / \partial \overline{t}$ can be set to zero and the equation becomes an ordinary second-order differential equation, which can be solved using MADONNA.

Again the entrance and exit boundary conditions must be considered. Thus the two boundary conditions at Z=0 and Z=L are used for solution, as shown in Fig. 4.15. Note that these boundary conditions refer to the inner side of the tubular reactor. A discontinuity in concentration at Z=0 is apparent in Fig. 4.16.

Balancing the material flows at the inlet gives a relation for the boundary conditions at $Z=0$

$$FC_{A0} = (FC_A)_{Z=0} - DA_c \left(\frac{dC_A}{dZ}\right)_{Z=0}$$

The zero flux condition at the closed outlet requires a zero gradient, thus

$$\left(\frac{dC_A}{dZ}\right)_{Z=L} = 0$$

According to the boundary conditions, the concentration profile for A must change with a discontinuity in concentration from C_{A0} to $(C_{A0})_{Z=0}$ occurring at the reactor entrance, as shown in Fig. 4.16.

In dimensionless form the boundary condition at $\bar{Z}=0$ is represented by

$$\bar{C}_A - \frac{D}{Lv}\frac{d\bar{C}_A}{d\bar{Z}} = 1$$

and at $\bar{Z}=1$ by $\frac{d\bar{C}_A}{d\bar{Z}} = 0$.

The solution requires two integrations as shown in Fig. 4.17.

Referring to Fig. 4.15, it is seen that the concentration and the concentration gradient are unknown at $Z=0$. The above boundary condition relation indicates that if one is known, the other can be calculated. The condition of zero gradient at the outlet ($Z=L$) does not help to start the integration at $Z=0$, because, as Fig. 4.17 shows, two initial conditions are necessary. The procedure to solve this split-boundary value problem is therefore as follows:

1. Guess $(\bar{C}_A)_{Z=0}$ and calculate $\left(\frac{d\bar{C}_A}{d\bar{Z}}\right)_{\bar{Z}=0}$ from the boundary condition relation.
2. Integrate to $\bar{Z}=1$ and check whether $\left(\frac{d\bar{C}_A}{d\bar{Z}}\right)_{\bar{Z}=1}$ equals zero.
3. Vary the guess and iterate between $\bar{Z}=0$ and $\bar{Z}=1$ until convergence is obtained.

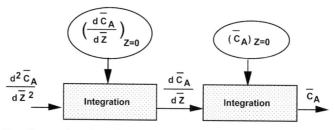

Fig. 4.17 Flow diagram for solving the second-order differential equation from the axial dispersion model.

4.4
Differential Mass Transfer

This section concerns the modelling of countercurrent flow, differential mass transfer applications, for both steady-state and non-steady-state design or simulation purposes. For simplicity, the treatment is restricted to the case of a single solute, transferring between two inert phases, as in the standard treatments of liquid–liquid extraction or gas absorption column design.

4.4.1
Steady-State Gas Absorption with Heat Effects

Figure 4.18 represents a countercurrent-flow, packed gas absorption column, in which the absorption of solute is accompanied by the evolution of heat. In order to treat the case of concentrated gas and liquid streams for which total flow rates of both gas and liquid vary throughout the column, the solute concentrations in the gas and liquid are defined in terms of mole ratio units and related to the molar flow rates of solute free gas and liquid respectively, as discussed previously in Section 3.3.2. By convention, the mass transfer rate equation is however expressed in terms of mole fraction units. In Fig. 4.18, G_m is the molar flow of solute free gas (kmol/m²s) and L_m is the molar flow of solute free liquid (kmol/m²s), where both L_m and G_m remain constant throughout the column. Y is the mole ratio of solute in the gas phase (kmol solute/kmol solute free gas), X is the mole ratio of solute in the liquid phase (kmol solute/kmol solute free liquid), y is the mole fraction of solute in the gas phase, x is the mole fraction of solute in the liquid phase and T is temperature (K).

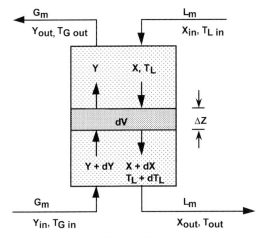

Fig. 4.18 Steady-state gas absorption with heat effects.

Mole ratio and mole fraction contents are related by

$$Y = \frac{y}{1-y} \quad \text{and} \quad X = \frac{x}{1-x}$$

$$x = \frac{Y}{1+Y} \quad \text{and} \quad y = \frac{X}{1+X}$$

Subscripts L and G refer to the liquid and gas phases, respectively, and subscripts 'in' and 'out' refer to the inlet and outlet streams.

4.4.1.1 Steady-State Design

In the steady-state design application, the flow rates L_m and G_m and concentrations Y_{in}, X_{in}, Y_{out} and X_{out} will either be specified or established by an overall steady-state solute balance, where

$$L_m X_{in} + G_m Y_{in} = L_m X_{out} + G_m Y_{out}$$

Temperatures T_{Lin} and T_{Gin} will also be known. The problem then consists of determining the height of packing required to obtain the above separation.

Component Material Balance Equations

For a small element of column volume dV

$$\begin{pmatrix} \text{Rate of loss} \\ \text{of solute from} \\ \text{the gas} \end{pmatrix} = \begin{pmatrix} \text{Rate of gain} \\ \text{of solute in} \\ \text{the liquid} \end{pmatrix} = \begin{pmatrix} \text{Rate} \\ \text{of solute} \\ \text{transfer} \end{pmatrix}$$

$$-G_m A_c dY = L_m A_c dX = K_{Lx} a(x^* - x) dV$$

Here $K_{Lx}a$ (kmol/m^{3}s) is the overall mass transfer coefficient for the liquid phase, based on mole fraction in the L-phase, x^* is the equilibrium liquid phase mole fraction, and A_c is the cross-sectional area of the column (m^2). Hence with $dV = A_c dZ$

$$\frac{dY}{dZ} = \frac{K_{Lx} a(x^* - x)}{G_m}$$

and

$$\frac{dX}{dZ} = \frac{K_{Lx} a(x^* - x)}{L_m}$$

Energy Balance

It is assumed that there are no heat losses from the column and that there is zero heat exchange between the gas and liquid phases. Consequently the gas phase temperature will remain constant throughout the column. A liquid phase heat balance for element of volume dV is given by

$$\begin{pmatrix} \text{Rate of gain of heat} \\ \text{by the liquid} \end{pmatrix} = \begin{pmatrix} \text{Rate of generation of} \\ \text{heat by absorption} \end{pmatrix}$$

or

$$L A_c c_p dT_L = K_{Lx} a (x^* - x) dV \Delta H_{abs}$$

where L is the total mass flow rate of liquid (kg/m^2 s), c_p is the specific heat capacity of the liquid (kJ/kg K) and ΔH_{abs} is the exothermic heat of absorption (kJ/kmol solute transferred). Hence

$$\frac{dT_L}{dZ} = \frac{K_{Lx} a (x^* - x) \Delta H_{abs}}{L c_p}$$

The temperature variation throughout the column is important, since this affects the equilibrium concentration x^*, where

$$x^* = f_{eq}(y, T_L)$$

Solution of the required column height is achieved by integrating the two component balance equations and the heat balance equation down the column from the known conditions x_{in}, y_{out} and T_{Lin}, until the condition that either Y is greater than Y_{in} or X is greater than X_{out} is achieved. In this solution approach, variations in the overall mass transfer capacity coefficient both with respect to temperature and to concentration, if known, can also be included in the model as required. The solution procedure is illustrated by the simulation examples AMMONAB and BIOFILT.

Using the digital simulation approach to steady-state design, the above design calculation is shown to proceed naturally from the defining component balance and energy balance equations, giving a considerable simplification to conventional text book approaches.

4.4.1.2 Steady-State Simulation

In this case, the flow rates L_m and G_m, concentrations Y_{in} and X_{in}, temperatures T_{Gin} and T_{Lin}, are known and in addition the height of packing Z is also known. It is now, however, required to establish the effective column performance by determining the resulting steady-state concentration values, Y_{out} and X_{out}, and also temperature T_{Lout}. The problem is now of a split-boundary type and must be solved by assuming a value for Y_{out}, integrating down the column for a column distance Z and comparing the calculated value for Y_{in} with the known inlet gas concentration condition. A revised guess for the starting value Y_{out} can then be taken and the procedure repeated until convergence is achieved. This is easy using MADONNA's split boundary tool.

4.4.2
Dynamic Modelling of Plug-Flow Contactors: Liquid–Liquid Extraction Column Dynamics

A plug-flow, liquid–liquid, extraction column is represented in Fig. 4.19. For convenience, it is assumed that the column operates under low concentration conditions, such that the aqueous and organic flow rates, L and G, respectively are constant. At low concentration, mole fraction x and y are identical to mole ratios X and Y, which are retained here in the notation for convenience. This however leads to a more complex formulation than when concentration quantities are used, as in the example AXDISP.

Consider a differential element of column volume ΔV, height ΔZ and cross-sectional area, A_c, such that $\Delta V = A_c \Delta Z$. Component material balance equations can be written for each of the liquid phases, where

$$\begin{pmatrix} \text{Rate of} \\ \text{accumulation} \\ \text{of solute} \end{pmatrix} = \begin{pmatrix} \text{Convective} \\ \text{flow of} \\ \text{solute in} \end{pmatrix} - \begin{pmatrix} \text{Convective} \\ \text{flow of} \\ \text{solute out} \end{pmatrix} + \begin{pmatrix} \text{Rate of} \\ \text{solute} \\ \text{transfer} \end{pmatrix}$$

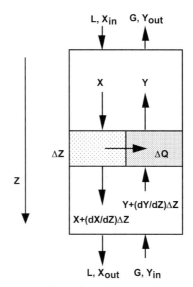

Fig. 4.19 Liquid–liquid extraction column dynamic representation.

4.4 Differential Mass Transfer

The actual volume of each phase in element ΔV is that of the total volume of the element, multiplied by the respective fractional phase holdup. Hence considering the direction of solute transfer to occur from the aqueous or feed phase into the organic or solvent phase, the material balance equations become:

For the aqueous phase

$$\rho_L h_L \Delta V \frac{\partial X}{\partial t} = -LA_c \frac{\partial X}{\partial t} \Delta Z - K_{LX} a (X - X^*) \Delta V$$

for the organic phase

$$\rho_G h_G \Delta V \frac{\partial Y}{\partial t} = GA_c \frac{\partial Y}{\partial Z} \Delta Z + K_{LX} a (X - X^*) \Delta V$$

where each term in the equations has units of kmol solute/s and where the symbols are as follows:

- a is the specific interfacial area related to the total volume (m^2/m^3)
- A_c is the column cross-sectional area (m^2)
- G is the molar flow rate of the light, organic phase per unit area $(kmol/m^2 s)$
- h_G is the volumetric holdup fraction of the light organic phase (–)
- h_L is the volumetric holdup fraction of the heavy aqueous phase (–)
- $K_{LX} a$ is the overall mass transfer capacity coefficient based on the aqueous phase mole ratio X $(kmol/m^3 s)$
- L is the molar flow rate of the heavy aqueous phase per unit area $(kmol/m^2 s)$
- X is the aqueous phase mole ratio (kmol solute/kmol water)
- X^* is the equilibrium mole ratio in the heavy phase, corresponding to light phase mole ratio Y (kmol solute/kmol water)
- Y is the organic phase mole ratio (kmol solute/kmol organic)
- Z is the height of the packing (m)
- ΔV is the total volume of one column segment with length ΔZ (m^3)
- ρ_G is the density of the solute-free light phase $(kmol/m^3)$
- ρ_L is the density of the solute-free heavy phase $(kmol/m^3)$

The above balance equations simplify to

$$\rho_L h_L \frac{\partial X}{\partial t} = L \frac{\partial X}{\partial Z} - K_{LX} a (X - X^*)$$

$$\rho_G h_G \frac{\partial Y}{\partial t} = -G \frac{\partial Y}{\partial Z} + K_{LX} a (X - X^*)$$

with the equilibrium relationship represented by

$$X^* = f_{eq}(Y)$$

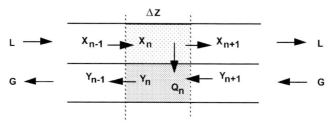

Fig. 4.20 Finite-difference element for the two-phase transfer system.

Thus the system is defined by two coupled partial differential equations, which can be solved by finite-differencing.

Consider a finite-difference element of length ΔZ as shown in Fig. 4.20.

Approximating the concentrations entering and leaving each section by an arithmetical mean of the neighbouring concentrations, as shown in Section 4.3.5, the component balance equations for stage n become

$$\rho_L h_L A_c \Delta Z \frac{dX_n}{dt} = L A_c \left(\frac{(X_{n-1} + X_n)}{2} - \frac{(X_n + X_{n+1})}{2} \right) - Q_n$$

and

$$\rho_G h_G A_c \Delta Z \frac{dY_n}{dt} = G A_c \left(\frac{(Y_{n+1} + Y_n)}{2} - \frac{(Y_n + Y_{n-1})}{2} \right) + Q_n$$

where Q_n (kmol solute/s) is rate of solute transfer in element n given by

$$Q_n = K_{LX} \, a \, (X - X^*) \Delta V$$

Hence with $\Delta V = A_c \Delta Z$

$$\rho_L h_L \frac{dX_n}{dt} = -L \frac{(X_{n+1} - X_{n-1})}{2 \Delta Z} - K_{LX} \, a \, (X - X^*)$$

$$\rho_G h_G \frac{dY_n}{dt} = -G \frac{(Y_{n+1} - Y_{n-1})}{2 \Delta Z} + K_{LX} \, a \, (X - X^*)$$

The boundary conditions are formulated with the help of Figs. 4.21 and 4.22 and in accordance with the methodology of Franks (1967).

This gives for stage 1

$$\rho_L h_L \frac{dX_1}{dt} = \frac{L}{2 \Delta Z} (2X_0 - X_1 - X_2) - K_{LX} \, a \, (X_1 - X_1^*)$$

$$\rho_G h_G \frac{dY_1}{dt} = \frac{G}{2 \Delta Z} (Y_2 + Y_1 - 2Y_0) + K_{LX} \, a \, (X_1 - X_1^*)$$

4.4 Differential Mass Transfer

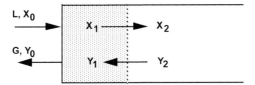

Fig. 4.21 Aqueous and organic phase streams at the inlet.

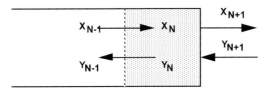

Fig. 4.22 Aqueous and organic phase streams at the outlet.

The balances for the end stage N thus become

$$\rho_L h_L \frac{dX_N}{dt} = \frac{L}{2\Delta Z}(X_{N-1} - X_N - 2X_{N+1}) - K_{LX} a (X_N - X_N^*)$$

$$\rho_G h_G \frac{dY_N}{dt} = \frac{G}{2\Delta Z}(Y_{N+1} - Y_N - Y_{N-1}) + K_{LX} a (X_N - X_N^*)$$

The representation of the boundary conditions for both the top and bottom of the column are really more mathematical than practical in nature and fail to take into account the actual geometry and construction of the upper and lower parts of the column and the relative positioning of the inlet and outlet connections. They may therefore require special modelling appropriate to the particular form of construction of the column, as discussed previously in Section 3.3.1.10.

4.4.3
Dynamic Modelling of a Liquid–Liquid Extractor with Axial Mixing in Both Phases

Axial mixing is known to have a very significant effect on the performance of agitated liquid–liquid extraction columns, and any realistic description of column performance must take this into account. Figure 4.23 represents a small differential element of column volume ΔV and height ΔZ. Here the convective flow rates, as in Fig. 4.19, are shown by the solid arrows, and the additional dispersion contributions, representing axial mixing, are shown by dashed arrows. It is assumed that axial mixing in both phases can be described by analogy to Fick's Law, but using an effective eddy dispersion coefficient appropriate to the respective liquid phase.

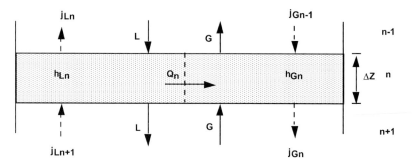

Fig. 4.23 Differential element of height, ΔZ, for a liquid–liquid extractor with axial mixing in both phases.

Writing unsteady-state component balances for each liquid phase results in the following pair of partial differential equations which are linked by the mass transfer rate and equilibrium relationships

$$\rho_L h_L \frac{\partial X}{\partial t} = -L\frac{\partial X}{\partial Z} + \rho_L h_L D_L \frac{\partial^2 X}{\partial Z^2} - K_{LX} a\,(X - X^*)$$

$$\rho_G h_G \frac{\partial Y}{\partial t} = G\frac{\partial Y}{\partial Z} + \rho_G h_G D_G \frac{\partial^2 Y}{\partial Z^2} + K_{LX} a\,(X - X^*)$$

Here the nomenclature is the same as in Section 4.4.2. In addition, D_G is the effective eddy dispersion coefficient for the organic or extract phase (m²/s) and D_L is the effective eddy dispersion coefficient for the aqueous or feed phase (m²/s). The above equations are difficult to solve analytically (Lo et al., 1983) but are solved with ease using digital simulation.

Referring to Fig. 4.23, the extractor is again divided into N finite-difference elements or segments of length ΔZ. The convective terms are formulated for each segment using average concentrations entering and leaving the segment, as shown in Section 4.4.2. The backmixing terms j are written in terms of dispersion coefficients times driving-force mole-ratio gradients. The resulting equations for any segment n are then for the aqueous feed phase with each term in kmol solute/s.

$$\rho_L h_{Ln} \Delta Z A_c \frac{dX_n}{dt} = L A_c \left(\frac{X_n + X_{n-1}}{2} - \frac{X_{n+1} + X_n}{2} \right) +$$

$$+ \rho_L D_L h_{Ln} A_c \left(\frac{X_{n+1} - X_n}{\Delta Z} - \frac{X_n - X_{n-1}}{\Delta Z} \right) - K_{LXn}\, a_n\, (X_n - X_n^*) \Delta Z A_c$$

Rearranging

$$\frac{dX_n}{dt} = -\frac{L}{2h_{Ln}\rho_L \Delta Z}(X_{n+1} - X_{n-1}) + \frac{D_L}{\Delta Z^2}(X_{n+1} - 2X_n + X_{n-1}) - \frac{K_{LXn}\,a_n}{h_{Ln}\rho_L}(X_n - X_n^*)$$

Similarly for the light solvent–extract phase

$$\Delta Z A_c h_{Gn} \rho_G \frac{dY_n}{dt} = A_c G \left(\frac{Y_n + Y_{n+1}}{2} - \frac{Y_{n-1} + Y_n}{2}\right) +$$

$$+ \rho_G D_G h_{Gn} A_c \left(\frac{Y_{n-1} - Y_n}{\Delta Z} - \frac{Y_n - Y_{n+1}}{\Delta Z}\right) + K_{LXn}\,a_n (X_n - X_n^*)\Delta Z A_c$$

Rearranging

$$\frac{dY_n}{dt} = \frac{G}{2h_{Gn}\rho_G \Delta Z}(Y_{n+1} - Y_{n-1}) + \frac{D_G}{\Delta Z^2}(Y_{n+1} - 2Y_n + Y_{n-1}) - \frac{K_{LXn}\,a_n}{h_{Gn}\rho_G}(X_n - X_n^*)$$

Note that the above formulation includes allowance for the fractional phase holdup volumes, h_L and h_G, the phase flow rates, L and G, the diffusion coefficients D_L and D_G, and the overall mass transfer capacity coefficient K_{LX} a, all of which may vary with position along the extractor.

Boundary Conditions

The column end sections require special treatment to allow for the fact that zero diffusive flux enters through the end wall of the column. The equations for the end section are derived by setting the diffusion flux leaving the column to zero. In addition, the liquid phase outlet concentrations leaving the respective end sections of the column are approximated by an extrapolation of the concentration gradient from the preceding section. The resulting model equations give the concentrations of each segment in both phases as well as the outlet concentrations as a function of time. The resulting model formulation is shown in the simulation example AXDISP.

4.4.4
Dynamic Modelling of Chromatographic Processes

Preparative chromatographic processes are of increasing importance particularly in the production of fine chemicals. A mixture of compounds is introduced into the liquid mobile phase, and this then flows through a packed column containing the stationary solid phase. The contacting scheme is thus differential, but since the adsorption characteristics of the compounds in the mixture are similar, many equivalent theoretical stages are required for their separation. Chromatographic processes are mostly run under transient conditions, such that concentration variations occur with respect to both time and space, but steady-state and quasi-steady-state systems are also being applied increasingly to overcome the inherent disadvantages of batch operation. The transient operational mode is essentially a scaled-up version of the usual analytical chromatography, but whereas analytical systems are usually run with low concentrations to avoid non-linearities, preparative industrial systems are highly loaded to increase productivity. Countercurrent chromatography is still not available, but simulated moving bed chromatography with switching between a series of columns is used increasingly industrially. An important feature of chromatographic process behaviour is that it is generally governed more by the adsorption equilibrium than by the kinetics of adsorption.

The modelling of chromatographic processes is treated in great detail by Ruthven and Ching (1993) and by Blanch and Clark (1996), with two alternative approaches being available. In a most rigorous approach the chromatographic separation column is considered as a plug flow contactor with axial dispersion analogous to previous descriptions in this chapter (Section 4.3.6). The second approach is to represent the column as a large number of well mixed stages, with a treatment similar to that shown in Chapter 3.

The interaction of the two phases can be accomplished either through the assumption of equilibrium or through a transfer rate that will eventually reach equilibrium. The transfer rate approach is closer to the real process and simplifies the calculations for nonlinear equilibrium. This is similar to the modelling of extraction columns with backmixing as found in Section 4.4.3. For linear equilibrium, simplifications in the models are possible. In the following section, the dispersion model is developed and is presented as a simulation example CHROMDIFF. A further simulation example, CHROMPLATE, considers the stagewise model for linear equilibrium. Dynamic modelling and simulation of simulated moving bed chromatography has been studied by Storti et al. (1993) and Strube et al. (1998).

4.4.4.1 Axial Dispersion Model for a Chromatography Column

Generally for modelling chromatograph systems, component mass balances are required for each component in each phase. The differential liquid phase component balances for a chromatographic column with non-porous packing take the partial differential equation form

$$\frac{\partial C_L}{\partial t} = -v\frac{\partial C_L}{\partial Z} + D\frac{\partial^2 C_L}{\partial Z^2} - \frac{1-\varepsilon}{\varepsilon}r_{ads}$$

where C_L is the liquid phase concentration of each component and D is the axial dispersion coefficient.

The linear superficial flow velocity in the packing voids v is calculated from volumetric flow rate L_{in}, voidage fraction of the adsorbant bed ε and column diameter d as

$$v = \frac{4L_{in}}{\varepsilon \pi d^2}$$

The transfer rate of the sorbate from the liquid phase to the solid adsorbant r_{ads} can be written as

$$r_{ads} = k(C_S^* - C_S)$$

Here the rate is specific to a unit volume of solids (g/cm^3 s) and k is a mass transfer rate coefficient (1/s).

The solid phase concentrations are influenced only by the rate of mass transfer, with convection and dispersion effects both being zero for this phase.

$$\frac{\partial C_S}{\partial t} = \frac{1-\varepsilon}{\varepsilon} r_{ads}$$

Equilibrium relations are required to calculate the values of C_S^*, the solid phase equilibrium concentrations, for each component. For very dilute systems these relations may be of linear form

$$C_S^* = KC_L$$

where K is the equilibrium constant for the particular component.

For concentrated systems the Langmuir adsorption form may be appropriate and for an interacting two-component system (A and B) may take the form

$$C_{SA}^* = \frac{K_A C_{LA}}{1 + b_A C_{LA} + b_B C_{LB}}$$

$$C_{SB}^* = \frac{K_B C_{LB}}{1 + b_A C_{LA} + b_B C_{LB}}$$

Here the constants b_A and b_B account for the competitive adsorption effects between components A and B.

Writing the model in dimensionless form, the degree of axial dispersion of the liquid phase will be found to depend on a dimensionless group vL/D or Peclet number. This is completely analogous to the case of the tubular reactor with axial dispersion (Section 4.3.6).

4.4.4.2 Dynamic Difference Equation Model for Chromatography

Instead of the partial differential equation model presented above, the model is developed here in dynamic difference equation form, which is suitable for solution by dynamic simulation packages, such as MADONNA. Analogous to the previous development for tubular reactors and extraction columns, the development of the dynamic dispersion model starts by considering an element of tube

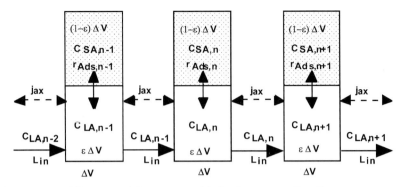

Fig. 4.24 Finite difference axial dispersion model of a chromatographic column.

length ΔZ, with a cross-sectional area A_c, a superficial flow velocity v and an axial dispersion coefficient or diffusivity D. Convective and diffusive flows of component A enter and leave the liquid phase volume of any element n, as indicated in Fig. 4.24 below. Here j represents the diffusive flux, L the liquid flowrate and C_{SA} and C_{LA} the concentration of any species A in both the solid and liquid phases, respectively.

For each element, the material balance in the liquid phase, here for component A, is

$$\begin{pmatrix} \text{Rate of} \\ \text{acumulation} \\ \text{of component} \\ \text{A} \end{pmatrix} = \begin{pmatrix} \text{Convective} \\ \text{flow of} \\ \text{A in} \end{pmatrix} - \begin{pmatrix} \text{Convective} \\ \text{flow of} \\ \text{A out} \end{pmatrix} + \begin{pmatrix} \text{Diffusive} \\ \text{flow of} \\ \text{A in} \end{pmatrix} -$$

$$- \begin{pmatrix} \text{Diffusive} \\ \text{flow of} \\ \text{A out} \end{pmatrix} + \begin{pmatrix} \text{Rate of} \\ \text{loss of A due} \\ \text{to transfer} \end{pmatrix}$$

As before, the concentrations are taken as the average in each segment and the diffusion fluxes are related to the concentration gradients at the segment boundaries.

The concentrations of reactant entering and leaving any section n are

$$C_{LA,in} = \frac{C_{LA,n-1} + C_{LA,n}}{2}$$

and

$$C_{LA,out} = \frac{C_{LA,n} + C_{LA,n+1}}{2}$$

4.4 Differential Mass Transfer

The concentration gradients at the inlet and outlet of the section are

$$\left(\frac{dC_{LA}}{dZ}\right)_{in} = \frac{C_{LA,n-1} - C_{LA,n}}{\Delta Z}$$

and

$$\left(\frac{dC_{LA}}{dZ}\right)_{out} = \frac{C_{LA,n} - C_{LA,n+1}}{\Delta Z}$$

The convective mass flows in and out of the segment are calculated by multiplying the respective concentrations by the constant volumetric flow rate, L_{in}. The diffusive mass flows are calculated from Fick's Law, using the inlet and outlet concentration gradients and the area εA_c.

The transfer rate of A, Tr_A (g/s), from liquid to solid is given by

$$Tr_A = r_{Aads}\Delta V_S = k_{eff}\, a_p\, (C_{SA}^* - C_{SA})(1 - \varepsilon)\Delta Z A_c$$

where k_{eff} (cm/s) is a transfer coefficient, a_p is the specific area of the spherical packing $(6/d_p)$ and ΔV_S is the volume of the solid phase. C_{SA}^* is given by the equilibrium relation

$$C_{SA}^* = f_{equil}(C_{LA})$$

The component balance for reactant A in the liquid phase, in any section n, becomes

$$\varepsilon A_c \Delta Z \frac{dC_{LA,n}}{dt} = L_{in}\left(\frac{C_{LA,n-1} + C_{LA,n}}{2} - \frac{C_{LA,n} + C_{LA,n+1}}{2}\right) +$$

$$+ \varepsilon A_c D\left(\frac{C_{LA,n-1} - C_{LA,n}}{\Delta Z} - \frac{C_{LA,n} - C_{LA,n+1}}{\Delta Z}\right) - (1-\varepsilon)r_{Aads}\Delta Z A_c$$

Here the specific transfer rate r_{ads} is related to the solid phase volume.

Dividing by $\varepsilon A_c \Delta Z$ gives the defining component material balance equation for segment n as

$$\frac{dC_{LA,n}}{dt} = \frac{L_{in}}{\varepsilon A_c \Delta Z}\left(\frac{C_{LA,n-1} - C_{LA,n+1}}{2}\right) +$$

$$+ D\frac{(C_{LA,n-1} - 2C_{LA,n} + C_{LA,n+1})}{\Delta Z^2} - \frac{1-\varepsilon}{\varepsilon}r_{Aads,n}$$

A dimensionless form of the balance equation can be obtained in the same way as described for the tubular reactor (Section 4.3.6).

The balance equation for component A in the solid phase balance for any element n is

$$(1 - \varepsilon) A_c \Delta Z \frac{dC_{SA,n}}{dt} = (1 - \varepsilon) A_c \Delta Z \, r_{Aads,n}$$

or simply

$$\frac{dC_{SA,n}}{dt} = r_{Aads,n}$$

The formulation of the end section balances needs special attention as already discussed in Section 4.3.6.1.

The axial dispersion coefficient may be calculated from a knowledge of the Peclet number, where

$$D = \frac{4 L_{in} D_p}{Pe \, \varepsilon \, \pi \, d^2}$$

The Reynolds number for a particle with diameter D_p is defined as

$$Re = \frac{\rho v D_p}{\eta}$$

and this is used to determine the Peclet number from a suitable correlation, such as

$$Pe = \frac{0.2}{\varepsilon} + \frac{0.011}{\varepsilon} \left(\frac{Re}{\varepsilon} \right)^{0.48}$$

These equations are applied in the simulation example CHROMDIFF to the case of a two-component separation with linear equilibrium. The situation of a non-linear equilibrium is considered as an exercise in the example.

The simulation program CHROMPLATE uses the plate model for the same column conditions as the simulation model CHROMDIFF. The results obtained are very similar in the two approaches, but the stagewise model is much faster to calculate.

With high concentrations, heat effects in the chromatographic column may be important. This would require the simultaneous application of an energy balance and the introduction of a term reflecting the influence of temperature on the adsorption equilibrium.

4.5
Heat Transfer Applications

4.5.1
Steady-State Tubular Flow with Heat Loss

Here a steady-state formulation of heat transfer is considered (Pollard, 1978). A hot fluid flows with linear velocity v, through a tube of length L, and diameter D, such that heat is lost via the tube wall to the surrounding atmosphere. It is required to find the steady-state temperature profile along the tube length.

Consider an element of tube of length ΔZ, distance Z from the tube inlet as shown in Fig. 4.25. If the temperature at the inlet to the tube element is T, then the temperature at the outlet of the element can be written as $T+(dT/dZ)\,\Delta Z$.

The energy balance for the element of tube length can be stated as

$$\begin{pmatrix}\text{Rate of} \\ \text{accumulation} \\ \text{of enthalpy} \\ \text{in the element}\end{pmatrix} = \begin{pmatrix}\text{Heating of} \\ \text{inlet stream} \\ \text{to element} \\ \text{temperature}\end{pmatrix} - \begin{pmatrix}\text{Rate of} \\ \text{heat loss} \\ \text{to the wall}\end{pmatrix}$$

As shown in Section 1.2.5 the heat balance equation, assuming constant fluid properties, becomes

$$Mc_p \frac{dT}{dt} = Wc_p \left(T - \left(T + \frac{dT}{dZ}\Delta Z \right) \right) - UA(T - T_w)$$

where

A is the heat transfer surface area for the element = $2\pi D\,\Delta Z$ (m²)
D is the tube diameter (m)
M is mass of fluid in the element = $(\pi D^2/4)\rho\Delta Z$ (kg)
T_s is the wall temperature (K)
T_w is the temperature of the wall (K)
U is the heat transfer coefficient between the fluid and the wall (kJ/m² s)
v is the linear velocity of the fluid (m/s)
W is the mass flow rate along the tube = $(\pi D^2/4)\rho v$ (kg/s)

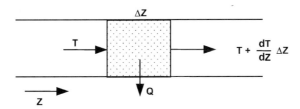

Fig. 4.25 Tubular flow with heat loss.

Simplifying the above equation gives

$$\frac{dT}{dt} = -v\frac{dT}{dZ} - \frac{4U}{\rho c_p D}(T - T_w)$$

Under steady-state conditions, $\frac{dT}{dt} = 0$, and the resulting temperature profile along the tube is given by

$$v\frac{dT}{dZ} + \frac{4U}{\rho c_p D}(T - T_s) = 0$$

Assuming constant coefficients, both the dynamic and steady-state equations describing this system can be solved analytically, but the case of varying coefficients requires solution by digital simulation.

4.5.2
Single-Pass, Shell-and-Tube, Countercurrent-Flow Heat Exchanger

4.5.2.1 Steady-State Applications

Figure 4.26 represents a steady-state, single-pass, shell-and-tube heat exchanger. For this problem W is the mass flow rate (kg/s), T is the temperature (K), c_p is the specific heat capacity (kJ/m² s), A (= $\pi D Z$) is the heat transfer surface area (m²), and U is the overall heat transfer coefficient (kJ/m² s K). Subscripts c and h refer to the cold and hot fluids, respectively.

Heat balances on a small differential element of heat transfer surface area, ΔA, give

$$\begin{pmatrix} \text{Rate of} \\ \text{accumulation} \\ \text{of enthalpy} \\ \text{in the element} \end{pmatrix} = \begin{pmatrix} \text{Rate of} \\ \text{heat transfer} \\ \text{to the element} \end{pmatrix}$$

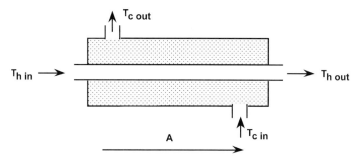

Fig. 4.26 Steady-state, countercurrent flow heat exchange.

Thus for the hot fluid

$$(W\,c_p)_h \Delta T_h = -U(T_h - T_c)\Delta A$$

and for the countercurrent cold fluid

$$(W\,c_p)_c \Delta T_c = -U(T_h - T_c)\Delta A$$

In the limit, the defining model equations for countercurrent flow become

$$\frac{dT_h}{dA} = -\frac{U\,(T_h - T_c)}{(Wc_p)_h}$$

and

$$\frac{dT_c}{dA} = -\frac{U\,(T_h - T_c)}{(Wc_p)_c}$$

where for cocurrent flow the sign in cold-fluid equation would be positive.

For design purposes the two equations can be integrated directly starting for the known temperature conditions at one end of the exchanger and integrating towards the known conditions at the other end, hence enabling the required heat exchange surface to be determined. This procedure is very similar to that of the steady-state mass transfer column calculation of Section 4.4.1.1. The design approach for a steady-state two-pass exchanger is illustrated by simulation example SSHEATEX.

However, the simulation of the steady-state performance for a heat exchanger with a known heat transfer surface area will demand an iterative split boundary solution approach, based on a guessed value of the temperature of one of the exit streams, as a starting point for the integration.

4.5.2.2 Heat Exchanger Dynamics

The modelling procedure is again based on that of Franks (1967). A simple, single-pass, countercurrent flow heat exchanger is considered. Heat losses and heat conduction along the metal wall are assumed to be negligible, but the dynamics of the wall (thick-walled metal tube) are significant.

Figure 4.27 shows the temperature changes over a small differential element of exchanger length ΔZ.

In this problem W is the mass flow rate (kg/s), T is temperature (K), c_p is the specific heat capacity (kJ/kg K), D is the diameter (m), U is the heat transfer coefficient (kJ/m^2 s K), Q is the rate of heat transfer (kJ/s), V is the volume (m^3), ρ is the density (kg/m^3) and A is the heat transfer area (m^2). The subscripts are as follows: t refers to tube conditions, s to shellside conditions, and m to the metal wall.

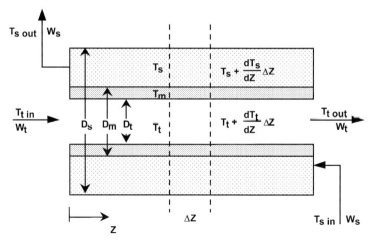

Fig. 4.27 Shell-and-tube heat exchanger: differential model.

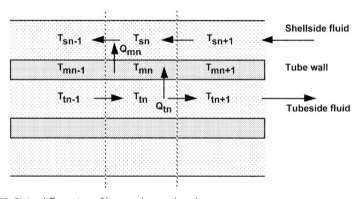

Fig. 4.28 Finite-differencing of heat exchanger length.

Heat balance equations on the element of heat exchanger length ΔZ according to enthalpy balance relationship

$$\begin{pmatrix} \text{Rate of} \\ \text{accumulation} \\ \text{of enthalpy} \\ \text{in the element} \end{pmatrix} = \begin{pmatrix} \text{Flow rate} \\ \text{of enthalpy} \\ \text{into} \\ \text{the element} \end{pmatrix} - \begin{pmatrix} \text{Flow rate} \\ \text{of enthalpy} \\ \text{out of} \\ \text{the element} \end{pmatrix} + \begin{pmatrix} \text{Rate of} \\ \text{heat transfer} \\ \text{to the element} \end{pmatrix}$$

lead to three coupled first-order partial differential equations, which can be converted into difference equations for simulation language solution using standard finite-difference formulae as mentioned in Section 4.6.

4.5 Heat Transfer Applications

Alternatively, the difference-equation model form can be derived directly by dividing the length of the heat exchanger into N finite-difference elements or segments, each of length ΔZ, as shown in Fig. 4.28.

The heat balance equation can now be applied to segment n of the heat exchanger. The heat transfer rate equations are given by the following terms:

Rate of heat transfer from tube contents to the metal wall

$$Q_{tn} = U_t \Delta A_t (T_{tn} - T_{mn})$$

where U_t is the tubeside film heat transfer coefficient and ΔA_t is the incremental tubeside area

$$\Delta A_t = \pi D_t \Delta Z$$

Rate of heat transfer from the metal wall to the shellside contents

$$Q_{mn} = U_m \Delta A_m (T_{mn} - T_{sn})$$

where U_m is the film heat transfer coefficient from the wall to the shell and ΔA_m is the incremental metal wall outside area

$$\Delta A_m = \pi D_m \Delta Z$$

Using a similar treatment as described previously in Section 4.3.5, the resulting finite difference form of the enthalpy balance equations for any element n become

$$\frac{dT_n}{dt} = \frac{W_t c_{pt} (T_{tn-1} - T_{tn+1}) - Q_{tn}}{2 \Delta V_t c_{pt} \rho_t}$$

$$\frac{dT_{mn}}{dt} = \frac{(Q_{tn} - Q_{mn})}{\Delta V_m c_{pm} \rho_m}$$

$$\frac{dT_{sn}}{dt} = \frac{W_s c_{ps} (T_{sn+1} - T_{sn-1}) + Q_{mn}}{2 \Delta V_s c_{ps} \rho_s}$$

where

$$\Delta V_s = \frac{\Delta Z \pi (D_s^2 - D_m^2)}{4}$$

$$\Delta V_t = \frac{\Delta Z \pi D_t^2}{4}$$

$$\Delta V_m = \frac{\Delta Z \pi (D_m^2 - D_t^2)}{4}$$

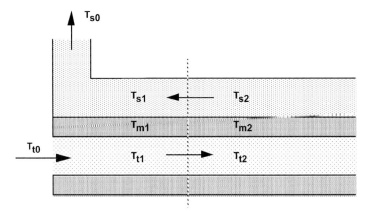

Fig. 4.29 Tube inlet and shell outlet segment.

Boundary Conditions

The consideration of the boundary conditions again follows Franks (1967). The position at the tube inlet and shell outlet section, segment number 1 is shown in Fig. 4.29.

Considering segment 1, the temperature of the entering shell side fluid is $(T_{s2}+T_{s1})/2$. The outlet shellside fluid temperature can also be approximated, either as

$$T_{s0} = T_{s1}$$

or

$$T_{s0} = T_{s1} + \frac{T_{s1} - T_{s2}}{2}$$

The heat balance equations for end segment 1 thus become

$$c_{ps}\rho_s \Delta V_s \frac{dT_{s1}}{dt} = W_s c_{ps} \left(\frac{T_{s2} + T_{s1}}{2} - T_{s0} \right) + Q_{m1}$$

and

$$c_{pt}\rho_t \Delta V_t \frac{dT_{t1}}{dt} = W_t c_{pt} \left(T_{t0} - \frac{T_{t1} + T_{t2}}{2} \right) - Q_{t1}$$

Similar reasoning for the tube outlet and shell inlet segment, number N, give for the tubeside fluid

$$\Delta V_t c_{pt} \rho_t \frac{dT_{tN}}{dt} = W_t c_{pt} \left(\frac{T_{tN-1} - T_{tN}}{2} - T_{tN+1} \right) - Q_{tN}$$

and for the shellside fluid

$$\Delta V_s c_{ps} \rho_s \frac{dT_{sN}}{dt} = W_s c_{ps} \left(T_{sN+1} - \frac{T_{sN} - T_{sN-1}}{2} \right) + Q_{mN}$$

Note that the outlet approximations must be consistent with a final steady-state heat balance. Note also that it is easy to allow in the simulation for variations in the heat transfer coefficient, density and specific heats as a function of temperature. The modelling methods demonstrated in this section are applied in the simulation example HEATEX.

4.6
Difference Formulae for Partial Differential Equations

As shown in this chapter for the simulation of systems described by partial differential equations, the differential terms involving variations with respect to length are replaced by their finite-differenced equivalents. These finite-differenced forms of the model equations are shown to evolve as a natural consequence of the balance equations, according to Franks (1967), and as derived for the various examples in this book. The approximation of the gradients involved may be improved, if necessary, by using higher order approximations. Forward and end-sections can be better approximated by the forward and backward differences as derived in the previous examples. The various forms of approximation based on the use of central, forward and backward differences have been listed by Chu (1969).

a) First-Order Approximations

Central difference as extensively used in this chapter

$$\left(\frac{\partial U}{\partial X} \right)_n = \frac{U_{n+1} - U_{n-1}}{2\Delta X}$$

$$\left(\frac{\partial^2 U}{\partial X^2} \right)_n = \frac{U_{n+1} - 2U_n + U_{n-1}}{\Delta X^2}$$

Forward difference

$$\left(\frac{\partial U}{\partial X} \right)_n = \frac{U_{n+1} - U_n}{\Delta X}$$

$$\left(\frac{\partial^2 U}{\partial X^2} \right)_n = \frac{U_{n+2} - 2U_{n+1} + U_n}{\Delta X^2}$$

Backward difference

$$\left(\frac{\partial U}{\partial X}\right)_n = \frac{U_n - U_{n-1}}{\Delta X}$$

$$\left(\frac{\partial^2 U}{\partial X^2}\right)_n = \frac{U_n - 2U_{n-1} + U_{n-2}}{\Delta X^2}$$

b) Second-Order Central Difference Approximations

$$\left(\frac{\partial U}{\partial X}\right)_n = \frac{-U_{n+2} + 8U_{n+1} - 8U_{n-1} + U_{n-2}}{12\Delta X}$$

$$\left(\frac{\partial^2 U}{\partial X^2}\right)_n = \frac{-U_{n+2} + 16U_{n+1} - 30U_n + 16U_{n-1} - U_{n-2}}{12\Delta X^2}$$

4.7
References Cited in Chapters 1 to 4

Aris, R. (1989) Elementary Chemical Reactor Analysis, Butterworth Publ., Stoneham.

Bird, R. B., Stewart, W. E. and Lightfoot, E. N. (1960) Transport Phenomena, Wiley.

Biwer, A., Heinzle, E. (2004) Environmental Assessment in Early Process Development. J. Chem. Technol. Biotechnol. 79, 597–609.

Blanch, H. W. and Clark, D. S. (1996) Biochemical Engineering, Marcel Dekker, New York.

Brodkey, R. S. and Hershey, H. C. (1988) Transport Phenomena, McGraw-Hill.

Carslaw, H. S. and Jaeger, J. C. (1959) Conduction of Heat in Solids, Clarendon Press.

Chu, Y. (1969) Digital Simulation of Continuous Systems, McGraw-Hill.

Coughanowr, D. R. and Koppel, L. B. (1965) Process Systems Analysis and Control, McGraw-Hill.

Dunn, I. J., Heinzle, E., Ingham, J. and Prenosil, J. E. (2003) Biological Reaction Engineering: Principles, Applications and Modelling with PC Simulation, 2nd edition, VCH.

Dunn, I. J. and Mor, J. R. (1975) Variable Volume Continuous Cultivation. Biotechnol. Bioeng. 17, 1805–1822.

Draper, N. R. and Smith, H. (1981) Applied Regression Analysis, Wiley.

Fogler, H. S. (2005) Elements of Chemical Reaction Engineering, 4th edition, Prentice-Hall.

Franks, R. G. E. (1967) Mathematical Modeling in Chemical Engineering, Wiley.

Franks, R. G. E. (1972) Modelling and Simulation in Chemical Engineering, Wiley-Interscience.

Froment, G. F. and Bischoff, K. B. (1990) Chemical Reactor Analysis and Design, Wiley.

Grewer, T. (1994) Thermal Hazards of Chemical Reactions, Elsevier Science.

Heinzle, E., Biwer, A. P. and Cooney, C. L. (2007) Development of Sustainable Bioprocesses – Modelling and Assessment, Wiley.

Heinzle, E. and Hungerbühler, K. (1997) Integrated Process Development: The Key to Future Production of Chemicals, Chimia, 51, 176–183.

Heinzle, E., Weirich, D., Brogli, F., Hoffmann, V., Koller, G., Verduyn, M. A. and Hungerbühler, K. (1998) Ecological and Economic Objective Functions for Screening in Integrated Development of Fine Chemical Processes. 1. Flexible and Expandable Framework Using

Indices. Ind. Eng. Chem. Res. 37, 3395–3407.

Ingham, J. and Dunn, I. J. (1974) Digital Simulation of Stagewise Processes with Backmixing. The Chem. Eng., June, 354–365.

Joshi, J. B., Pandit, A. B. and Sharma, M. M. (1982) Mechanically Agitated Gas-Liquid Reactors. Chem. Eng. Sci. 37, 813.

Kapur, J. N. (1988) Mathematical Modelling, Wiley.

Keller, A., Stark, D., Fierz, H., Heinzle, E. and Hungerbühler, K. (1997) Estimation of the Time to Maximum Rate Using Dynamic DSC Experiments. J. Loss Prev. Process Ind. 10, 31–41.

Keller, R. and Dunn, I. J. (1978) Computer Simulation of the Biomass Production Rate of Cyclic Fed Batch Continuous Culture. J. Appl. Chem. Biotechnol. 28, 508–514.

Koller, G., Weirich, D., Brogli, F., Heinzle, E., Hoffmann, V., Verduyn, M. A. and Hungerbühler, K. (1998) Ecological and Economic Objective Functions for Screening in Integrated Development of Fine Chemical Processes. 2. Stream Allocation and Case Studies. Ind. Eng. Chem. Res. 37, 3408–3413.

Levenspiel, O. (1999) Chemical Reaction Engineering, Wiley.

Luyben, M. L. and Luyben, W. L. (1997) Essentials of Process Control, McGraw-Hill.

Luyben, W. L. (1973) Process Modeling, Simulation, and Control for Chemical Engineers, McGraw-Hill

Luyben, W. L. (1990) Process Modelling, Simulation, and Control for Chemical Engineers, 2nd edition, McGraw-Hill.

Maria, G. and Heinzle, E. (1998) Kinetic System Identification by Using Short-Cut Techniques in Early Safety Assessment of Chemical Processes, J. Loss Prev. Process Ind. 11, 187–206.

Moser, A. (1988) Bioprocess Technology: Kinetics and Reactors, Springer.

Nash, J. C. and Walker-Smith, M. (1987) Nonlinear Parameter Estimation: An Integrated System in Basic, Marcel Dekker.

Noye, J. Ed. (1984) Computational Techniques for Differential Equations, North-Holland.

Oosterhuis, N. M. G. (1984) PhD Thesis, Delft University, The Netherlands.

Perlmutter, D. D. (1972) Stability of Chemical Reactors, Prentice-Hall.

Pollard, A. (1978) Process Control, Heinemann Educational Books.

Prenosil, J. E. (1976) Multicomponent Steam Distillation: A Comparison between Digital Simulation and Experiment. Chem. Eng. J. 12, 59–68.

Press, W. H., Flannery, B. P., Teukolsky, S. A. and Vetterling, W. T. (1992) Numerical Recipes: The Art of Scientific Computing, Cambridge University.

Ramirez, W. F. (1976) Process Simulation, Lexington Books.

Ruthven, D. M. and Ching, C. B. (1993) Modeling of chromatographic processes. Preparative and production scale chromatography. In: Chromatographic Science Series, Vol. 61. Eds. Ganetos, G., Barker, P. E. Marcel Dekker, pp. 629–672.

Seborg, D. E., Edgar, T. F., Mellichamp, D. A. (1989) Process Dynamics and Control, Wiley.

Shaw, J. A., The PID Control Algorithm: How It Works and How To Tune It., 2nd edition, eBook, Process Control Solutions (www.jashaw.com).

Sheldon, R. A. (1994) Consider the environmental quotient. Chem. Tech. 24/3, 38–47.

Shinskey, F. G. (1996) Process Control Systems: Application, Design, and Tuning 4th edition, McGraw-Hill.

Smith, C. L., Pike, R. W. and Murrill, P. W. (1970) Formulation and Optimisation of Mathematical Models, Intext.

Smith, R. (1995) Chemical Process Design, McGraw-Hill

Stephanopoulos, G. (1984) Chemical Process Control, Prentice-Hall.

Stoessel, F. (1995) Design thermally safe semibatch reactors. Chem. Eng. Prog. 9, 46–53.

Storti, G., Masi, M., Morbidelli, M. (1993) Modeling of countercurrent adsorption processes. Preparative and production scale chromatography. In: Chromatographic Science Series, Vol. 61. Eds. Ganetos, G., Barker, P. E. Marcel Dekker, New York, 673–700.

Strube, J., Schmidt-Traub, H., Schulte, M. (1998) Chem.-Ing.-Tech. 70, 1271–1279.

Sweere, A. P. J., Luyben, K. Ch. A. M. and Kossen, N. W. F. (1987) Regime Analysis and Scale-Down: Tools to Investigate the Performance of Bioreactors. Enzyme Microb. Technol., 9, 386–398.

Ullmann's Encyclopedia of Industrial Chemistry (1995) Production-integrated environmental protection, B8, 213–309, VCH.

Walas, S. M. (1989) Reaction Kinetics for Chemical Engineers, McGraw-Hill, reprinted by Butterworths.

Walas, S. M. (1991) Modelling with Differential Equations in Chemical Engineering, Butterworth-Heinemann Series in Chemical Engineering.

Welty, J. R., Wicks, C. E. and Wilson, R. E. (1976) Fundamentals of Momentum, Heat and Mass Transfer, Wiley.

Wilkinson, W. L. and Ingham, J. (1983) In: Handbook of Solvent Extraction, Eds. Lo, T. C., Baird, M. H. I., and Hanson, C., Wiley, pp. 853–886.

4.8
Additional Books Recommended

Al-Khafaji, A. W. and Tooley, J. R. (1986) Numerical Methods in Engineering Practice, Holt-Rinehart-Winston.

Aris, R. (1978) Mathematical Modelling Techniques, Pitman.

Aris, R. and Varma A. (1980) The Mathematical Understanding of Chemical Engineering Systems: Collected Papers of Neal R. Amundson, Pergamon.

Aris, R. (1999) Mathematical Modeling A Chemical Engineer's Perspective, Academic Press.

Babatunde, A., Ogunnaike, W., Harmon Ray (1994) Process Dynamics, Modelling and Control, Oxford University Press.

Beltrami, E. (1987) Mathematics for Dynamic Modelling, Academic Press.

Bender, E. A. (1978) An Introduction to Mathematical Modelling, Wiley-Interscience.

Bequette, B. W. (1998) Process Dynamics: Modeling, Analysis and Simulation, Prentice Hall.

Bequette, B. W. (2003) Process Control: Modeling, Design and Simulation, Prentice Hall.

Bronson, R. (1989) 2500 Solved Problems in Differential Equations, McGraw-Hill.

Burghes, D. N. and Borris, M. S. (1981) Modelling with Differential Equations, Ellis Horwood.

Butt, J. B. (2000) Reaction Kinetics and Reactor Design (Chemical Industries), Marcel Dekker.

Carberry, J. J. (1976) Chemical and Catalytic Reaction Engineering, McGraw-Hill.

Carberry, J. J. and Varma, A. (Eds.) (1987) Chemical Reaction and Reactor Engineering, Marcel Dekker.

Champion, E. R. and Ensminger, J. M. (1988) Finite Element Analysis with Personal Computers, Marcel Dekker.

Chapra, S. C. and Canale, R. P. (1988) Numerical Methods for Engineers, McGraw-Hill.

Constantinides, A. (1987) Applied Numerical Methods with Personal Computers, McGraw-Hill.

Courriou, J.-P. (2004) Process Control, Springer.

Cross, M. and Mascardini, A. O. (1985) Learning the Art of Mathematical Modelling, Ellis Horwood.

Darby, R. (2001) Chemical Engineering Fluid Mechanics, CRC Press.

Davis, M. E. (1984) Numerical Methods and Modeling for Chemical Engineers, Wiley.

Denn, M. M. (1987) Process Modelling, Longman Scientific and Technical Publishers.

Douglas, J. M. (1972) Process Dynamics and Control, Vols. 1 and 2, Prentice Hall.

Douglas, J. M. (1988) Conceptual Design of Chemical Processes, McGraw-Hill.

Finlayson, B. A. (1980) Nonlinear Analysis in Chemical Engineering, McGraw-Hill.

Fishwick, P. A. and Luker, P. A. (Eds.) (1991) Qualitative Simulation, Modelling and Analysis, Springer.

Gates, B. C. (1992) Catalytic Chemistry, Wiley.

Geankoplis, C. J. (1983) Transport Processes and Unit Operations, 2nd edition, Allyn and Baker.

Guenther, R. B. and Lee, J. W. (1988) Partial Differential Equations of Mathematical Physics and Integral Equations, Prentice-Hall.

Hannon, B. and Ruth, M. (1994) Dynamic Modeling, Springer.

Hayes, R. E. (2001) Introduction to Chemical Reactor Analysis, Gordon and Breach Science Publishers.

Himmelblau, D. M. (1985) Mathematical Modelling. In: Bisio, E. and Kabel, R. L. (Eds.) Scaleup of Chemical Processes, Wiley.

Hines, A. L. and Maddox, R. N. (1985) Mass Transfer, Prentice-Hall.

Holland, C. D. and Anthony, R. G. (1989) Fundamentals of Chemical Reaction Engineering, 2nd edition, Prentice-Hall.

Holland, C. D. and Liapis, A. I. (1983) Computer Methods for Solving Dynamic Separation Problems, McGraw-Hill.

Huntley, I. D. and Johnson, R. M. (1983) Linear and Nonlinear Differential Equations, Ellis Horwood.

Hussam, A. (1986) Chemical Process Simulation, Halstead/Wiley.

Kaddour, N. (Ed.) (1993) Process Modeling and Control in Chemical Engineering, Marcel Dekker.

Kocak, H. (1989) Differential and Difference Equations Through Computer Experiments, with Diskette, Springer.

Korn, G. A. and Wait, J. V. (1978) Digital Continuous System Simulation, Prentice Hall.

Lapidus, L. and Pinder, G. F. (1982) Numerical Solution of Partial Differential Equations in Science and Engineering, Wiley-Interscience.

Leesley, M. E. (Ed.) (1982) Computer Aided Process Plant Design, Gulf Publishing Company.

Lo, T. C., Baird, M. H. I., and Hanson, C. (1983) Handbook of Solvent Extraction, Wiley.

Loney, N. W. (2001) Applied Mathematical Methods for Chemical Engineers, CRC Press.

Luyben, W. L. (2002) Plantwide Dynamic Simulators in Chemical Processing and Control, Marcel Dekker.

Marlin, T. E. (2000) Process Control: Designing Processes and Control Systems for Dynamic Performance, 2nd edition, McGraw-Hill College.

Missen, R. W., Mims, C. A. and Saville, B. A. (1999) Introduction to Chemical Reaction Engineering and Kinetics, John Wiley & Sons, Inc.

Morbidelli, M and Varma, A. (1997) Mathematical Methods in Chemical Engineering, Oxford University Press.

Naumann, E. B. (1987) Chemical Reactor Design, Wiley.

Naumann, E. B. (1991) Introductory Systems Analysis for Process Engineers, Butterworth-Heinemann.

Naumann, E. B. (2001) Chemical Reactor Design, Optimization and Scaleup, McGraw-Hill.

Neelamkavil, F. (1987) Computer Simulation and Modelling, Wiley.

Ogunnaike, B. A. and Ray, W. H. (1994) Process Dynamics, Modeling, and Control, Oxford Univ. Press.

Palazoglu, A. and Romagnoli, J. A. (2005) Introduction To Process Control, Marcel Dekker.

Peters, M. S., Timmerhaus, K. D., and West, R. E. (2003) Plant Design and Economics for Chemical Engineers, 5th edition, McGraw-Hill.

Ramirez, W. F. (1989) Computational Methods for Process Simulation, Butterworth-Heinemann.

Rase, H. F. (1977) Chemical Reactor Design for Process Plants, Vol. 11: Case Studies and Design Data, Wiley.

Richardson, J. F. and Peacock, D. G. (1994) Coulson & Richardson's Chemical Engineering, Vol. 3: Chemical & Biochemical Reactors & Process Control, Pergamon.

Riggs, J. B. (1988) An Introduction to Numerical Methods for Chemical Engineers, Texas Techn. Univ. Press.

Robinson, E. R. (1975) Time-Dependent Chemical Processes, Applied Science.

Russell, T. W. F. and Denn, M. M. (1972) Introduction to Chemical Engineering Analysis, Wiley.

Seider, W. D., Seader, J. D., Lewin, D. R. (2004) Product and Process Design Principles, Synthesis, Analysis, and Evaluation, 2nd edition, Wiley.

Schmidt, L. D. (1998) The Engineering of Chemical Reactions, Oxford Univ. Press.

Shuler, M. L. and Kargi, F. (2002) Bioprocess Engineering. Basic Concepts; Prentice Hall: USA

Smith, C. A. and Corripio, A. B. (1985) Principles and Practice of Automatic Process Control, Wiley.

Smith, J. M. (1981) Chemical Engineering Kinetics, McGraw-Hill.

Smith, J. M. (1987) Mathematical Modelling and Digital Simulation for Engineers and Scientists 2nd edition, Wiley-Interscience.

Stewart, W. E., Ray, W. H. and Conley, C. C. (1980) Dynamics and Modeling of Reactive Systems, Academic Press.

Szekely, J. and Themelis, N. J. (1971) Rate Phenomena in Process Metallurgy, Wiley.

Tarhan, M. O. (1983) Catalytic Reactor Design, McGraw-Hill.

Turton, R., Baillie, R. C., Whiting, W. B. and Shaeiwitz, J. A. (2002) Analysis, Synthesis, and Design of Chemical Processes, 2nd edition, Prentice Hall.

Ulrich, G. D. and Vasudevan, P. T. (2004) Chemical Engineering Process Design and Economics: A Practical Guide, 2nd edition, Process Publishing, Durham.

Vemuri, V. and Karplus, W. J. (1981) Digital Computer Treatment of Partial Differential Equations, Prentice-Hall.

Wankat, P. C. (1990) Rate Controlled Separations, Elsevier Applied Science.

Woods, D. R. (1995) Process Design and Engineering Practice, Prentice-Hall.

Zwillinger, D. (1989) Handbook of Differential Equations, Academic Press.

5
Simulation Tools and Examples of Chemical Engineering Processes

In this chapter the simulation examples are presented. They are preceded by a short description of simulation tools and the MADONNA program in particular. As seen from the Table of Contents, the examples are organised according to thirteen application areas: Batch Reactors, Continuous Tank Reactors, Tubular Reactors, Semi-Continuous Reactors, Mixing Models, Tank Flow Examples, Process Control, Mass Transfer Processes, Distillation Processes, Heat Transfer, Biological Process Examples and Environmental Process Examples. There are aspects of some examples that make them relevant to more than one application area, and this is usually apparent from their titles. Within each section, the examples are listed in order of their degree of difficulty.

Each simulation example is identified by a section number, a file name and a title, and each comprises the qualitative physical description with a drawing, the model equation development, the nomenclature, exercises, sample graphical results and literature references. A few selected examples have the BERKELEY-MADONNA program in the text. The CD in the pocket at the back of the book contains all the programs and the MADONNA software versions for Windows and MAC OSX. Guidance for running a MADONNA program is given in the Appendix, in the form of both a simplified manual and a screenshot guide. The Help section of MADONNA should also be consulted.

The objective of the exercises with each example is, of course, to aid in understanding the example physical and mathematical models. Usually changes in parameters are suggested, which can either be accomplished by the use of overlay plots and sliders or by multiple runs. Some exercises involve making minor changes in a program statement, some require adding a few lines to the program and some even require writing a new self-standing program. Since the MADONNA version on the CD is not registered, it is not possible to save any modified programs or new programs directly. However, any modified programs can still be run. The CD contains registering information for the full modestly priced shareware to be made available to the reader.

The reader-simulator should try to relate the simulation results to the parameters used and to the model equations. For this understanding, it is necessary to refer back to the model equations and to the program. Most users will find it useful to make notes of the parameters investigated and to sketch the expected graphical results before they are obtained by simulation.

Chemical Engineering Dynamics: An Introduction to Modelling and Computer Simulation, Third Edition
J. Ingham, I.J. Dunn, E. Heinzle, J.E. Prenosil, J.B. Snape
Copyright © 2007 WILEY-VCH Verlag GmbH & Co. KGaA, Weinheim
ISBN: 978-3-527-31678-6

5.1
Simulation Tools

5.1.1
Simulation Software

Many different digital simulation software simulation software packages are available on the market. Modern tools are numerically powerful, highly interactive and allow sophisticated types of graphical and numerical output. Many packages also allow optimisation and parameter estimation.

A small selection of available software is given in Table 5.1. MADONNA is very user-friendly and is used in this book. This recent version has a facility for parameter estimation and optimisation. MODELMAKER is also a more recent powerful and easy to use program, which also allows optimisation and parameter estimation. ACSL has quite a long history of application in the control field, and also for chemical reaction engineering.

Both MATLAB and ASCL include powerful algorithms for non-linear optimisation, which can also be applied for parameter estimation. Optimisation and parameter estimation are also discussed in greater detail in Sections 2.4.1 and 2.4.2.

Software Links

A large compilation of information on the many simulation software tools is available on the Internet at www.idsia.ch/%7Eandrea/sim/simtools.html.

Table 5.1 Software Examples

Program	Computer	Characteristics
BERKELEY MADONNA	PC and Mac	Easy to use, powerful, highly interactive, easy and effective parameter estimation and optimisation.
MODELMAKER	PC	Easy to use, graphical model representation, parameter estimation and optimisation.
ACSL-OPTIMIZE	PC	Powerful mathematics, building blocks especially for control purposes, includes powerful and user-friendly optimisation and parameter estimation.
MATLAB-SIMULINK	All types	Powerful mathematical package especially for linear problems includes optimisation routines, powerful integration routines, model building blocks and graphical interface.
HYSYS and SPEEDUP	PC and larger	Dynamic and steady-state simulation of large processes with data base.

For the software of particular interest here, refer to the following websites: BERKELEY MADONNA, www.berkeleymadonna.com; MODELMAKER, www.modelkinetix.com/modelmaker/; MATLAB, http://www.mathworks.com; ACSL, http://www.aegisxcellon.com/; HYSYS and SPEEDUP http://www.aspentec.com

5.1.2
Teaching Applications

The solution technique, based on the use of a simulation language, is particularly beneficial in teaching, because it permits easy programming, easily comprehensible program listings, a rapid and efficient solution of the model equations, interaction with the model and a convenient graphical output of the computed results. In our experience, digital simulation has proven itself to be the most effective way of introducing and reinforcing new concepts which involve multiple interactions.

5.1.3
Introductory MADONNA Example: BATSEQ-Complex Reaction Sequence

The complex chemical reaction, shown below, is carried out in an isothermal, constant-volume, batch reactor. All the reactions follow simple first-order kinetic rate relationships, in which the rate of reaction is directly proportional to concentration (Fig. 1).

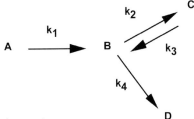

Fig. 1 Example of a complex reaction sequence.

The model equations for each component of this system are

$$\frac{dC_A}{dt} = -k_1 C_A$$

$$\frac{dC_B}{dt} = k_1 C_A - k_2 C_B + k_3 C_C - k_4 C_B$$

$$\frac{dC_C}{dt} = k_2 C_B - k_3 C_C$$

$$\frac{dC_D}{dt} = k_4 C_B$$

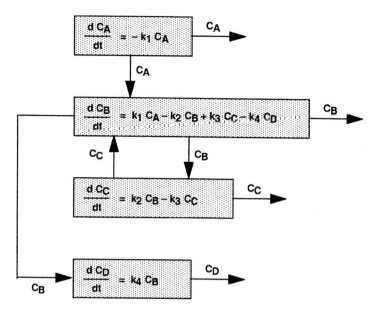

Fig. 2. Information flow diagram for the complex reaction model equations.

where C_A, C_B, C_C and C_D are the concentrations of components, A, B, C and D, respectively and k_1, k_2, k_3 and k_4 are the respective kinetic rate coefficients.

If the number of equations is equal to the number of unknowns, the model is complete and a solution can be obtained. The easiest way to demonstrate this is via an information flow diagram as shown in Fig. 2.

It is seen that all the variables, required for the solution of any one equation block are obtained as results from other blocks. The information flow diagram thus emphasises the complex interrelationship of this very simple problem.

BERKELEY-MADONNA Program BATSEQ

MADONNA provides an effective means of solving very large and complicated sets of simultaneous non-linear ordinary differential equations. The above complex reaction problem is solved with considerable ease by means of the following MADONNA program, which is used here to illustrate some of the main features of solution.

{Example BATSEQ}

{COMPLEX BATCH REACTION SEQUENCE}

METHOD AUTO
DT = 0.1
STOPTIME= 500 {Time period of simulation, sec.}

K1 = 0.01 {First order rate constant, 1/s}
K2 = 0.01 {First order rate constant, 1/s}
K3 = 0.05 {First order rate constant, 1/s}
K4 = 0.02 {First order rate constant, 1/s}

INIT CA=1 {Initial concentration, kmol/m3}
INIT CB=0 {Initial concentration, kmol/m3}
INIT CC=0 {Initial concentration, kmol/m3}
INIT CD=0 {Initial concentration, kmol/m3}

{Balance equations for components A, B, C and D}
d/dt (CA) = rA
d/dt (CB) = rB
d/dt (CC) = rC
d/dt (CD) = rD

{Kinetic rate terms for components A, B, C and D }
rA = -K1*CA
rB = K1*CA-K2*CB+K3*CC-K4*CB
rC = K2*CB-K3*CC
rD = K4*CB

MOLE = CA+CB+CC+CD {Sum of moles}

Note that the program does not have to have a pre-specified structure, although it is useful for purposes of program clarity, to group like items together. Here the first two program lines are used to describe the file name and title of the program and are bounded by the brace brackets to show that they represent non-executable program statements. The next three lines concern the integration specifications and there are discussed in more detail later. The following four lines are used to give the values of the kinetic rate constants K1 to K4. These are then followed by the initial conditions for each of the four model variables, CA, CB, CC, and CD, using the Madonna prefix of INIT.

The initial structure of the program is then followed by statements reflecting the dynamic model equations. These are also provided with comment lines with surrounding braces to distinguish them from the executable program lines. Note that the kinetic rate equations are expressed separately apart from the balance equations, to provide additional simplicity and additional flexibility. The kinetic rates are now additional variables in the simulation and the rates can also be plotted for further study. Note also the direct relationship between each individual program statement and the corresponding model equation.

Finally, an algebraic model relationship is included in order to check on the total component material balance achieved in the simulation. The last lines specify the chemical reaction rate terms and calculate the total number of moles present at any time during the reaction.

All the variables appearing on the left-hand side of the program equations are evaluated by the computer, during the simulation and are thus all available for graphing. The derivative notation d/dt(CA) denotes a differential equation, requiring the application of a numerical integration procedure for its solution. Here line 3 of the program specifies the Madonna integration method AUTO, as the means of solution. Numerical integration of all the differential equations comprising the model commences from their initial conditions, as specified by lines 10 to 13, and from an initial time. This may be specified by STARTTIME, but is usually taken by default as being equal to 0. Note that the numerical integration proceeds in parallel with all the model variables being updated simultaneously, at the end of each integration step length, with the integration continuing up to the final time of the simulation, set by STOPTIME as specified by line 5. Line 4 determines the increment of time DT within some of the integration routines. Each routine has its own set of parameters, as seen in the **Parameter/Parameter Window**. A too high value of DT can result in a failure of the integration. Too low a value of DT may slow the speed of integration and also generate an unduly high quantity of numerical data, which can be compensated by setting an appropriate value for DTOUT.

The program BATSEQ is available on the CD, with this book, but can be copied to an alternative hard disk location, or placed on diskette. The following steps represent a usual run procedure:

1. Open the file BATSEQ and examine the model equations, using the **Model/Equations** menu.
2. Specify a new graph with **Graph/New Window** and the variables to be used with **Graph/Choose Variables**. Click on **RUN** and the plot will be made.
3. If the run fails to produce satisfactory values, this may be due to the choice of an inappropriate integration method or an inappropriate choice of the integration parameters of DT, DTMAX or DTMIN. Changes in these parameters can then be best achieved by use of the **Parameters/Parameter Window**. Here any other parameter values for the problem can also be changed. Note that when viewing the value of the parameters in this window, a * before a particular value indicates a changed value from that specified in the program. Clicking on **Reset** will restore the original value.
4. Changing the value of say K1 and clicking on **RUN** will enable a second simulation, based on the changed value of the parameter to be obtained. Note that it may also be necessary to change the value of STOPTIME, to accord either with a shorter or longer run time.

5. Run a few changes again but first click the **O Button** on the graph to obtain an overlay plot. In this way, each run for each new value of K1 will be plotted on top of the previous runs and thus providing an easy means of comparing the effects of system parameter changes.
6. Sliders can also be used to conveniently change parameters. This can be done by going to **Parameters/Define Sliders** and choosing the parameter and its desired range. Each time the value of the slider parameter is slid to a new value, a run will be made and the results will be graphed, which can be done either separately or as an overlay plot.
7. A series of runs can be made automatically for a range of parameter values by choosing **Parameters/Batch Runs.** The parameter, to be changed, is then defined together with the number of runs and the range of values required. All the results of the runs will then be plotted together.
8. For a multiple run graph, click on the **Lock Button** above the graph window. Otherwise the new multiple run will not ask for a new window and the first runs will be lost.
9. Other useful buttons on the graph window are **Table** output, **Legend**, **Parameters**, **Colours**, **Dashed Lines**, **Data Points**, **Grid**, **Readout** and **Zoom**.
10. The axis scales can be rearranged to plotting variables on either the left and right hand side of the graph, as chosen in **Graph/Choose Variables**. TIME is always displayed on the X axis as the default condition, but any calculated value can be chosen.

The Screenshot Guide in the Appendix provides a full explanation for the major features of BERKELEY-MADONNA and the mechanics of running the simulation examples. The Help menu of MADONNA should also be consulted.

5.2
Batch Reactor Examples

5.2.1
BATSEQ – Complex Batch Reaction Sequence

System

The following complex reaction sequence can be used to study various reaction kinetics of interest simply by varying the rate constants.

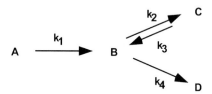

Fig. 1. A complex reaction sequence.

Model

For batch kinetics the model equations become

$$\frac{dC_A}{dt} = -k_1 C_A$$

$$\frac{dC_B}{dt} = k_1 C_A - k_2 C_B + k_3 C_C k_4 C_B$$

$$\frac{dC_C}{dt} = k_2 C_B - k_3 C_C$$

$$\frac{dC_D}{dt} = k_4 C_B$$

For this batch reaction case, the sum of all the mols must equal the initial mols charged.

Program

Part of the program is given below.

```
{Batch mass balances for components A, B, C and D }
d/dt (CA) = rA
d/dt (CB) = rB
d/dt (CC) = rC
d/dt (CD) = rD
{Kinetic rate terms for components A, B, C and D}
rA=-K1*CA
rB=K1*CA-K2*CB+K3*CC-K4*CB
rC=K2*CB-K3*CC
rD=K4*CB
```

Nomenclature

Symbols

C	Component concentration	kmol/m^3
k	First-order reaction rate constant	1/s
t	Time	s

Indices

1, 2, 3, 4	Refer to reactions
A, B, C, D	Refer to components

Exercises

1. Study the effect of varying rate coefficients k_1, k_2, k_3, k_4 in the range (0.001 to 0.1) on the time-dependent concentrations of A, B, C and D. Note that C_B and C_C pass through maximum values.

2. Set $k_3 = k_4 = 0$, to simulate a simple consecutive reaction sequence.

$$A \xrightarrow{k_1} B \xrightarrow{k_2} C$$

3. Set $k_2 = k_3 = k_4 = 0$, to simulate a simple first-order reaction.

$$A \xrightarrow{k_1} B$$

4. Set the initial values for $C_A = C_C = C_D = 0$, $C_B = 1$ and $k_1 = k_4 = 0$, to simulate a simple first-order reversible reaction.

$$B \underset{k_3}{\overset{k_2}{\rightleftarrows}} C$$

5. Note that when using the initial conditions of Exercise 4 but with $k_1 = k_3 = 0$, the program models a simple first-order parallel reaction sequence.

$$B \overset{k_2}{\underset{k_4}{\diagdown}} \overset{C}{\underset{D}{}}$$

6. Write the model and program in terms of mole fractions.

7. BATSEQ is also one of the first examples that is also of interest to study from the waste minimisation viewpoint. As suggested in Section 2.5, this can be achieved by means of incorporating a simple weighting or cost factor for each material to account for positive sales value, negative cost values representing unused or excess reactants and negative penalty cost values representing the potential damage of the waste to the environment. Regarding B and C to be useful products and D to be harmful waste, modify the program to define appropriate cost values ($/kmol) for each compound and define a total environmental cost function. Hence find the optimum reaction time producing the minimum environmental cost and compare this value to that simply based on the maximum concentration of B or C. Vary the relative cost values and study the sensitivity of the process to the harmful nature of the waste.

Results

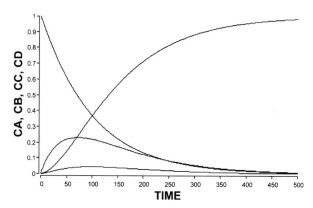

Fig. 2. Profiles of concentrations versus time as obtained from the rate constants in the program.

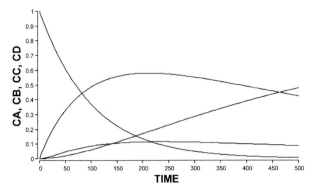

Fig. 3. Changing K4 from 0.02 to 0.002 resulted in these profiles, otherwise the same constants as Fig. 2.

5.2.2
BATCHD – Dimensionless Kinetics in a Batch Reactor

System

An nth-order homogeneous liquid phase reaction is carried out in a batch tank reactor.

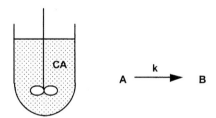

Fig. 1. Batch reactor with simple n-th order reaction

Model

The batch balance for the reactant A with nth-order kinetics is

$$V\frac{dC_A}{dt} = -k\, C_A^n\, V$$

Dimensionless variables are defined as

$$\overline{C_A} = \frac{C_A}{C_{A0}}$$

$$\bar{t} = t k\, C_{A0}^{n-1}$$

where C_{A0} is the initial concentration.

The dimensionless balance equation is then

$$\frac{d\overline{C_A}}{d\bar{t}} = -\overline{C_A}^{\,n}$$

In this form the results are independent of k.

Nomenclature

Symbols

C_A	Concentration of A	kmol/m^3
$\overline{C_A}$	Dimensionless concentration of A	–
C_{A0}	Initial concentration of A	kmol/m^3
k	Reaction rate constant	m$^{3(n-1)}$/kmol$^{(n-1)}$ s
n	Reaction order	–
t	Time	s
$\bar{t}$	Dimensionless time	–
V	Volume	m^3

Exercises

1. For a first-order reaction $n = 1$, the dimensionless time is given by the value of [kt]. Make a series of runs with different initial concentrations and compare the results, plotting the variables in both dimensional and dimensionless terms.
2. For $n = 2$ the dimensionless time is [k t C_{A0}]. Make the runs and study the results as in Exercise 1. Verify that a change in k and a change in C_{A0} have the same influence as long as the product term $k\,C_{A0}$ is maintained constant.
3. The sensitivity of the reaction is given by the order of reaction, n. Experiment with this, changing n in the range 0 to 3.

Results

Dimensionless and normal output variables are plotted below.

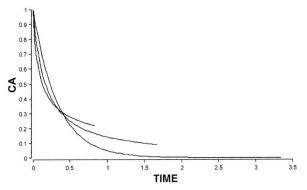

Fig. 2. Profile of dimensionless concentration of A versus dimensionless time for n=1,2, and 3. CA0=2.

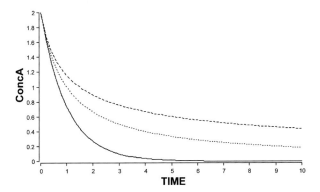

Fig. 3. Profile of concentration A versus time for the same parameters as in Fig. 2.

5.2.3
COMPREAC – Complex Reaction

System

The complex batch reaction between formaldehyde, A, and sodium p-phenol sulphonate, B, proceeds in accordance with the following complex reaction scheme. All the reactions follow second-order kinetics. Components C, D and F are intermediates, and E is the final product.

Reaction	Rate Coefficient
A + B → C	k_1
A + C → D	k_2
C + D → E	k_3
B + D → F	k_4

$$C + C \rightarrow F \quad k_5$$
$$C + B \rightarrow G \quad k_6$$
$$A + G \rightarrow F \quad k_7$$
$$A + F \rightarrow E \quad k_8$$

Model

For a constant volume batch reaction, the balance equations for each component lead to

$$\frac{dC_A}{dt} = -k_1 C_A C_B - k_2 C_A C_C - k_7 C_A C_G - k_8 C_A C_F$$

$$\frac{dC_B}{dt} = -k_1 C_A C_B - k_4 C_B C_D - k_6 C_C C_B$$

$$\frac{dC_C}{dt} = k_1 C_A C_B - k_2 C_A C_C - k_3 C_C C_D - 2 k_5 C_C^2 - k_6 C_C C_B$$

$$\frac{dC_D}{dt} = k_2 C_A C_C - k_3 C_C C_D - k_4 C_B C_D$$

$$\frac{dC_E}{dt} = -k_3 C_C C_D + k_8 C_A C_F$$

$$\frac{dC_F}{dt} = k_4 C_B C_D + k_5 C_C^2 + k_7 C_A C_G - k_8 C_A C_F$$

$$\frac{dC_G}{dt} = k_6 C_C C_B - k_7 C_A C_G$$

Nomenclature

Symbols

C	Concentration	$kmol/m^3$
k	Rate coefficient	$m^3/kmol\ min$
R	Rate	$kmol/m^3\ min$

Indices

A, B, …, G Refer to components
1, 2, …, 8 Refer to reactions

Exercises

1. Vary the initial concentrations of the reactants A and B and observe their effect on the relative rates of the individual reactions and on the resulting product distribution. Note how the distribution varies with time.
2. Write the model in dimensionless form, program it and compare the results with those from Exercise 1.

Results

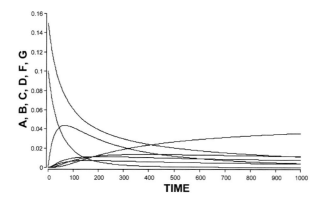

Fig. 1. Concentration-time profiles using the parameters as given in the program

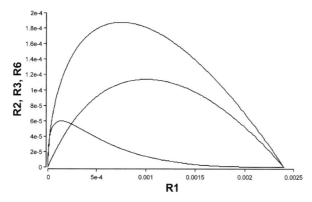

Fig. 2. Phase-plane plot showing variation of R_2, R_3 and R_6 versus R_1.

Reference

Stults, F.C., Moulton, R.W. and McCarthy, J.L. (1952) Chem. Eng. Prog. Symp. Series, Vol. 4, 38.

5.2.4
BATCOM – Batch Reactor with Complex Reaction Sequence

System

Product B is manufactured by the liquid phase batch reaction, whose complexity gives a range of byproducts.

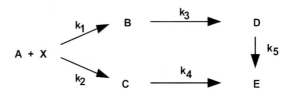

Fig. 1. This complex reaction can be altered to fit various situations.

The batch reactor can be run in the temperature range 180 to 260 C. Reactant X is in large excess, and C, D, and E are undesired byproducts.

The reactions follow first-order relationships, where,

$$r = -k\,C$$

and

$$k = Z e^{-E_a/RT} K$$

Model

For the batch reactor with these complex kinetics, the batch balances are as follows

$$\frac{dC_A}{dt} = -(k_1 + k_2)C_A$$

$$\frac{dC_B}{dt} = k_1 C_A - k_3 C_B$$

$$\frac{dC_C}{dt} = k_2 C_A - k_4 C_C$$

$$\frac{dC_D}{dt} = k_3 C_B - k_5 C_D$$

$$\frac{dC_E}{dt} = k_4 C_C + k_5 C_D$$

Program

The program can be used to generate repeated runs for differing values of reactor temperature T_C. The range of interest is 180 to 260 °C.

Nomenclature

Symbols

C	Concentration	kmol/m³
E_a	Activation energy	kJ/kmol
k	Reaction rate constant	1/s
R	Gas constant	kJ/kmol K
t	Time	s
T_K	Reactor temperature	K
T_C	Reactor temperature	°C
Z	Arrhenius constant	1/s

Indices

A, B, C, D, E Refer to components
1, 2, 3, 4, 5 Refer to reactions

Exercises

1. Determine the optimum batch time and batch temperature giving maximum yield of B.
2. Under optimum conditions, determine the concentrations of A, B, C, D and E obtained.
3. Check your results with the following analytical solution

$$C_A = e^{-(k_1+k_2)t}$$

$$C_B = \frac{k_1}{k_1 + k_2 - k_3} \left(e^{-k_3 t} - e^{-(k_1+k_2)t} \right)$$

$$C_C = \frac{k_2}{k_1 + k_2 - k_4} \left(e^{-k_4 t} - e^{-(k_1+k_2)t} \right)$$

Note: These equations can be easily programmed using MADONNA.

4. Introduce a new variable r_B describing the net formation rate of B. Use parametric runs to determine the batch time and batch temperature for maximum B formation.

Results

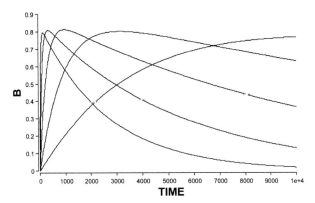

Fig. 2. Profiles of B versus time for a range of reactor temperatures, TC, from 180 to 260 K, curves A to E. The optimum corresponds to TR = 220 (middle curve).

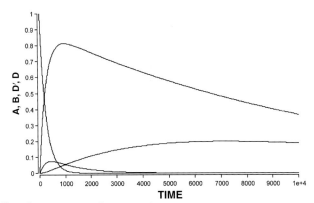

Fig. 3. Profiles of concentrations for TC = 220.

Reference

Binns, D.T., Kantyka, K.A. and Welland, R.C. (1969) Trans. Inst. Chem. Engrs. 47, 53.

5.2.5
CASTOR – Batch Decomposition of Acetylated Castor Oil

System

The batch decomposition of acetylated castor oil to drying oil proceeds via the following reaction:

Acetylated Castor Oil $\xrightarrow{\text{Heat}}$ Drying Oil + Acetic Acid Vapour

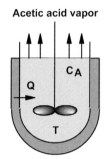

Fig. 1. The reactor for CASTOR, showing acetic acid vapour escaping.

The reaction is endothermic, and heat transfer to the reactor is required in order to accomplish the decomposition of the acetylated oil and to liberate acetic acid vapour. The example has been considered previously by Smith (1972), Cooper and Jeffreys (1971) and Froment and Bischoff (1990). Data values are based on those used by Froment and Bischoff.

Model

With respect to the material balances, the solution here treats the process as occurring with variable reactor mass because of the loss of acetic acid vapour.

The rate of production of acetic acid (kg acid/kg charge s)

$$r_p = k C_A$$

Here C_A is the equivalent mass fraction of acetyl, and the rate constant varies with T as

$$k = \frac{1}{60} \exp\left(35.2 - \frac{22450}{T}\right)$$

The acetic acid vaporizes immediately, and the total balance is

$$\frac{dM}{dt} = -m_{vap}$$

where

$$m_{vap} = r_p M$$

The acetyl component balance for variable total mass conditions is given by

$$\frac{d(MC_A)}{dt} = -r_p M$$

where

$$C_A = \frac{MC_A}{M}$$

and the fraction conversion is

$$X_A = 1 - \frac{MC_A}{M_0 C_{A0}}$$

According to the derivation given in Section 1.2.5 the energy balance equation is

$$M c_p \frac{dT}{dt} = -r_p M \Delta H + Q$$

where the heat of reaction, ΔH, includes the vaporisation of acetic acid.

Program

The simplicity of the Madonna program should be compared with the solutions in the literature. PID control has been added to the program to control the heat to the reactor. To turn off the control set $K_p = 0$. Then Q will be equal to Q1, i.e., the base heating rate.

Nomenclature

Symbols

C_A	Equivalent acetic acid content	kg/kg
C_{A0}	Initial equivalent acetic acid content	kg/kg
c_p	Specific heat	kJ/kg K
ΔH	Heat of reaction with vaporisation	kJ/kg

k	Kinetic rate constant	1/s
M	Total mass of castor oil charge	kg
M_0	Initial total mass of castor oil charge	kg
Q	Heat input	kJ/s
r_p	Rate of production of acetic acid	kg/kg s
T	Reactor temperature	K
t	Time	s

Exercises

1. Study the effect of differing rates of heat input on the fractional conversion X and reactor temperature T, for the range Q = 0 to 200 kJ/s, and note particularly the effect on the reactor time required to obtain a given conversion efficiency.
2. Repeat Exercise 1 for differing batch starting temperatures.
3. In your simulations, plot temperature and conversion versus time and also plot temperature versus fractional conversion.
4. Calculate the temerature drop for adiabatic operation from the equation.

$$M\ C_{A0}\ \Delta H = M\ c_p\ \Delta T_{adiab}.$$

Verify this by simulation.

5. Try to operate the reactor isothermally with the PID-control as programmed to control the heat input rate.
6. Add a first order time delay function to the heater to give a more realistic situation. Experiment with the control settings and compare the results with and without the heater delay.

Results

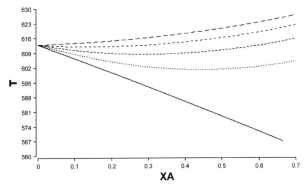

Fig. 2. Temperature versus fractional conversion obtained by setting Q = 0, 40, 80, 120 and 160 kJ/s.

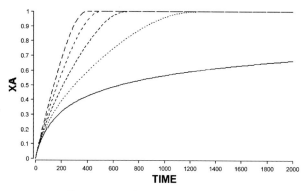

Fig. 3. Conversion-time profiles corresponding to Fig. 2.

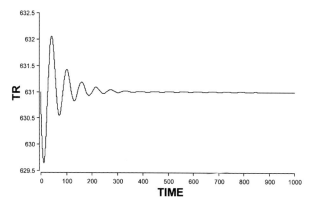

Fig. 4. Temperature profile with PID-control. Kp = 4.35, TAUINT = 0.21, TAUDER = 6.2, DT = 0.01

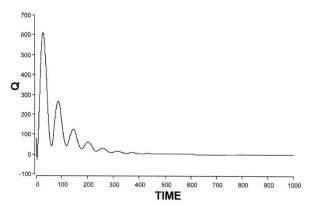

Fig. 5. Heat input with PID-control corresponding to Fig. 4

References

Smith (1972) Chemical Engineering Kinetics., McGraw-Hill, Vol. 2.
Cooper and Jeffreys (1971) Chemical Kinetics and Reactor Design, Chem Engr. Texts.
Froment, G. F. and Bischoff, D. B. (1990) Chemical Reactor Analysis and Design. 2nd edition, Wiley.

5.2.6
HYDROL – Batch Reactor Hydrolysis of Acetic Anhydride

System

The liquid phase hydrolysis reaction of acetic anhydride to form acetic acid is carried out in a constant volume, adiabatic batch reactor. The reaction is exothermic with the following stoichiometry

$$(CH_3CO)_2O + H_2O \rightarrow 2CH_3COOH + Heat$$

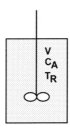

Fig. 1. The adiabatic batch reactor with variables of concentration and temperature.

Model

The acetic anhydride balance can be formulated as

$$\begin{pmatrix} \text{Rate of accumulation} \\ \text{in reactor} \end{pmatrix} = \begin{pmatrix} \text{Rate of loss due to} \\ \text{chemical reaction} \end{pmatrix}$$

hence, for the first-order reaction

$$\frac{d(VC_A)}{dt} = -k\,C_A\,V$$

and for constant volume conditions

$$\frac{dC_A}{dt} = -k\,C_A$$

The fractional conversion for the batch is

$$X_A = \frac{C_{A0} - C_A}{C_{A0}}$$

The temperature influence on the rate is given by

$$k = Z e^{-E/RT} R$$

The heat balance under adiabatic conditions is as follows:

$$\begin{pmatrix} \text{Rate of accumulation} \\ \text{of heat in the reactor} \end{pmatrix} = \begin{pmatrix} \text{Rate of heat generation} \\ \text{by chemical reaction} \end{pmatrix}$$

$$M c_p \frac{dT_R}{dt} = -k C_A V \Delta H$$

or

$$\frac{dT_R}{dt} = -k C_A V \frac{\Delta H}{M c_p}$$

Nomenclature

Symbols

C_A	Concentration of A	kmol/m³
C_{A0}	Initial concentration of A	kmol/m³
c_p	Specific heat	kJ/kg K
E	Activation energy	kJ/kmol
ΔH	Heat of reaction	kJ/kmol
k	Rate constant	1/s
M	Mass in tank	kg
R	Gas constant	kJ/kmol K
T_R	Temperature in reactor	K
V	Volume of charge	m³
X	Fractional conversion	–
ρ	Solution density	kg/m³

Exercises

1. Study the effect of differing initial batch charge temperatures on the fractional conversion XA, the reactor temperature TR, and the required batch time.
2. What initial temperature is required to give 99% conversion if the batch temperature must not exceed 300 K? What is the corresponding batch time required?

3. What is the batch time requirement for the reactor operating isothermally at 20 C. How must the heat removal requirement vary during the batch in order to maintain the batch temperature constant?

Results

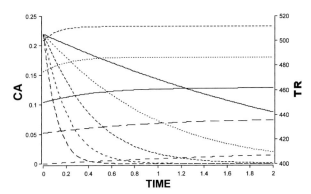

Fig. 2 Temperature and concentration variations with time for a range of starting temperature, TR0.

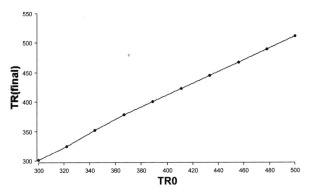

Fig. 3. Variation of fraction conversion during the runs in Fig. 2.

Reference

Cooper, A. R. and Jeffreys, G. V. (1971) Chemical Kinetics and Reactor Design, Chem. Engng. Texts, 133.

5.2.7
OXIBAT – Oxidation Reaction in an Aerated Tank

System

To be investigated are the influences of gassing rate and stirrer speed on an oxidation reaction in an aerated batch reactor. The outlet gas is assumed to be of normal air composition, which eliminates the need of a balance for the well-mixed gas phase, since the gas oxygen composition does then not change.

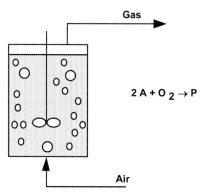

Fig. 1. Schematic drawing of the batch enzymatic oxidation reactor.

Model

The reaction kinetics are described by a second-order relation in reactant A and oxygen

$$r_A = -k_r\, C_A\, C_O$$

The stoichiometry gives

$$r_O = \frac{1}{2} \cdot r_A$$

$$r_P = -\frac{1}{2} \cdot r_A$$

The batch mass balances are

$$V\frac{dC_A}{dt} = r_A V$$

$$V\frac{dC_O}{dt} = K_L a(C_O^* - C_O)V + r_O V$$

$$V\frac{dC_P}{dt} = r_P V$$

$K_L a$ varies with stirring speed (N) and aeration rate (G) according to

$$K_L a = k_t N^3 G^{0.5}$$

where $k_t = 23.8$ with N in 1/s, G in m³/s and $K_L a$ in 1/s.

Nomenclature

Symbols

C_O	Dissolved oxygen concentration	kmol/m³
C_O^*	Saturation oxygen concentration	kmol/m³
G	Aeration rate	m³/s
$K_L a$	Transfer coefficient	1/s
k	Second-order rate constant	m³/kmol s
k_t	Constant in $K_L a$ correlation	$s^{2.5}/m^{1.5}$
N	Stirring rate	1/s
P	Product concentration	kmol/m³
r	Reaction rate	kmol/m³s

Indices

0	Refers to feed
A	Refers to reactant
O	Refers to oxygen
P	Refers to product

Exercises

1. Vary the stirrer speed N and aeration rate G and observe the response in dissolved oxygen C_O and product concentration C_P.
2. Vary the system operating parameters in the very low C_O range ($2 \cdot 10^{-6}$ kmol/m³) and note the influence on the rates.
3. Graph the oxygen transfer rate and verify that the rate of reaction is limited by oxygen transfer.
4. Calculate the value of $K_L a$ required to give $C_O = C_O^*/2$ when $C_A = 0.2$. Verify by simulation.
5. Reduce the aeration rate until the reaction becomes oxygen-limited. Maintain this and then reduce the catalyst amount by 1/2 (reduce k_r by one-half). Does the initial rate decrease by 1/2? Is the reaction still oxygen-limited?

6. Run the process oxygen-limited and observe the increase of C_O at the end of the reaction.
7. Increase N by 25% and notice the change of K_La and C_O.
8. Vary N and G under non-oxygen-limiting conditions. What is the effect on the reaction rate?

Results

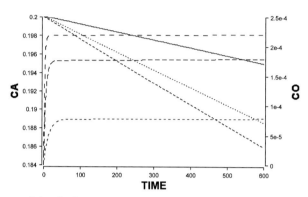

Fig. 2. Response of dissolved oxygen and reactant concentration for three stirrer speeds.

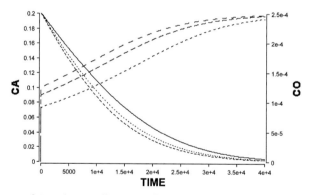

Fig. 3. Response of C_A and C_O profiles with different gassing rates until the reactant C_A is consumed.

5.2.8
RELUY – Batch Reactor of Luyben

System

The first-order consecutive exothermic reaction sequence, A → B → C, is carried out in a thick-walled, jacketed batch reactor, provided with both jacketed heating and cooling, as shown below.

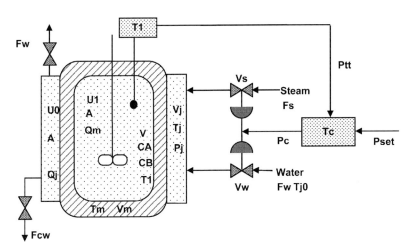

Fig. 1. Batch reactor with temperature control, adding steam and water via a split-range valve to the reactor jacket.

The reactants are charged and the contents heated by steam up to the starting batch temperature via the split range valve V_s. Steam to the jacket is supplied until the reactor temperature, T_1, reaches the maximum allowable temperature T_{max} when the steam flow is then stopped. With cooling control, T_{max} is held for a time period and then a linear ramp cooling follows. The cooling water is supplied to the jacket via the split range valve V_w at a controlled rate, set in accordance with a prescribed temperature–time profile, by a controller cam operating on the set point.

Component B is the desired product of the reaction, and the aim is to find the optimum batch time and temperature to maximise the selectivity for B. Saturated steam density data are taken from steam tables and fitted to a polynomial. The model and data for this example are taken from Luyben (1973).

Model

The balances for the reactor liquid are as follows:

Total mass balance, assuming constant density

$$\frac{d(\rho V)}{dt} = 0$$

Thus V = constant.

Component balances,

$$V\frac{dC_A}{dt} = -V k_1 C_A$$

$$V\frac{dC_B}{dt} = V k_1 C_A - V k_2 C_B$$

where

$$k_1 = Z_1 e^{-E_1/RT}$$

and

$$k_2 = Z_2 e^{-E_2/RT}$$

The reactor energy balance is

$$\rho V c_p \frac{dT}{dt} = -V k_1 C_A (-\Delta H_1) + V k_2 C_B (-\Delta H_2) - U_1 A(T - T_m)$$

The energy balance for the metal wall at constant c_p is

$$\rho_m V_m c_{pm} \frac{dT_m}{dt} = U_1 A(T_1 - T_m) - U_s A(T_m - T_j)$$

For both heating and cooling, the energy balance remains the same except that differing values of the overall heat transfer coefficient now apply. Thus for the coolant U_s is replaced by U_c.

A mass balance for the jacket is used to calculate the temperature of the saturated steam.

$$V_j \frac{d\rho_s}{dt} = F_s - F_{cw}$$

From steam table relations

$$T_j = f(\rho_s)$$

$$P_j = (1/14.5) * \exp\left(\frac{K_A}{T_j + K_B}\right)$$

During the cooling period, the energy balance for the jacket is

$$V_j\, c_{pw}\, \rho_w\, \frac{dT_j}{dt} = F_w\, c_{pw}\, \rho_w\, (T_{j0} - T_j) + U_c\, A(T_m - T_j)$$

The steam valve is initially opened slowly by increasing X_s from 0 to 1 and then kept open until the reactor temperature, T_1, reaches T_{1max}. Then it is closed and the temperature is controlled at T_{1max} for time $TIME_{hold}$. Then the cooling ramp is started.

The temperature transmitter has a range of 283 to 390 K and an output 0.2 to 1 bar. Hence,

$$P_{tt} = 0.2 + (T_1 - 283) K_{PT}$$

where KPT is a constant.

For the proportional feedback controller

$$P_c = P_0 + K_c (P_{set} - P_{tt})$$

The set point for the water ramp cooling is given by

$$P_{set} = P_{set0} - K_{ramp}(TIME - (TIME_{st} + TIME_{hold}))$$

where $TIME_{st}$ is the time when the reactor temperature first reaches a temperature T_{max}, and $TIME_{hold}$ is the time period to maintain T_{max} constant.

Program

The Madonna program makes use of conditional statements to distinguish between the heating and cooling periods: This is done by switching values of the conditional variables HEAT and COOL, to take values of either 0 or 1 in order to switch the heating or cooling on or off. Heating is accomplished with steam at a mass flow rate of F_s. The material balance for the steam jacket determines the steam density ρ_s, which, toether with a steam table function, determines the temperature of the steam in the jacket T_j. A root finding routine in Madonna is used to set the initial density of the steam in the jacket.

When the reactor temperature T_1 becomes greater than the maximum, Tmax, the values of the conditional variables switch to HEAT = 0 and COOL = 1, such that the program turns the cooling water on with a mass flow rate of F_w. This flow is controlled by a proportional controller with control constant K_c. The set point for the controller (P_{set}) is varied according to a time ramp function with a constant K_{ramp}. The output of the controller to the control valve is Pc. The ramp set point remains constant until time period $TIME_{hold}$ has passed. Then the setpoint is decreased linearly. The temperature is sensed using a pressure transmitter with output P_{tt}. The conditional logic is further described in the program.

Nomenclature

A	Heat transfer area of jacket	m^2
C	Concentration	$kmol/m^3$
c_p	Specific heat of vessel contents	$kJ/kg\ K$
c_{pm}	Specific heat of metal wall	$kJ/kg\ K$
c_{pw}	Specific heat of cooling water	$kJ/kg\ K$
C_{vs}	Valve constant for steam	$kg/s\ (bar)^{0.5}$
C_{vw}	Valve constant for water	$m^3/s\ (bar)^{0.5}$
E_i	Activation energy of reaction	kJ/mol
F_s	Steam flow rate	kg/s
F_w	Cooling water rate	m^3/s
F_{cw}	Condensed water flow rate	m^3/s
k_i	Reaction rate constant	$1/s$
K_A	Constant in steam pressure relation	K
K_B	Constant in steam pressure relation	–
K_C	Controller constant	–
K_{ramp}	Ramping constant	bar/s
P_0	Controller pressure	bar
P_c	Controller pressure output	bar
P_j	Steam vapour pressure	bar
P_{j0}	Steam supply pressure	bar
P_{Set}	Set point pressure	bar
P_{tt}	Pressure of temperature transmitter	bar
P_{w0}	Water supply pressure	bar
R	Gas constant	$kJ/kmol\ K$
T_1	Temperature in reactor	K
T_{1abs}	Absolute reactor temperature	K
T_j	Temperature of jacket contents	K
T_{j0}	Steam supply temperature	K
T_m	Temperature of metal wall	K
T_{w0}	Water supply temperature	K
T_{1max}	Temperature for batch start	K
$TIME_{st}$	Time when T_{max} is first reached	s
$TIME_{hold}$	Time period to hold T_{max}	s
U_1	Heat transfer coeff. contents to wall	$kJ/s\ m^2\ K$
U_w	Heat transfer coeff. wall to coolant	$kJ/K\ s\ m^2$
U_s	Heat transfer coeff. wall to steam	$kJ/K\ s\ m^2$
V	Vessel volume	m^3
V_j	Jacket volume	m^3
V_m	Wall volume	m^3
X_s	Steam valve position	–
X_w	Water valve position	–
Z_i	Arrhenius rate constant	$1/s$

ΔH_i	Reaction enthalpy	kJ/mol
λ	Latent heat of steam	kJ/kg
ρ	Density of vessel contents	kg/m^3
ρ_j	Density of steam in jacket	kg/m^3
ρ_s	Steam supply density	kg/m^3
ρ_m	Density of wall	kg/m^3
ρ_w	Density of cooling water	kg/m^3

Indices

A	Refers to component A
B	Refers to component B
0	Refers to inlet
1	Refers to reaction A —> B
2	Refers to reaction B —> C

Exercises

1. Vary the temperature of the heating period limit, T_{max}, and note the influence on the concentration profiles.
2. Study the effects of the parameters of the cooling water ramp function (TIME$_{hold}$ and Kramp) on the selectivity defined in the program as

$$\text{SEL} = \frac{C_B - C_{B0}}{C_{A0} - C_A}$$

3. Investigate the influence of the wall mass and the heat transfer coefficients on the maximum yield of B.

Results

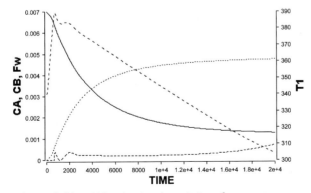

Fig. 2. A run with TIMEhold = 3000 s showing the variation of temperature, cooling water flowrate, and concentrations of A and B.

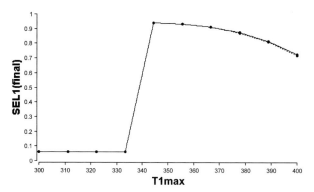

Fig. 3. A "parametric run" varying T1max to obtain the selectivity at the end of the run.

Reference

Luyben, W. L. (1973) Process Modeling, Simulation, and Control for Chemical Engineers, McGraw-Hill.

5.2.9
DSC – Differential Scanning Calorimetry

System

Differential scanning calorimetry (DSC) is a method for measuring the heat effects and temperature stability of reactants with a small scale experiment. Keller et al. (1997) used dynamic simulation to investigate screening methods for the thermal risk of chemical reactions using this method. In this example, the measurement of heat production rate of a consecutive reaction (A→B→C) using DSC is modelled. The sample is filled into the reaction crucible, while a second crucible serves as reference. The experiment is started by slowly increasing the oven temperature. During the process of decomposition caused by the high temperatures, the temperature difference between the two crucibiles is measured. As seen by the simulation results, this temperature difference closely follows the total heat produced by the reaction.

Model

The total mass is constant, and it is assumed that the volume is constant as well. For the consecutive reactions the two component balances are

$$\frac{dC_A}{dt} = -k_1 C_A$$

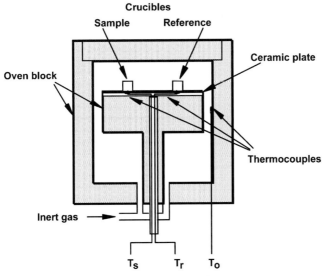

Fig. 1. Differential Scanning Calorimeter (DSC). T_s = sample temperature, T_r = reference temperature, T_o = oven temperature.

$$\frac{dC_B}{dt} = k_1 C_A - k_2 C_B$$

The rate constants k_1 and k_2 are functions of the temperature

$$k_1 = k_{01} \exp\left(-\frac{E_{a1}}{R\, T_S}\right)$$

$$k_2 = k_{02} \exp\left(-\frac{E_{a2}}{R\, T_S}\right)$$

The heat production rates are given by

$$q_1 = -k_1 C_A \Delta H_1$$

$$q_2 = -k_2 C_B \Delta H_2$$

The heat balance for the reference crucible is

$$\frac{dT_r}{dt} = \frac{U A (T_o - T_r)}{C_{pc} m_c}$$

The heat balance of the sample crucible yields

$$\frac{dT_r}{dt} = \frac{UA(T_o - T_s) + (q_1 + q_2)V_s}{C_{pc} m_c + C_{ps} m_s}$$

Exercises

1. Vary the activation energies of the two reactions and note for which values the reaction can be distinguished from a single decomposition reaction. Compare these results by "turning off" the second reaction and enhancing the first by setting $k_{02} = 0$ and setting $\Delta H_1 = -500000$ J/mol.
2. Study the influence of the oven heating rate on the resolution of the two measured peaks.
3. Modify the program to study isothermal differential calorimetry by operating at constant temperature for a series of runs. Estimate the activation energy with Arrhenius plots from the rates observed at constant conversion.
4. Modify the program to study an auto-catalytic reaction (A→B and A+B→2B).
5. Modify the existing program to study adiabatic decomposition. Compare the results with those for an adiabatic decomposition of an auto-catalytic reaction.

Nomenclature

Symbols

C	Concentration	mol/m^3
c_p	Heat capacity	J/K kg
E_a	Activation energy	J/mol
k	Rate constants	1/s
k_0	Preexponential factors	1/s
m	Mass	kg
q	Heat production rate	J/m^3 s
r	Reaction rate	mol/m^3 s
T	Temperature	h
UA	Heat transfer coefficient	J/ K s
V	Volume	m^3
ΔH	Reaction enthalpy	J/ mol
ρ	Density	kg/m^3
τ	Measurement delay	1/s

Subscripts

A	Refers to reactant A
B	Refers to intermediate B
c	Refers to crucible
o	Refers to oven
r	Refers to reference
S	Refers to sample
1	Refers to the first reaction
2	Refers to the second reaction

Results

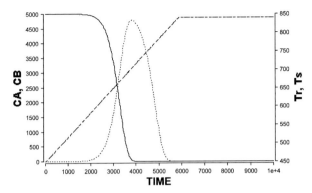

Fig. 2. Concentration and temperature profiles in a DSC measurement of a consecutive reaction.

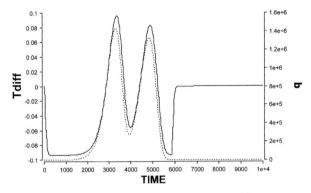

Fig. 3. Temperature difference between sample and reference crucibles (solid line) and the total heat production (dashed line).

Reference

Keller, A., Stark, D., Fierz, H., Heinzle, E., Hungerbühler, K. (1997) Estimation of the time to maximum rate using dynamic DSC experiments. J. Loss Prev. Process Ind. 10, 31–41.

5.2.10
ESTERFIT – Esterification of Acetic Acid with Ethanol. Data Fitting

System

In this example, the determination of rate constants by curve-fitting the model to real experimental data is demonstrated.

The process involves ethanol and acetic acid reacting reversibly to ethyl acetate, using a catalyst, ethyl hydrogen sulfate, which is prepared by reaction between sulfuric acid and ethanol. After equilibration of sulphuric acid with ethanol, acetic acid is added to the batch reactor in equimolar amounts, equivalent to the remaining free ethanol.

$$\text{Acetic acid} + \text{Ethanol} \rightleftharpoons \text{Ethyl acetate} + \text{water}$$

$$A + B \underset{k_2}{\overset{k_1}{\rightleftharpoons}} C + D$$

Model

The rate of batch reaction for reactant A (acetic acid) is modelled as

$$\frac{dC_A}{dt} = r_A = -k_1 C_A C_B + k_2 C_C C_D$$

The progress of the reaction is followed by taking samples at regular time intervals and titrating the remaining free acid with alkali (mL).

Program

The program is on the CD.

Experimental Data

The file of experimental data (ESTERdat) giving time (min.) versus mL titrated is on the disk.

Nomenclature

Symbols

C	Component concentration	$kmol/m^3$
k	Second-order reaction rate constant	$m^3/kmol\ s$
t	Time	min

Indices

1, 2, 3, 4	Refer to reactions
A, B, C, D	Refer to components

Exercises

1. Study the effect of varying rate coefficients k_1, k_2 in the range (0.001 to 0.1) on the time-dependent concentrations of A, B, C and D. Starting from the default values (use RESET in the Parameter Menu) make one run each with a lower and higher k_1 value. Repeat the same for k_2. Describe the effects of these changes. Use Overlay-Plots to document the changes.
2. Starting from the default values, decrease both k_1 and k_2 by half simultaneously. Then double both values starting from default. Describe the resulting effects. Again use overlay-plots.
3. Start again from default values. Modify the initial amount of water (CD_init), using values about 2 and 4 times the default values.
4. Import the table of measured data, which actually originate from a student experiment, into the program from the external file ESTERdat.txt (Menu Model/Datasets). Data are listed as time (min) versus titrated volume (mL). Modify the plot window correspondingly (Double click. Now, adjust Integration Time, STOPTIME, to see all data points. Modify k1 and k2 manually to obtain an optimal fit of measured data (MLtitrated) to experimental data (#ESTERdat).
5. Use the Curve Fit tool to obtain optimal results automatically. Document the differences to the manually obtained results. Clicking CURVE-FIT will allow the selection of the variable to be fit to the experimental data and the parameters to be determined. For each parameter, two initial guessed values and the maximum and minimum allowable values can be entered. Limit k_1 and k_2 to positive values. On running under CURVE-FIT, the values of the required parameters are repeatedly updated until the final converged values are obtained. The updated values can be found in the Parameter Window.

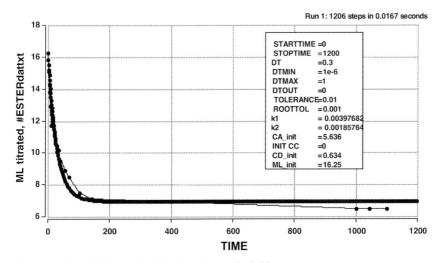

Fig. 1. Results of the Curve Fit. The simulation is the bold curve.

6. One student wanted to get a quick result. Instead of including the values obtained at equilibrium, he(she) just used the values until 180 min for parameter estimation. Compare the results with those of exercise 5 and comment.
7. It is assumed that the initial titration to yield ML_init is quite erroneous. Therefore, ML_init is also adjusted to improve the fit. What are the changes resulting?

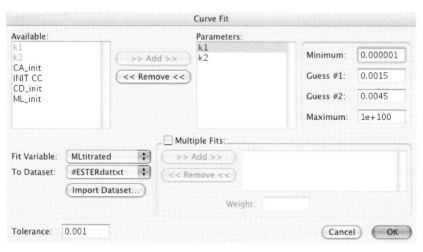

Fig. 2. Set-up of the Curve Fit window.

5.3
Continuous Tank Reactor Examples

5.3.1
CSTRCOM – Isothermal Reactor with Complex Reaction

System

Complex reactions can be conveniently investigated by simulation techniques. In this example from Russell and Denn (1972), the characteristics of the following complex reaction can be investigated.

$$A + B \xrightarrow{k_1} X$$

$$B + X \xrightarrow{k_2} Y$$

$$B + Y \xrightarrow{k_3} Z$$

This sequential-parallel reaction has sequential (A→X→Y→Z) as well as parallel characteristics (B→X, B→Y, B→Z).

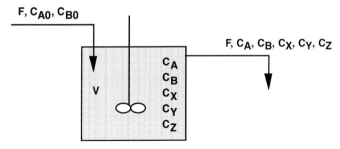

Fig. 1 Stirred tank with series-parallel reaction.

Model

For each component

$$\begin{pmatrix} \text{Rate of} \\ \text{accumulation} \\ \text{in reactor} \end{pmatrix} = \begin{pmatrix} \text{Rate of} \\ \text{flow} \\ \text{in} \end{pmatrix} - \begin{pmatrix} \text{Rate of} \\ \text{flow} \\ \text{out} \end{pmatrix} + \begin{pmatrix} \text{Rate of} \\ \text{formation} \\ \text{by reactor} \end{pmatrix}$$

All three reactions are assumed to follow simple second-order kinetics, and thus the mass balances become

$$V\frac{dC_A}{dt} = F(C_{A0} - C_A) - V k_1 C_A C_B$$

$$V\frac{dC_B}{dt} = F(C_{B0} - C_B) - V k_1 C_A C_B - V k_2 C_B C_X - V k_3 C_B C_Y$$

$$V\frac{dC_X}{dt} = F(C_{X0} - C_X) + V k_1 C_A C_B - V k_2 C_B C_X$$

$$V\frac{dC_Y}{dt} = F(C_{Y0} - C_Y) + V k_2 C_B C_X - V k_3 C_B C_Y$$

$$V\frac{dC_Z}{dt} = F(C_{Z0} - C_Z) + V k_3 C_B C_Y$$

Dividing by V, the equations can be written in terms of the mean residence time of the reactor, $\tau = F/V$.

Nomenclature

Symbols

C	Concentration	kmol/m^3
F	Flow rate	m^3/s
k	Rate constant	m^3/kmol s
V	Volume	m^3

Indices

A, B, X, Y, Z	Refer to components
1, 2, 3	Refer to the reaction steps

Exercises

1. Compare the steady-state values of A, B, X, Y, Z with calculated values obtained by solving the steady-state component balance equations for this problem.
2. Study the effect of varying residence time τ (by changing feed rate), feed concentration, and rate constants on reactor performance. This can be done using "Parametric Runs" to obtain plots of the steady state concentrations as final values versus the corresponding change in parameter.
3. Note that when $k_1 = k_2 = k_3 = 0$, $C_{A0} = 1$, $C_{B0} = 0$, the program solves the case of a step input of tracer solution, which can be used to generate the typical F-diagram for a single perfectly mixed tank. Compare this result with the analytical solution.

Results

The concentration levels are very sensitive to the kinetic parameters. Using the parameters in the program the following results were obtained

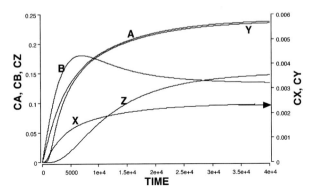

Fig. 2 Start up of the reactor for the F = 0.0001, i.e. residence time τ = 10 000 s. The components X and Y are present at very small concentrations (right hand scale).

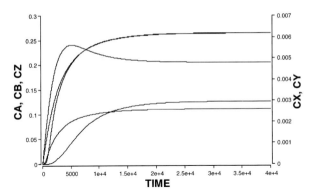

Fig. 3 Start up of the reactor for the F = 0.0002, i.e. residence time τ = 5000 s. The components X and Y are present at very small concentrations (right hand scale).

Reference

Russell, T.W.F. and Denn, M.M. (1972) Introduction to Chemical Engineering Analysis, McGraw-Hill.

5.3.2
DEACT – Deactivating Catalyst in a CSTR

System

Solid catalysts can be conveniently studied in loop reactors, which allow measuring the reaction rates by concentration difference measurement across the catalyst bed. When operated continuously with high recycle rate, the entire loop usually can be modelled as a single well-stirred tank. Here the case of catalyst deactivation is studied.

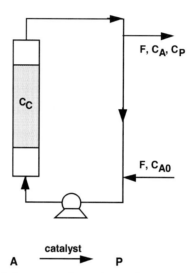

Fig. 1 A continuous recycle-loop reactor for catalyst studies.

Model

The model mass balances are formulated in mol quantities. The kinetics are assumed to be

$$r_A = -\frac{k C_C C_A}{1 + K_A C_A}$$

and

$$r_P = -r_A$$

The mass balance for component A for the entire loop is

$$\frac{dC_A}{dt} = \frac{C_{A0} - C_A}{\tau} = r_P$$

where τ is calculated as the volume of the entire loop system divided by the entering flow rate, V/F.

The mass balance for product P is

$$\frac{dC_P}{dt} = -\frac{C_P}{\tau} + r_P$$

The batch balance for the catalyst with first-order deactivation is

$$\frac{dC_C}{dt} = -k_D C_C$$

Nomenclature

Symbols

C	Concentration	mol/m³
C_C	Catalyst concentration	kg/m³
F	Flow rate	m³/h
k	Reaction rate constant	m³/kg h
K_A	Kinetic constant	m³/mol
K_D	Deactivation constant	1/h
r	Reaction rate	mol/m³ h
τ	Residence time in loop	h
V	Total loop reactor volume	m³

Indices

A, C, P Refer to reactant, catalyst and product
0   Refers to initial value

Exercises

1. Observe the influence of the catalyst decay constant, K_D, on the outlet product concentration. Make runs for a range of K_D values.
2. Design and carry out simulated experiments to evaluate the proposed reaction kinetics expression.
3. Assuming P can be measured, devise and program a control system to keep the product level constant.

Results

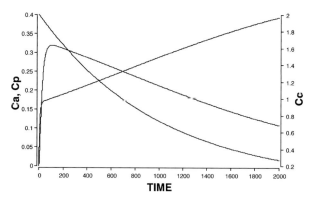

Fig. 2 Long time effect of the catalyst deactivation, Cc, on the concentrations Ca and Cp.

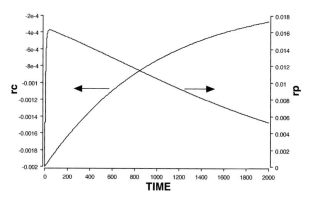

Fig. 3 The influence of rc changes on the production rate, rp.

5.3.3
TANK and TANKDIM – Single Tank with Nth-Order Reaction

System

An nth-order reaction is run in a continuous stirred-tank reactor. The model and program are written in both dimensional and dimensionless forms. This example provides experience in the use of dimensionless equations.

5.3 Continuous Tank Reactor Examples

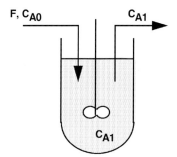

Fig. 1 A continuous tank reactor with nth-order reaction.

Model

The balance for the reactant A is

$$V \frac{dC_{A1}}{dt} = F(C_{A0} - C_{A1}) + r_A V$$

and the kinetics are

$$r_A = -k\,C^{n,A1}$$

Dimensionless variables are defined as

$$\overline{C}_{A1} = \frac{C_{A1}}{C_{A0}}, \quad \overline{C}_{A0} = \frac{C_{A0}}{C_{A0}} = 1 \quad \text{and} \quad \bar{t} = \frac{t}{\tau}$$

where τ, the residence time, is V/F.

The dimensionless balance equation is then

$$\frac{d\overline{C}_{A1}}{d\bar{t}} = (1 - \overline{C}_{A1}) + \frac{\tau}{C_{A0}} r_A$$

and the rate equation becomes

$$\overline{r}_A = -k\,C_{A0}{}^n\,\overline{C}_{A1}{}^n$$

Thus the governing dimensionless parameter is $[k\,\tau\,C_{A0}{}^{n-1}]$.

Program

Program TANK solves the normal dimensional model equation for the problem, whereas TANKDIM is formulated in terms of the dimensionless model equations.

Nomenclature

Symbols

C_{A0}	Feed concentration	kg/m³
C_{A1}	Concentration in reactor	kg/m³
F	Feed	m³/s
k	Reaction constant	$m^{3(n-1)}/kg^{n-1}s$
n	Reaction order	–
τ	Residence time	s
t	Time	s
V	Volume	m³

Exercises

Starting with program TANK:

1. Run the simulation for a range of residence times. When are the steady states reached?
2. Change the initial concentration and rerun the simulations.
3. Turn off the flow by setting the value of TAU initially very high, and start the reactor as a batch system. Change over to continuous operation using the an IF-THEN-ELSE statement to define a value of TAU.
4. Run for several values of n and plot rA versus CA1 for each case. Calculate the steady-state concentrations for n = 0, n = 1 and n = 2.
5. Compare the results from TANK and TANKDIM for a range of parameters.
6. Calculate the steady state for n = 1 by hand. Run the simulation and compare. Change C_{A0} for this case and run. Compare the effects on $\overline{C}_{A1}$ and C_{A1}.
7. Investigate the time to reach steady state for various parametric values.
8. With zero initial conditions, run for n = 0, n = 1, n = 1.5 and n = 2. Plot r_A versus C_{A1} and compare.

Results

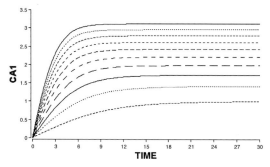

Fig. 2 Startup for a range of flow rates using the program TANK.

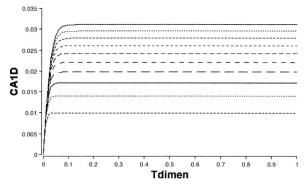

Fig. 3 Dimensionless results from TANKDIM corresponding to the same flow rates as the run in Fig. 2.

5.3.4
CSTRPULSE – Continuous Stirred-Tank Cascade Tracer Experiment

System

A system of N continuous stirred-tank reactors is used to carry out a first-order isothermal reaction. A simulated pulse tracer experiment can be made on the reactor system, and the results can be used to evaluate the steady state conversion from the residence time distribution function (E-curve). A comparison can be made between reactor performance and that calculated from the simulated tracer data.

$$A \xrightarrow{k} Products$$

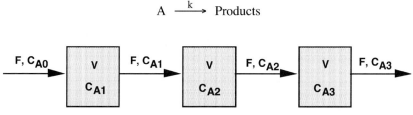

Fig. 1 Tanks-in-series with first-order reaction.

Model

Component balances for the three tanks give

Tank 1 $\qquad V\dfrac{d(C_{A1})}{dt} = F(C_{A0} - C_{A1}) - V k C_{A1}$

Tank 2 $\qquad V\dfrac{d(C_{A2})}{dt} = F(C_{A1} - C_{A2}) - V k C_{A2}$

Tank 3
$$V\frac{d(C_{A3})}{dt} = F(C_{A2} - C_{A3}) - V k C_{A3}$$

or dividing by V

$$\frac{d(C_{A1})}{dt} = \frac{C_{A0} - C_{A1}}{\tau_1} - k C_{A1}$$

$$\frac{d(C_{A2})}{dt} = \frac{C_{A1} - C_{A2}}{\tau_2} - k C_{A2}$$

$$\frac{d(C_{A3})}{dt} = \frac{C_{A2} - C_{A3}}{\tau_3} - k C_{A3}$$

where the mean residence time τ in tank n is given by

$$\tau_n = \frac{V_n}{F}$$

Thus for tank n the balances have the form

$$\frac{C_{An}}{dt} = \frac{C_{An-1} - C_{An}}{\tau_n} - kC_{An}$$

where for the tracer experiments, the reaction rate term is, of course, set to zero.

Information from tracer experiments can be used to calculate the steady state conversion according to the following relationship (Levenspiel, 1999)

$$\frac{C_{Ass}}{C_{A0}} = \int_0^\infty E(t)e^{-kt}dt$$

The E-curve is evaluated by measuring experimentally the outlet tracer concentration versus time curve from a tracer pulse input and applying the following defining equation

$$E(t) = \frac{C}{\int C dt}$$

Program

The program is formulated in vector form, allowing for any number of tanks N in the cascade but with all N tanks having an equal volume, V. The function PULSE is used in the program in relation to the tracer experiments, and the calculation of the actual conversion based on the E curve data is made by the program. Instructions on its use are found within the program.

Nomenclature

Symbols

C_{An}	Concentration of A in tank n	kmol/m³
C_{tr}	Concentration of tracer leaving system	kmol/m³
F	Flow rate	m³/s
k	Rate constant	1/s
V	Volume of all tanks	m³
τ_n and Tau	Residence time in each tank	s
E	E-curve	1/s

Indices

0, 1, 2, 3, n   Refer to tank number

Exercises

1. Study the effect of varying the total volume V, the number of tanks, and the rate constant on the performance of the reaction cascade.
2. Compare the steady state concentration values obtained with those from the solution of the steady state balance equations.
3. Setting k = 0, simulate the response to a pulse of tracer for 3 perfectly-stirred tanks in series. Repeat this for various numbers of tanks and plot E versus dimensionless time for these on an overlay graph.
4. Use the E-curve from a simulated tracer experiment to calculate the conversion for 3 tanks. Compare this result with that obtained by a simulation with reaction.
5. Compare the steady-state values obtained with the solution of the steady-state algebraic equations.
6. Derive the dimensionless form of the balances to show $k\tau$ to be the governing parameter. Run simulations to confirm this.
7. Change the program to compare the conversion from various combinations of tank volumes such that the total volume is constant. For a total volume of 10 L this could be a tank of 2.5 L followed by 2.5 L and 5 L, or this sequence could be reversed.

Results

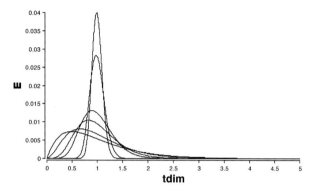

Fig. 2 Residence time distribution curves for N = 2, 3, 6, 10, 50 and 100.

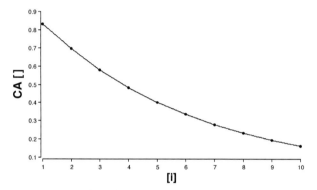

Fig. 3 Concentrations in 10 tanks with reaction at steady state.

Reference

O. Levenspiel (1999) Chemical Reaction Engineering, 3rd edition. Wiley.

5.3.5
CASCSEQ – Cascade of Three Reactors with Sequential Reactions

System

A cascade of 3 tanks in series is used to optimise the selectivity of a complex sequential-parallel reaction. Depending on the kinetics, distributing the feed of one reactant among the tanks may lead to improved selectivity.

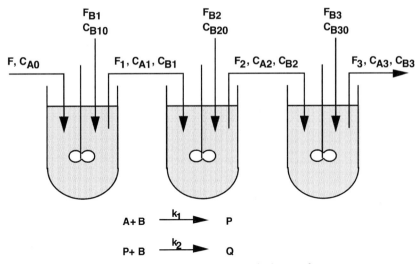

Fig. 1 Stirred tanks in series with feeding to accomplish the best performance with this complex reaction.

Model

The kinetics are

$$r_A = -k_1 C^{n_{A,A}} C^{n_{B1,B}}$$

$$r_B = -k_1 C^{n_{A,A}} C^{n_{B1,B}} - k_2 C^{n_{P,P}} C^{n_{B2,B}}$$

$$r_P = -k_1 C^{n_{A,A}} C^{n_{B1,B}} - k_2 C^{n_{P,P}} C^{n_{B2,B}}$$

$$r_Q = +k_2 C^{n_{P,P}} C^{n_{B2,B}}$$

The component balances for tank 2 are

$$\frac{V dC_{A2}}{dt} = F_1 C_{A1} - F_2 C_{A2} + r_A V$$

$$\frac{V dC_{B2}}{dt} = F_1 C_{B1} - F_{B2} C_{B20} F_1 C_{B2} + r_B V$$

$$\frac{V dC_{P2}}{dt} = F_1 C_{P1} - F_2 C_{P2} + r_P V$$

$$\frac{V dC_{Q2}}{dt} = F_1 C_{Q1} - F_2 C_{Q2} + r_Q V$$

The other tanks have analogous balances.

Note that the flow rates vary from tank to tank

$$F_1 = F + F_{B1}$$

$$F_2 = F + F_{B1} + F_{B2}$$

$$F_3 = F + F_{B1} + F_{B2} + F_{B3}$$

Nomenclature

Symbols

V_1, V_2, V_3	Initial tank volume	m^3
C_{A0}	Concentration of A in feed	kmol/m^3
C_{B10}	Concentration of B in feed to tank 1	kmol/m^3
C_{B20}	Concentration of B in feed to tank 2	kmol/m^3
C_{B30}	Concentration of B in feed to tank 3	kmol/m^3
F	Flow rates	m^3/s
k_1	Rate constant reaction 1	depends on n
k_2	Rate constant reaction 2	depends on n
n	Reaction oeder	–

Subscripts

A,B,C,P,Q	Refer to components
1,2,3	Refer to tanks
0	Refers to inlet
Init	Refers to initial

Exercises

1. Set $n_{B1} = n_{B2}$, and set equal feed concentrations of B and A to tank 1. Keeping $F = F_{B1}$, vary the volumetric flows to the reactor cascade and study the effect of total holdup time on the maximum concentration of the product C_{P3}.
2. Feed B at differing rates into each tank, but maintain the same total molar flow rate of B, as in Exercise 1. Carry out simulations for the kinetic case $n_{B1} < n_{B2}$ and optimise the resultant selectivity. Show by simulation that high values for C_{P3} are given by low values of C_B.
3. Repeat Exercises 1, 2 for the kinetic case $n_{B1} > n_{B2}$. Now high C_{P3} will be obtained with high C_B. Explain the results obtained with reference to Section 3.2.11.
4. By varying k_2 relative to k_1, investigate the influence of the kinetic constant ratio on the selectivity.

Program

The program calculates some quantities, such as selectivity, that are valid only at the steady state condition.

Results

Startup and steady states are shown for the kinetic case $k_1 = 0.1$, $k_2 = 0.05$, $n_A = 1$, $n_P = 1$, $n_{B1} = 1$ and $n_{B2} = 1$ or 2.

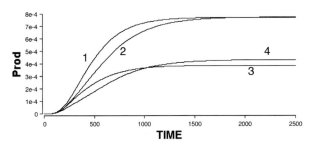

Fig. 2 When all of reactant B was fed into tank 1 with flow rate $F_{B1} = 0.003$ (curves 1, 3). The productivity was slightly lower than in the case for equal distribution of reactant B to all tanks (2, 4). Higher reaction order with respect to B gave higher productivity ($n_{B2} = 2$: curves 1, 2). The maximum productivity possible was 0.001.

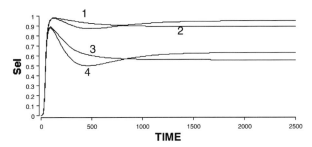

Fig. 3 Distributing reactant B with feed rates $F_{B1} = F_{B2} = F_{B3} = 0.001$ (curves 2, 4) gave lower concentrations of B in all tanks and consequently a higher selectivity. The productivity was also slightly higher (0.13 compared to 0.12). A higher reaction order with respect to B gave high selectivity ($n_{B2} = 2$: curves 1, 2).

5.3.6
REXT – Reaction with Integrated Extraction of Inhibitory Product

System

Consider the necessity to extract a product, which impedes a reaction in a continuous tank system. This can be accomplished with an integrated extraction unit, here a liquid–liquid extraction system.

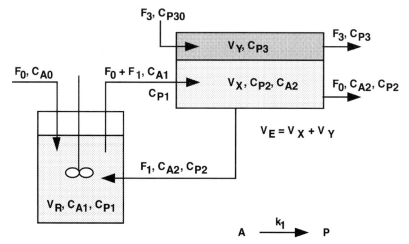

Fig. 1 Integrated reaction and liquidliquid extraction.

Model

The kinetics exhibit inhibition by the product,

$$-r_A = r_P = \frac{k_1 C_{A1}}{1 + \frac{C_{P1}}{K_I}}$$

The balances for reactant and product in the reactor are for A

$$V_R \frac{dC_{A1}}{dt} = F_0 C_{A0} + F_1 C_{A2} - (F_0 + F_1) C_{A1} - V_R r_P$$

and for P

$$V_R \frac{dC_{P1}}{dt} = F_1 C_{P2} - (F_0 + F_1) C_{P1} + V_R r_P$$

The single-stage extraction unit is modelled by balances for P in both phases X and Y as follows:

$$\frac{d(V_X C_{P2})}{dt} = (F_0 + F_1)C_{P1} - F_1 C_{P2} - F_0 C_{P2} - K_a V_E (C_{P3}^* C_{P3})$$

$$\frac{d(V_Y C_{P3})}{dt} = F_3 C_{P30} - F_3 C_{P3} + K_a V_E (C_{P3}^* - C_{P3})$$

where the equilibrium is given by

$$C_{P3}^* = m C_{P2}$$

The reactant is not removed by the extraction. Thus,

$$\frac{d(V_X C_{A2})}{dt} = (F_0 + F_1)C_{A1} - F_0 C_{A2} - F_1 C_{A2}$$

The recycle flow, F_1, can be written in terms of fraction recycled R as

$$F_1 = R F_0$$

Thus,

$$F_0 + F_1 = (1 + R) F_0$$

Program

The program also calculates the production rate of the extractor and the conversion fraction, product produced/reactant fed.

Nomenclature

Symbols

F_0	Feed rate of A	m³/s
C	Concentration	kmol/m³
F_3	Extraction solvent flow	m³/s
C_{P3}	Product concentration in solvent	kmol/m³
R	Recycle ratio	–
V_R	Reactor volume	m³
V_Y	Solvent volume	m³
V_X	Extract volume of solvent	m³
k_1	Reaction constant	1/s
K_I	Inhibition constant	kmol/m³
m	Saturation constant	–
K_a	Specific transfer coefficient	1/s
V_E	Total extractor volume	m³

Indices

A, P   Refer to reactant and product
0, 1, 2, 3   Refer to steams
X, Y   Refer to extractor phases

Exercises

1. Verify that the component balances for the system are correct at steady state.
2. Vary the main operating variables for the reactor system and note their influence on the conversion efficiency for differing values of the inhibition coefficient, K_I.
3. Study the effect of the extraction stage on reactor performance by varying the magnitudes of the the mass transfer coefficient K_a, the equilibrium distribution ratio m, the recycle ratio R, the relative reactor and extraction volumes and solvent flowrate.
4. The introduction of a solvent extraction process and an increasing flow of solvent in order to improve extraction efficiency constitute an increased environmental loading on the system. How might one take account of such effects in the simulation program?

Results

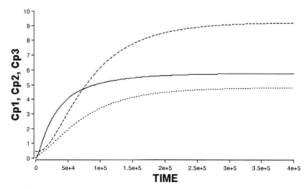

Fig. 2 Startup with recycle ratio R = 0.5 showing the product concentrations reaching steady-state.

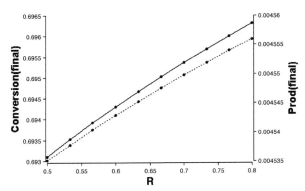

Fig. 3 Steady-state fraction conversion and productivity of the process as a function of recycle ratio.

5.3.7
THERM and THERMPLOT – Thermal Stability of a CSTR

System

A first-order, exothermic reaction occurs within a continuous stirred-tank reactor, equipped with jacket cooling, where the kinetics and reactor schemes are

$$A \rightarrow \text{products} + \text{heat}$$

$$-r_A = kC_A \quad k = Ze^{-E/RT_R}$$

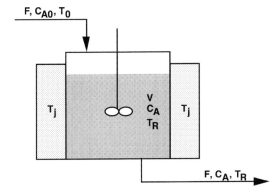

Fig. 1 Continuous tank with heat transfer to jacket.

Model

The dynamic model involves a component mass balance, an energy balance, the kinetics and the Arrhenius relationship. Hence

$$V \frac{dC_A}{dt} = F C_{A0} - F C_A - k V C_A$$

$$V \rho c_p \frac{dT_R}{dt} = F \rho c_p (T_0 - T_R) - k V C_A (-\Delta H) - UA(T_R - T_j)$$

The steady-state equations allow evaluation of the possible steady-state operating points. For the component balance:

$$F C_{A0} = F C_A + k V C_A$$

Giving

$$C_A = \frac{C_{A0}}{1 + (kV/F)}$$

Separating the energy balance into heat loss and heat generation terms:

Rate of heat loss (H_L) = Rate of heat generation (H_G)

$$F \rho c_p (T_R - T_0) + UA(T_R - T_j) = -k V C_A (-\Delta H)$$

which with the mass balance gives

$$F c_p (T_R - T_0) + UA(T_R - T_j) = k V C_{A0} \Delta H / (1 + (k V/F))$$

Programs

Program **THERMPLOT** models the steady-state variation in H_L and H_G as a function of temperature. For the conditions of this reaction, there may be three possible steady-state operating conditions for which $H_G = H_L$. Only two of these points, however, are stable.

Program **THERM** solves the dynamic model equations. The initial values of concentration and temperature in the change with in multiple runs with Madonna or with sliders Note that for comparison, both programs should be used with the same parameter values.

Portions of the programs are shown below.

```
{Example THERMPLOT}
{TEMP IS CHANGED BY DUMMY INTEGRATION}
INIT TR=250
d/dt (TR) =1 {TR becomes TR+250}
```

```
K0=Z*EXP(-E/(R*273))
K=K0*EXP(-(E/R)*(1/TR-1/273)) {AVOIDS MATH OVERFLOW}
CA=CA0/(1+K*V/F)  {STEADY STATE MASS BALANCE}
HL = F*(TR-T0)/V+UA*(TR-TJ)/(V*RHO*CP) {HEAT LOSS}
HG = -K*CA*HR/(CP*RHO)             {HEAT GAIN}

{Example THERM}

K0=Z*EXP(-E/(R*273)){Rate constant at 273 K, 1/s}
K=K0*EXP(-(E/R)*(1/TR-1/273){Avoids math overflow}

{Dynamic mass balance, first order kinetics}
d/dt(CA) = F*(CA0-CA)/V-K*CA
Limit CA>=0

{Dynamic energy balance}
{Heat lost, K/s}
HL = F*(TR-T0)/V+UA*(TR-TJ)/(V*RHO*CP)
HG = -K*CA*HR/(CP*RHO)  {Heat gained, K/s}
d/dt(TR) = HG-HL        {Heat balance, K/s}
```

Nomenclature

Symbols

c_p	Specific heat	kJ/kg K
ρ	Density	kg/m^3
F	Volumetric feed rate	m^3/s
V	Reactor volume	m^3
UA	Heat transfer capacity coefficient	kJ/K s
ΔH	Exothermic heat of reaction	kJ/kmol
T	Feed temperature	K
T_R	Reactor temperature	K
C	Concentration	kmol/m^3
Z	Collision frequency	1/s
E	Activation energy	kJ/kmol
R	Gas constant	kJ/kmol K

Indices

0	Refers to feed condition
A	Refers to reactant
J	Refers to jacket

Exercises

1. Using THERMPLOT, locate the steady states. With the same parameters, verify the steady states using THERM. To do this, change the initial conditions CAI and TRI using multiple runs or sliders. Plot as a phase-plane and also as concentration and temperature versus time.
2. Make small changes in the operating and design values (V, F, CA0, T0, TJ, UA) separately and note the effect on the steady states, using THERMPLOT. Verify each with THERM.
3. Choose a multiple steady-state case and try to upset the reactor by changing CA0, F, T0 or TJ. Only very small changes are required to cause the reactor to move to the other steady state. Plot as time and phase-plane graphs.

Results

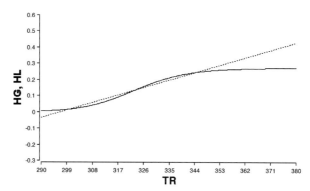

Fig. 2 Program THERMPLOT provides the graphical steady-state solution. The parameters in the program give two steady states and one unstable state.

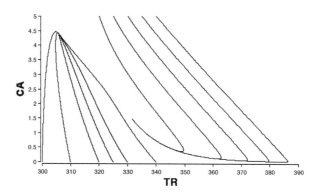

Fig. 3 This plot was made with THERM by changing the initial conditions TRI and CAI in two multiple runs on an overlay plot.

Reference

Bilous, O., and Amundson, N. R. (1956) Chem. Eng. Sci., 5, 81.

5.3.8
COOL – Three-Stage Reactor Cascade with Countercurrent Cooling

System

A cascade of three continuous stirred-tank reactors arranged in series, is used to carry out an exothermic, first-order chemical reaction. The reactors are jacketed for cooling water, and the flow of water through the cooling jackets is countercurrent to that of the reaction (Fig.1).

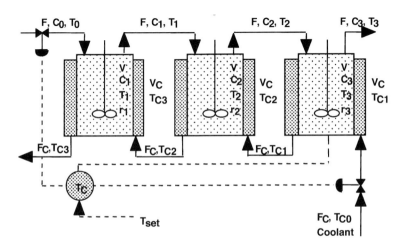

Fig. 1 A three-stage reactor system with countercurrent controlled cooling.

A variety of control schemes can be employed and are of great importance, since the reactor scheme shows a multiplicity of possible stable operating points. Obviously the aim in operating the reactor cascade is to ensure operation at the most favourable conditions, and for this both the startup policy and the control strategy are important. This example is taken from the paper of Mukesh and Rao (1977).

Model

The kinetics are

$$r_i = k_i C_i$$

where

$$k_i = Z^{-E/Rt_i}$$

The mass and heat balances in any reactor i are

$$V\frac{dC_i}{dt} = F(C_{i-1} - C_i) - V k_i C_i$$

$$\rho c_p V \frac{dT_i}{dt} = F \rho c_p (T_{i-1} - T_i) - V k_i C_i(-\Delta H) - UA(T_i - T_{ci})$$

For the cooling jackets of any reactor i the energy balance is

$$\rho_c c_{pc} V_C \frac{dT_{ci}}{dt} = F_C \rho_c c_{pc}(T_{ci-1} - T_{ci}) + UA(T_i - T_{ci})$$

Nomenclature

Symbols

A	Heat transfer area	m²
C_0	Inlet concentration of A	kmol/m³
c_p	Specific heat	kJ/kg K
c_{pc}	Coolant specific heat	kJ/kg K
E	Activation energy	kJ/kmol
F	Volumetric flow rate to reactor	m³/s
F_C	Coolant flow	m³/s
R	Gas constant	kJ/kmol K
ρ	Density	kg/m³
ρ_c	Coolant density	kg/m³
T_0	Inlet temperature	K
T_{co}	Coolant inlet temperature	K
U	Heat transfer coefficient	kJ/m² s K
V	Reactor volume	m³
V_C	Jacket volume	m³
Z	Frequency factor	1/s
ΔH and H_r	Heat of reaction	kJ/kmol

Exercises

1. Study the normal start up procedure with the reactors empty of reactant and cold ($CA_1 = CA_2 = CA_3 = 0$, $T_1 = T_2 = T_3 = 0$) and confirm that the system proceeds to a low yield condition.
2. By modifying the initial concentration and/or temperature profile along the cascade show that other stable operating conditions are realisable. This can be done by making a parametric runs to see how changes in any one parameter influence the steady state.
3. Investigate alternative start up policies to force the cascade to a more favourable, stable yield condition, as given below.
3a. Introduction of a feed-preheater and the effect of a short-burst preheat condition can be studied by starting with an increased value of the feed temperature, T_0, which is returned to normal once the reaction is underway. For this purpose a conditional operator is built into the program. Both the duration and temperature of the preheat can be defined.
3b. Temperature control of feed flow rate can be described by

    F = F0-KC * (TSET-T3)

    Proportional control can be based on the temperature of the third stage. Here F0 is the base flow rate, KC is the proportional controller gain, and TSET is the temperature set point. Note that in order to guard against the unrealistic condition of negative flow, a limiter condition on F should be inserted. This can be accomplished with MADONNA by the following statement

    LIMIT F >= 0

3c. Temperature control of coolant flow can be described by

    FC = FC0-KC * (TSET-T3)

    A limiter on FC should be used as in 3b.
4. The response to step changes in feed flow rate, feed concentration or temperature may also be simulated.

Results

A change in the feed temperature will move the system to another steady state, as these results show.

Reference

Mukesh, D. and Prasada Rao, C.D. (1977) Ind. Eng. Chem. Proc. Des. Dev. 16, 186.

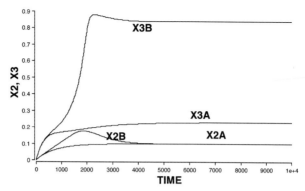

Fig. 2 A burst of warm feed at 310 K for the first 1400 seconds causes the system to shift into a higher temperature and higher conversion steady state. The fractional conversion curves A and C are for normal startup and B and D are for the startup with a warm feed period.

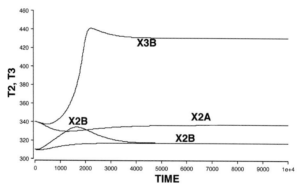

Fig. 3 Temperature profiles corresponding to the conditions for Fig. 2. The curve labels are the same.

5.3.9
OSCIL – Oscillating Tank Reactor Behaviour

System

A water-cooled, continuous stirred-tank reactor is used to carry out the following exothermic parallel reactions.

The parameters used in the program give a steady-state solution, representing, however, a non-stable operating point at which the reactor tends to produce natural, sustained oscillations in both reactor temperature and concentration. Proportional feedback control of the reactor temperature to regulate the coolant flow can, however, be used to stabilise the reactor. With positive feedback con-

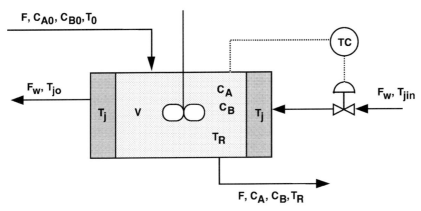

Fig. 1 Continuous tank reactor with coolant flow control.

trol, the controller action reinforces the natural oscillations and can cause complete instability of operation.

Model

The reaction kinetics are given by

$$r_A = -k_1 C_A^2 - k_2 C_A$$

and

$$r_B = k_1 C_A^2$$

The component mass balances become

$$V\frac{dC_A}{dt} = F(C_{A0} - C_A) - k_1 V C_A^2 k_2 V C_A$$

$$V\frac{dC_B}{dt} = F(C_{B0} - C_B) + k_1 V C_A^2$$

where

$$k_1 = Z_1 e^{-E1/RT_R}$$

and
$$k_2 = Z_2 \, e^{-E2/RT_R}$$

The heat balance is given by

$$V \rho c_p \frac{dT_R}{dt} = F \rho c_p (T_0 - T_R) - UA(T_R - T_j) - k_1 V C_A^2(-\Delta H_1) - k_2 V C_A(-\Delta H_2)$$

Note that this includes terms for the rates of heat generation of both reactions. As shown in Section 1.2.5, the heat balance equation is equivalent to

$$V \rho c_p \frac{dT_R}{dt} = F \rho c_p (T_0 - T_R) - U'(T_R - T_{jin}) - k_1 V C_A^2(-\Delta H_1) - k_2 V C_A(-\Delta H_2)$$

where

$$U' = \frac{UAKF_W}{1 + KF_W}$$

and

$$K = \frac{2 c_{pc} \rho_c}{UA}$$

Both the effects of coolant flow rate and coolant inlet temperature are thus incorporated in the model.

The effective controller action is represented by its effect on U' and hence on F_w, where

$$U' = U'_0 + K_c(T_R - T_{set})$$

Here $K_c = 0$ represents open loop conditions, $K_c < 0$ represents positive feed back conditions, and $K_c > 0$ represents conventional negative feedback control.

Program

The constant parameters for the program give a steady-state solution for the reactor, with the following conditions of

$$T = 333.4 \text{ K}, \quad C_A = 0.00314, \quad C_B = 0.000575 \text{ mol/cm}^3$$

but the system at this steady-state condition is unstable, and the simple open loop system of the reactor with no imposed controller action is such that the reactor itself generates sustained oscillations, which eventually form a limit cycle.

The program can be used to generate time dependent displays of CA, CB and TR for input values of the effective proportional gain KC and set point temperature TSET, for initial conditions CA0 = CB0 = 0, T = T0 at t = 0. Dimensionless variables are defined in the program, where

$$\overline{C_A} = \frac{C_A}{C_A[\text{steady} - \text{state}]}, \quad \overline{C_B} = \frac{C_B}{C_B[\text{steady} - \text{state}]} \quad \text{and} \quad \overline{T_R} = \frac{T_R}{T_R[\text{steady} - \text{state}]}$$

as these are useful in the following phase-plane analysis.

Nomenclature

Symbols

A	Heat transfer area	m^2
C	Concentration	kmol/m^3
c$_p$	Specific heat	kJ/kg K
ΔH	Exothermic heat of reaction	kJ/kmol
E	Activation energy	kJ/kmol K
F	Volumetric flow rate	m^3/s
k	Rate coefficient	1/s
K$_c$	Controller constant	kJ/m^2 K^2
ρ	Density	kg/m^3
T	Temperature	K
U	Heat transfer coefficient	kJ/m^2 K
U'	Heat transfer conductance factor	kJ/s K
V	Volume	m^3
Z	Frequency factor	1/s

Indices

A and B	Refer to components
1 and 2	Refer to reactions
0	Refers to inlet
j	Refers to jacket
in	Refers to inlet
O	Refers to outlet
R	Refers to reactor
w	Refers to coolant water

Exercises

1. Study the simple, open-loop (KC = 0) and closed-loop responses (KC = 1 to 5, TSET = TDIM, and 300 to 350 K) and the resulting yields of B. Confirm the oscillatory behaviour and find appropriate values of KC and TSET to give maximum stable and maximum oscillatory yield. For the open-loop response, show that the stability of operation of the CSTR is dependent on the operating variables by carrying out a series of simulations with varying T_0 in the range 300 to 350 K.
2. Make a dimensionless phase-plane display of BDIM versus ADIM, repeating the studies of Exercise 1. Note that the oscillatory behaviour tends to form stable limit cycles in which the average yield of B can be increased over steady-state operation.

Results

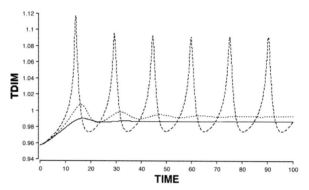

Fig. 2 Dynamic response of dimensionless temperature versus time for three setpoints; TSET = 325, 350 and 375 K.

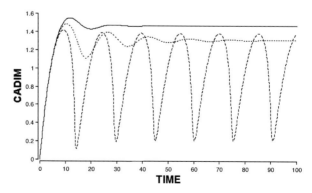

Fig. 3 Response of dimensionless concentration of A for the conditions of Fig. 2.

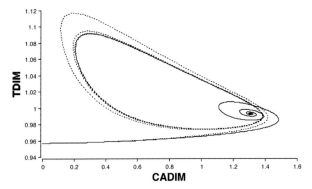

Fig. 4 A phase-plane plot of dimensionless CA versus dimensionless temperature for TSET = 325 and 350 K.

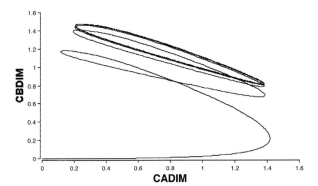

Fig. 5 A phase-plane plot of dimensionless concentrations of A and B for TSET = 350 K.

Reference

Dorawala, T.G. and Douglas, J.M. (1971) A.I.Ch.E.J. 17, 974.

5.3.10
REFRIG1 and REFRIG2 – Auto-Refrigerated Reactor

System

In the auto-refrigerated reactor shown below, an exothermic reaction A → B is carried out using a low-boiling solvent C. The heat of reaction is removed from the reactor by vapourising the solvent, condensing the vapour in the reflux condenser and returning the condensate as saturated liquid to the reactor. The total holdup of liquid in the reactor is maintained constant, but the temperature of the reactor is controlled by regulating the mass flow of vapour to the condenser. The example is taken from the paper of Luyben (1960).

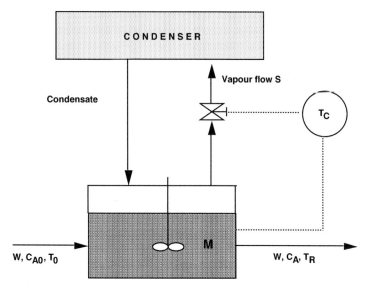

Fig. 1 A schematic drawing of the example REFRIG, showing control of the vapour rate.

Model

The balance equations are as follows:

Component balance equation

$$M\frac{dC_A}{dt} = W(C_{A0} - C_A) - k\,C_A\,M$$

Energy balance equation

$$Mc_p\frac{dT_R}{dt} = W\,c_p\,(T_0 - T_R) - M\,k\,C_A(-\Delta H_R) - S\,\Delta H_v$$

Arrhenius equation

$$k = Z\,e^{-E/R\,T_R}$$

Control equation

$$S = S_0 - K_P(T_{Set} - T_R)$$

At steady-state conditions

$$\frac{dC_A}{dt} = \frac{dT_R}{dt} = 0$$

$$\begin{pmatrix} \text{Rate of} \\ \text{generation} \\ \text{of heat} \end{pmatrix} = \begin{pmatrix} \text{Rate of} \\ \text{heat} \\ \text{removal} \end{pmatrix}$$

$$H_G = H_L$$

where

$$H_G = -M\, C_A\, k(-\Delta H_R)$$

and

$$H_L = W\, c_p (T_R - T_0) + S\, \Delta H_v$$

From the mass balance

$$C_A = \frac{C_{A0}}{1 + \frac{Mk}{W}}$$

Program

The program **REFRIG1** calculates the steady-state heat generation and heat loss quantities, HG and HL, as functions of the reactor temperature, TR, over the range 320 to 410 K. It is shown that, according to the van Heerden steady-state stability criterion that the simple loop control response, KP = 0 is unstable.

Program **REFRIG2** simulates the dynamic behaviour of the reactor and generates a phase-plane stability plot for a range of reactor concentrations and temperatures.

Nomenclature

Symbols

M	Mass of reactor contents	kg
C_A	Reactant concentration	kg/kg
t	Time	h
W	Mass flow rate	kg/h
k	Kinetic rate constant	1/h
c_p	Specific heat	kJ/kg K
T_0	Feed temperature	K
T_R	Reactor temperature	K
ΔH_R	Exothermic heat of reaction	kJ/kg
S	Vapour reflux rate	kg/h
ΔH_v	Latent heat of vapourisation	kJ/kg
Z	Frequency factor	1/h
E	Activation energy	J/mol

R	Ideal Gas Constant	J/mol K
S_0	Base level for vapour rate	kg/h
K_P	Controller gain	kg/h K
T_{Set}	Set point temperature	K
H_G	Net rate of heat generation	kJ/h
H_L	Net rate of heat loss	kJ/h

Exercises

1. Using program REFRIG1, find the magnitude of the steady-state vapourisation rate, S_0, required to give stable operation.
2. Using the value of S_0 = 330 kg/h, use program REFRIG2 to find the value of the proportional gain constant K_P required to give stability.

Results

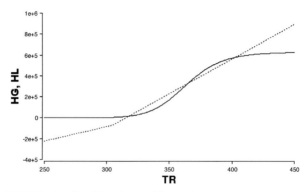

Fig. 2 Using REFRIG1, the plot of heat-loss and heat-gain terms versus temperature exhibits the stable state for KP = 6.

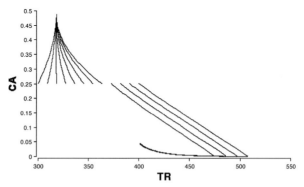

Fig. 3 Plots of the concentration-temperature phase-planes, corresponding to the conditions of Fig. 2 using REFRIG2.

Reference

Luyben, W. L. (1960) A.I.Ch.E.J, 12, 663.

5.3.11
REVTEMP – Reversible Reaction with Variable Heat Capacities

System

A reversible reaction $A \underset{k_2}{\overset{k_1}{\rightleftharpoons}} B$ is carried out in a well-mixed, tank reactor, which can be run in either batch or continuous mode. A high temperature at the beginning of the batch will increase the rate of approach to equilibrium, while a low temperature at the end of the run will give a favourable final equilibrium condition. After starting the reactor under adiabatic conditions, there will be therefore an optimal time to turn on the cooling water. A heat exchanger is used for temperature control. This example is important in understanding the application of the energy balance and related thermodynamic considerations in the simulation of complex reactions with associated heat effects. This simulation example is also used in Section 2.4.1.1, Case A, to demonstrate the application of optimization to dynamic models.

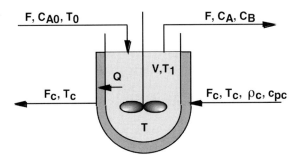

Fig. 1 Continuous stirred-tank reactor with cooling.

Model

Thermodynamics

The standard heat of reaction can be calculated from the heats of formation as

$$\Delta H_{RSt} = H_{FBSt} - H_{FASt}$$

The specific heats of reactant and product depend on temperature, according to

$$c_{pA} = a_A + b_A T$$

$$c_{pB} = a_B + b_B T$$

The equilibrium constant for the reaction at standard conditions is

$$K_{ESt} = \frac{C_{BE}}{C_{AE}}$$

and its temperature dependency is given by the van't Hoff Equation

$$\frac{d \ln K_E}{dT} = \frac{\Delta H_R(T)}{RT^2}$$

where ΔH_R at any temperature is calculated by

$$\Delta H_R = \Delta H_{RSt} + \int_{T_{ST}}^{T} (c_{pB} - c_{pB}) dT$$

Therefore

$$\Delta H_R = \Delta H_{RSt} + (a_B - a_A)(T - T_{st}) + \frac{b_B - b_A}{2}(T^2 - T_{st}^2)$$

or with

$$\Delta H_{RSt1} = \Delta H_{RSt} - (a_B - a_A)T_{st} - \frac{b_B - b_A}{2}T_{St}^2$$

this reduces to

$$\Delta H_R = \Delta H_{RSt1} + \Delta a T + \frac{\Delta b}{2} T^2$$

where $\Delta a = a_B - a_A$ and $\Delta b = b_B - b_A$.

Integration of the van't Hoff Equation then gives

$$\ln K_E = \ln K_{ESt} - \frac{\Delta H_{RSt1}}{R}\left(\frac{1}{T} - \frac{1}{T_{St}}\right) + \frac{\Delta a}{R} \ln \frac{T}{T_{St}} + \frac{\Delta b}{R}(T - T_{St})$$

Material Balance Equations

The total mass balance, assuming constant density is

$$\frac{d(\rho V)}{dt} = 0$$

Thus V is a constant under constant density conditions.

5.3 Continuous Tank Reactor Examples

The component balances are given by

$$V\frac{dC_A}{dt} = -F(C_{A0} - C_A) + r_A V$$

$$V\frac{dC_B}{dt} = -F C_B + r_B V$$

The kinetics are each assumed to be first order, where $r_A = k_1 C_A$ and $r_B = k_2 C_B$

At equilibrium

$$k_1 C_{AE} = k_2 C_{BE}$$

Since

$$K_E = \frac{C_{BE}}{C_{AE}}$$

then

$$k_2 = \frac{k_1}{K_E}$$

which allows the kinetics to be written as

$$r_A = -r_B = -k_1\left(C_A - \frac{C_B}{K_E}\right)$$

The kinetic rate constant k_1 is a function of temperature according to

$$k_1 = k_{A1}\, e^{-E_1/RT_1}$$

Energy Balance Equations

The reactor energy balance as derived in Section 1.2.5 is

$$\sum_{i=1}^{2} n_i c_{pi} \frac{dT_1}{dt} = F_0 \sum_{i=1}^{2} C_{i0} \int_{T_1}^{T_0} c_{pi} dT_1 - \Delta H_R\, r_A V + UA(T_c - T_1)$$

The individual terms are now

$$\alpha = \sum_{i=1}^{2} n_i c_{pi} = V(C_A(a_A + b_B T_1) + C_B(a_B + b_B T_1))$$

$$\beta = \sum_{i=1}^{2} C_{i0} \int_{T_1}^{T_0} c_{pi} dT_1 = C_{A0}\left(a_A(T_0 - T_1) + \frac{b_A}{2}(T_0^2 - T_1^2)\right)$$

$$\Delta H_R = \Delta H_{StR} + (a_B - a_A)(T_1 - T_{St}) + \frac{b_B - b_A}{2}(T_1^2 - T_{St}^2)$$

The energy balance for the reactor becomes

$$a\frac{dT_1}{dt} = F_0\beta + \Delta H_R\, r_A\, V + U\,A(T_c - T_1)$$

The energy balance for the jacket is

$$V_c\, c_{pc}\, \rho_c\, \frac{dT_c}{dt} = F_c\, c_{pc}\, \rho_c(T_{c0} - T_c) - U\,A(T - T_c)$$

Optimization

An objective function, SPTYB, is defined for batch operation

$$SPTYB = \frac{C_B^2}{T}$$

and for continuous operation

$$SPTYC = \frac{C_B^2 F}{V}$$

These functions represent space-time yields with special weightings for high conversion.

Program

The reactor is run initially as a batch reactor (F = 0) under adiabatic conditions (F_{C0} = 0) to obtain a high initial rates of reaction. At time TIMEON the cooling flow rate is set to FCON and the reaction temperature reduced to obtain favourable equilibrium conditions.

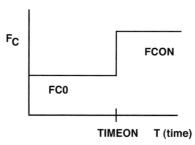

Fig. 2 Cooling flow rate profile.

Nomenclature

Symbols

A	Heat transfer area of jacket	m^2
C	Concentration	mol/m^3
c_p	Specific heat of vessel contents	kJ/kg K
c_{pc}	Specific heat of cooling water	kJ/kg K
E_i	Activation energy of reaction	kJ/mol
F	Flow rate to the reactor	kg/s
F_c	Coolant flow rate	m^3/s
k_{A1}	Arrhenius rate constant	1/s
K_E	Equilibrium constant	–
k_i	Reaction rate constant	1/s
N or n	Number of moles	mol
R	Gas constant	kJ/mol K
SPTYB	Objective function for batch operation	mol^2/m^6 s
T_1	Temperature in reactor	K
T_c	Temperature of jacket contents	K
T_{c0}	Coolant supply temperature	K
U	Heat transfer coefficient	kJ/s m^2
V	Vessel volume	m^3
V_c	Jacket volume	m^3
ΔH_R	Reaction enthalpy	kJ/mol
ρ	Density of vessel contents	kg/m^3
ρ_c	Density of cooling liquid	kg/m^3

Indices

A	Indicates component A
B	Indicates component B
1	Refers to reaction A → B
2	Refers to reaction B → A
E	Refers to equilibrium
0	Refers to reactor inlet
St	Refers to standard conditions (298 K)

Exercises

1. Compare the batch reactor performance at constant c_p with that for variable c_p. Do this by setting both c_p values constant at a temperature T_1 and calculating $c_{pi1} = a + b\, T_1$ for particular values of $a_A = c_{pA1}$, $a_B = c_{pB1}$, $b_A = 0$ and $b_B = 0$.
2. Optimise the reactor performance (space-time yield SPTB) by varying the value of TIMEON. Compare results obtained with those in Section 2.4

3. Vary both F0 and FCON to obtain the optimum SPTYB.
4. Run the reactor continuously with continuous feed to and from the reactor. Observe the effect of varying space-time on the reaction yield and on SPTYC. Set TIMEON = 0, and to give continuous cooling set FCON to the desired cooling flow rate.

Results

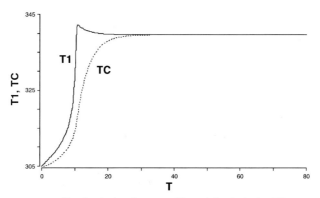

Fig. 3 Temperature profiles for the batch reactor, T1, and for the jacket TC, which lags behind. No cooling, TIMEON = 1000.

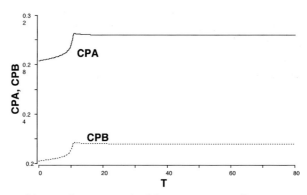

Fig. 4 Variation of the c_P values as a result of the temperature profiles, corresponding to Fig. 3.

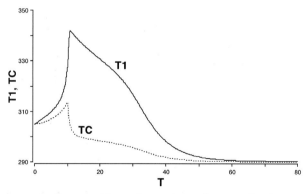

Fig. 5 Temperature profiles caused by turning on the cooling water flow by setting TIMEON to 10.

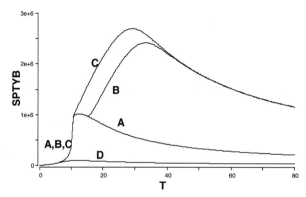

Fig. 6 Space-time yields, SPTYB, profiles in the batch reactor as a result of varying TIMEON (A – 1000, B – 15, C – 10, D – 5).

5.3.12
REVREACT – Reversible Reaction with Temperature Effects

System

A reversible reaction, $A \underset{k_2}{\overset{k_1}{\rightleftharpoons}} B$, takes place in a well-mixed tank reactor. This can be operated either batch-wise or continuously. It has a cooling jacket, which allows operation either isothermally or with a constant cooling water flowrate. Also without cooling it performs as an adiabatic reactor. In the simulation program the equilibrium constant can be set at a high value to give a first-order irreversible reaction.

This example allows the study of the interaction of chemical reaction with temperature influence. Various cooling control strategies can also be simulated.

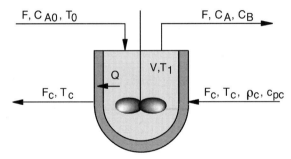

Fig. 1 Stirred-tank reactor with cooling

Model

Thermodynamics

The standard heat of reaction can be calculated from the heats of formation.

$$\Delta H_{RSt} = H_{FBSt} - H_{FASt}$$

Here the heat capacities are assumed to be constant.

At equilibrium there is a fixed ratio of product to reactant.

$$K_{ESt} = \frac{C_{BE}}{C_{AE}}$$

The temperature dependence on the equilibrium constant is described by the van't Hoff Equation

$$\frac{d \ln K_E}{dT} = \frac{\Delta H_R(T)}{RT^2}$$

where the heat of reaction is assumed constant over the temperature range.

Integration of the van't Hoff Equation gives

$$\ln K_E = \ln K_{ESt} - \frac{\Delta H_{RSt1}}{R}\left(\frac{1}{T} - \frac{1}{T_{St}}\right)$$

Mass Balances

Density ρ and volume V are constant. The component balances are then.

$$V\frac{dC_A}{dt} = -F(C_{A0} - C_A) + r_A V$$

$$V\frac{dC_B}{dt} = -F\,C_B + r_B V$$

The kinetics are first order for both reactions such that $r_1 = k_1 C_A$ und $r_2 = k_2 C_B$.

At equilibrium

$$k_1 C_{AE} = k_2 C_{BE}$$

Since

$$K_E = \frac{C_{Be}}{C_{AE}}$$

then

$$k_2 = \frac{k_1}{K_E}$$

giving

$$r_A = -r_B = -k_1 \left(C_A - \frac{C_B}{K_E} \right)$$

The kinetic rate constant k_1 is a function of temperature according to

$$k_1 = k_{A1} K_1 = K_{A1} \, e^{-\frac{E_1}{RT_1}}$$

Energy Balances

For the reactor

$$\rho \, c_p \, V \frac{dT_1}{dt} = F_0 \, \rho \, c_p (T_0 - T_1) - \Delta H_R \, r_A \, V + U \, A (T_c - T_1)$$

Giving

$$\frac{dT_1}{dt} = \frac{F_0}{V} \rho \, c_p (T_0 - T_1) - \frac{\Delta H_R \, r_A}{\rho \, c_p} + \frac{U \, A}{\rho \, c_p \, V} (T_c - T_1)$$

For the well-mixed cooling jacket

$$\rho_c \, c_{pc} \, V_c \frac{dT_c}{dt} = F_c \, \rho_c \, c_{pc} (T_{c0} - T_c) - U \, A (T_c - T_1)$$

Giving

$$\frac{dT_c}{dt} = \frac{F_c}{V} (T_{c0} - T_c) - \frac{U \, A}{\rho_c \, c_{pc} \, V} (T_c - T_1)$$

Program

The program provides for simple switching from one operation to another by setting the corresponding variables. Batch (1 batch, 0 continuous); Isothermal (1 isothermal, 0 non isothermal); Adiabatic (1 no heat transfer, 0 with heat transfer)

Nomenclature

Symbols

A	Heat transfer area of jacket	m^2
C	Concentration	mol/m^3
c_p	Specific heat of vessel contents	kJ/kg K
c_{pc}	Specific heat of cooling water	kJ/kg K
E_i	Activation energy of reaction	kJ/mol
F	Flow rate to the reactor	kg/s
F_c	Coolant flow rate	m^3/s
k_{A1}	Arrhenius rate constant	1/s
K_E	Equilibrium constant	–
k_i	Reaction rate constant	1/s
N or n	Number of moles	mol
R	Gas constant	kJ/mol K
SPTYB	Objective function for batch operation	mol^2/m^6 s
T_1	Temperature in reactor	K
T_c	Temperature of jacket contents	K
T_{c0}	Coolant supply temperature	K
U	Heat transfer coefficient	$kJ/s\ m^2$
V	Vessel volume	m^3
V_c	Jacket volume	m^3
ΔH_R	Reaction enthalpy	kJ/mol
ρ	Density of vessel contents	kg/m^3
ρ_c	Density of cooling liquid	kg/m^3

Indices

A	Indicates component A
B	Indicates component B
1	Refers to reaction A → B
2	Refers to reaction B → A
E	Refers to equilibrium
0	Refers to reactor inlet
St	Refers to standard conditions (298 K)

Exercises

1. Investigate an isothermal, batch reactor (Set batch=1 und isothermal=1) with an irreversible first-order ($k_1 \gg k_2$). For this purpose set KEST to a very high value, say 1e20. Determine the necessary reaction time to achieve a fraction conversion, XA, of 90, 95 und 99%. Determine also the cycle time and the productivity. For this assume the down-time between batches is 30 min (1800 s). Perform this for two different temperatures between 300K and 320K.
2. Investigate an isothermal, continuous reactor (Set batch = 0 und isothermal = 1) with an irreversible first-order ($k_1 \gg k_2$). For this purpose set KEST to a very high value, as in Example 1. Test whether at your temperature a steady-state is reached. If not then set the STOPTIME to a higher value. Determine what flow rate is required to give a 99% conversion. What residence times are required to give XA equal to 90, 95 und 99. Make a Parameter plot to show this. Using the same temperature as in Exercise 1 compare the residence times and productivities with the batch reactor case of Exercise 1.
3. Investigate the batch reactor for the case of an equilibrium reaction. Reset the equilibrium constant to the original value in the program. Run first the batch reactor isothermally in the range of 300 to 400 K and determine the equilibrium conversion. Use the Parameter Plot tool for this to obtain the values at each temperature. At low temperatures make sure the STOPTIME is always sufficiently long to reach equilibrium. Using the same temperatures as in Exercise 1, find the reaction times to achieve fraction conversions XA of 90, 95 und 99%.
4. Repeat the study of Exercise 3 but for a continous reactor, as in Exercise 2.
5. Operate the batch reactor of Exercise 1 with a constant cooling water flowrate (Parameter batch = 1 und isothermal = 0): Flowrates: FC = 0, 0.02, 0.025 and 0.03. For each value of FC observe the profiles of T1, TC, XA. Zoom in to make observations on the initial phase of the reaction.

Results

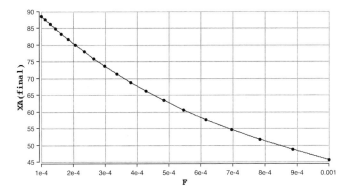

Fig. 2 Parameter plot showing variation of XA with F for the irreversible reaction run in an isothermal, continuous reactor at 305K

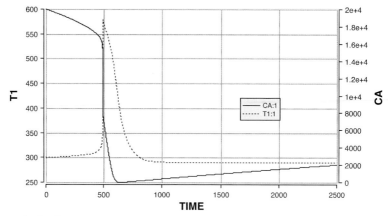

Fig. 3 The batch reactor with irreversible reaction and cooling water flow at 0.023 m^3/s gave these results.

5.3.13
HOMPOLY Homogeneous Free-Radical Polymerisation

System

A certain free-radical polymerisation reaction is described by the following sequence of initiation, addition and termination

Initiation $\qquad I \xrightarrow{k_d} 2 \bullet R$

Addition $\qquad \bullet R + M \longrightarrow \bullet M_1$

$$\bullet M_1 + M \xrightarrow{k_p} \bullet M_2$$

$$\bullet M_x + M \xrightarrow{k_p} \bullet M_{x+1}$$

Termination $\qquad \bullet M_x + \bullet M_y \xrightarrow{k_t} M_{x+y}$

The reaction is carried out in a non-cooled, continuous stirred-tank reactor (Fig. 1), and it is required to find the effect of changes in the reactor inlet conditions on the degree of polymerisation obtained. The model is that of Kenat, Kermode and Rosen (1967).

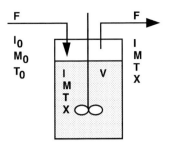

Fig. 1 The reactor for example HOMPOLY is a constant-volume stirred tank.

Model

Only the propagation reactions are significantly exothermic and the energy balance is given by

$$\frac{dT}{dt} = \frac{F}{V}(T_0 - T) + \frac{-\Delta H_P}{\rho\, c_P} r_p$$

the monomer balance is represented by

$$\frac{dM}{dt} = \frac{F}{V}(M_0 - M) - r_p$$

and the initiator balance by

$$\frac{dI}{dt} = \frac{F}{V}(I_0 - I) - r_i$$

The reaction kinetics are as follows

1) Initiation or initiation decomposition reaction

$$r_i = 2\, f\, k_d\, I$$

where f is the fraction of free-radicals, •R, initiating chains.

2) Termination $r_t = 2\, k_t\, M_0 2$

3) Propagation $r_p = k_p\, M\, M_0$

Assuming equal rates of initiation and termination

$$r_p = k_p \left(\frac{f k_d I}{k_t}\right)^{0.5}$$

The kinetic rate coefficients k_d, k_t are described by Arrhenius temperature dependencies.

Average chain length

$$X = \frac{f_p}{r_i} = \left(\frac{k_p M}{f k_d k_t I}\right)^{0.5}$$

The model equations are expressed in dimensionless form and defining new variables in terms of the fractional deviation from steady state.

Dimensionless perturbation variables are defined as

$$T' = \frac{T - \bar{T}}{\bar{T}} \qquad M' = \frac{M - \bar{M}}{\bar{M}} \qquad I' = \frac{I - \bar{I}}{\bar{I}}$$

$$F' = \frac{F - \bar{F}}{\bar{F}} \qquad X' = \frac{X - \bar{X}}{\bar{X}} \qquad \Theta = \frac{\bar{F}t}{V}$$

where the bar superscripts relate to steady-state conditions, and the dash superscript represents the perturbation from steady state.

The model equations can then be recast into dimensionless form as

$$\frac{dT'}{d\Theta} = \left((T'_0 + 1)\frac{\bar{T}_0}{\bar{T}} - (T' + 1)\right)(F' + 1) + \left(1 - \frac{\bar{T}_0}{\bar{T}}\right)(I' + 1)^{0.5}(M' + 1)\exp\left(E_2 \frac{T'}{1 + T'}\right)$$

$$\frac{dM'}{d\Theta} = \left((M'_0 + 1)\frac{\bar{M}_0}{\bar{M}} - (M' + 1)\right)(F' + 1) + \left(1 - \frac{\bar{M}_0}{\bar{M}}\right)(I' + 1)^{0.5}(M' + 1)\exp\left(E_2 \frac{T'}{1 + T'}\right)$$

$$\frac{dI'}{d\Theta} = \left((I'_0 + 1)\frac{\bar{I}_0}{\bar{I}} - (I' + 1)\right)(F' + 1) + \left(1 - \frac{\bar{I}_0}{\bar{I}}\right)(I' + 1)\exp\left(E_1 \frac{T'}{1 + T'}\right)$$

$$X' = \frac{(M' + 1)}{(I' + 1)^{0.5}}\exp\left((E_2 - E_1)\frac{T'}{1 + T'}\right) - 1$$

In the above model equations, the terms $\left(\frac{\bar{T}_0}{\bar{T}}\right)$, $\left(\frac{\bar{M}_0}{\bar{M}}\right)$, $\left(\frac{\bar{I}_0}{\bar{I}}\right)$, E_1 and E_2 now represent dimensionless constants, where

$$\left(\frac{\bar{M}_0}{\bar{M}}\right) = 1 + \frac{V A_p}{\bar{F}}\left(\frac{A_d \bar{I}}{A_t}\right)^{0.5}\exp(-E_2)$$

$$\left(\frac{\bar{T}_0}{\bar{T}}\right) = 1 - \frac{\bar{M}(-\Delta H_p)}{\bar{T}\rho c_p}\left(\frac{\bar{M}_0}{\bar{M}} - 1\right)$$

$$\left(\frac{\bar{I}_0}{\bar{I}}\right) = 1 + \frac{V A_d}{\bar{F}}\exp(-E_1)$$

$$E_1 = \frac{E_d}{R\bar{T}}, \quad E_2 = \frac{2E_d + E_d - E_t}{2R\bar{T}}$$

5.3 Continuous Tank Reactor Examples

For a full derivation of the above model equations, the reactor is referred to the original paper (Kenat et al., 1967).

Program

The dimensionless model equations are programmed into the MADONNA simulation program HOMPOLY, where the variables, M, I, X and TEMP are zero. The values of the dimensionless constant terms in the program are realistic values chosen for this type of polymerisation reaction. The program starts off at steady state, but can then be subjected to fractional changes in the reactor inlet conditions, M_0, I_0, T_0 and F, of between 2 and 5 per cent, using the IF-THEN-ELSE statement and parametric run facility. The value of TIME in the program, of course, refers to dimensionless time.

Nomenclature

Symbols

A	Arrhenius constant	1/s or kmol/m³s
c_p	Specific heat	kJ/kg K
E	Energy of activation	kJ/kmol
F	Flow rate	m³/s
I	Initiator's concentration	kmol/m³
M	Monomer's concentration	kmol/m³
•M	Chain radical concentration	kmol/m³
M_x	Dead polymer chain of x units	kmol/m³
•M_y	Growing polymer chain of y units	kmol/m³
•R	Initiator free radical	kmol/m³
T	Absolute temperature	K
V	Reactor volume	m³
X	Average degree of polymerisation or chain length	–
f	Reactivity fraction for free radicals	–
k	Rate constant	1/s or m³/kmol s
r	Rate of reaction	kmol/m³s
t	Time	s
Θ	Dimensionless time	–
r	Density	kg/m³
ΔH_p	Heat of reaction per monomer unit	kJ/kmol

Indices

d	Refers to decomposition of initiator
i	Refers to initiation process
p	Refers to propagation reaction
t	Refers to termination reaction
0	Refers to inlet

Exercises

1. Use the program to study the effects of both positive and negative fractional changes in feed temperature, monomer and initiator concentration and feed flow rate. Compare your observations with the results and explanations of Kenat et al. (1967).
2. One of the main conclusions of Kenat et al. (1967) was that the largest changes in polymer mean-chain length occur from the effect of inlet temperature changes and also that, therefore, controlling inlet temperature, rather than reactor temperature, is beneficial to reactor performance. Test this out by simulation.

Results

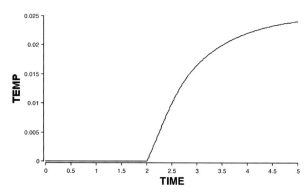

Fig. 2 An increase in the dimenionless feed temperature from 0 to 0.02 results in increased reactor temperature as shown.

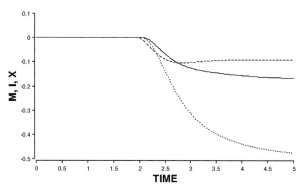

Fig. 3 The temperature change in Fig. 2 causes pronounced decreases in the dimensionless concentrations of monomer, initiator, and chain length.

Reference

Kenat, T.A., Kermode, R.I. and Rosen, S.L. (1967) Ind. Eng. Chem. Proc. Des. Devel, 6, 363.

5.4
Tubular Reactor Examples

5.4.1
TUBE and TUBEDIM – Tubular Reactor Model for the Steady State

System

The tubular reactor, steady-state design equation is of interest here. The dimensional and dimensionless forms are compared for the case of an nth-order reaction.

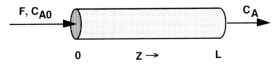

Fig. 1 Isothermal tubular reactor.

Model

The component mass balance at steady state is

$$\frac{dC_A}{dZ} = \frac{1}{v} r_A$$

where v is the linear flow velocity ($=F/A$).

The kinetics here are

$$r_A = -kC_A^n$$

Using the variables $\overline{C_A} = C_A/C_{A0}$ and $\overline{Z} = Z/L$, the dimensionless model is

$$\frac{d\overline{C_A}}{d\overline{Z}} = -\left[\frac{kL}{v} C_{A0}^{n-1}\right] \overline{C_A^n} = [PAR]\overline{C_A^n}$$

Note that the dimensionless parameter PAR is the ratio of the residence time, L/v, and the reaction time for an nth-order reaction, $1/kC_{A0}^{n-1}$.

Program

The results for both the separate dimensional TUBE and dimensionless TUBEDIM cases should be the same if the reaction order n and PAR are the same.

Nomenclature

Symbols

A	Area of tube cross-section	m^2
C_A	Concentration	$kmol/m^3$
F	Flow rate	m^3/s
k	Rate constant	various
n	Reaction order	–
r_A	Reaction rate	$kmol/m^3\ s$
t	Time	s
v	Flow velocity	m/s
Z	Length	m

Exercises

1. Run TUBE for various k, v and C_{A0} and compare with corresponding runs from TUBEDIM.
2. Graph $\overline{C_A}$ versus $\overline{Z}$ n=1 for and various values of PAR.
3. Repeat Exercise 2 for n=2.

Results

Using the same set of parameters, the results from both programs will be the same.

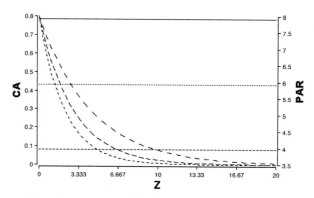

Fig. 2 These steady-state profiles were obtained from TUBE by setting three values of k.

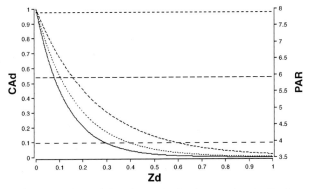

Fig. 3 These profiles were obtained from TUBEDIM by changing k as in Fig. 2, thus keeping the same values of PAR.

Reference

Levenspiel, O. (1999) Chemical Reaction Engineering, 3rd edition, Wiley.

5.4.2
TUBETANK – Design Comparison for Tubular and Tank Reactors

System

Steady-state conversions for both continuous tank and tubular reactors are compared for nth-order reaction kinetics.

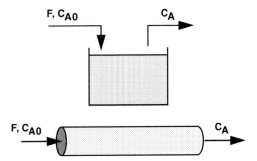

Fig. 1 Comparison of tank and tubular reactors.

The comparison is made by calculating the residence times and volumes required for a particular value of fraction conversion. This allows a comparison of the reactor volumes for a given flow rate. Levenspiel (1999) presents design graphs for the steady-state solution, similar to those generated by this program. One advantage of this method is an easy use of arbitrary kinetic functions.

Model

For a tank reactor with nth-order reaction at steady state

$$0 = FC_{A0} - FC_A - kC_A^n V$$

For n=1

$$C_{ATa} = \frac{FC_{A0}}{F - kV} = \frac{C_{A0}}{1 - k\tau_{Ta}}$$

where τ_{Ta} is the mean residence time for the tank.

For a plug-flow tubular reactor with nth-order reaction at steady state

$$\frac{dC_A}{dZ} = \frac{-1}{v} kC_A^n$$

For n=1

$$C_A = C_{A0} e^{-k\tau_{Tu}}$$

where the tubular reactor residence time τ_{Tu} is equal to Z/v.

For a given fraction conversion X_A, the residence time for n=1 can be calculated from

$$\tau_{Tu} = \frac{1}{k} \ln \frac{1}{1 - X_A}$$

For nth-order reaction in terms of fraction conversion X_A for a tank

$$\tau_{Ta} k C_{A0}^{n-1} = \frac{X_A}{(1 - X_A)^n}$$

and for a plug-flow tubular reactor (n≠1) the integration gives

$$\tau_{Tu} k C_{A0}^{n-1} = \frac{-1}{1 - n} \left((1 - X_A)^{1-n} - 1 \right)$$

Program

The problem is not one that would normally be solved with a program such as MADONNA. The values for X_A generated from an integration are used to calculate quasi backwards the residence time and the volume required from the analytical steady state solutions for tubular and tank reactors. The STOPTIME is renamed XaStop.

Nomenclature

Symbols

C_A	Concentration of A	kg/m^3
F	Flow rate	m^3/s
k	Reaction rate constant	m$^{3(n-1)}$/kg$^{(n-1)}$ s
n and n$_0$	Reaction order	
v	Flow velocity	m/s
V	Volume	m^3
V$_{tuta}$	Volume ratio, tube to tank	–
X$_A$	Fraction conversion	–
Z	Length variable	m
τ	Residence time	s

Indices

tu	Refers to tube
ta	Refers to tank
0	Refers to inlet conditions

Exercises

1. Set the volumetric flow rate and feed concentration for the tank and tubular reactors to desired values. Set also the order of reaction to n=1.01. Run for a range of fraction conversions from 0 to 0.99. Compare the required volumes for the two reactor types.
2. Rerun Exercise 1 for n=2 and compare the ratio of volumes V$_{tuta}$. Compare the required volumes for the two reactor types. Suppose a conversion of 90% is desired and the flow rate to the tank reactor is to be one-half that of the tubular reactor. What would be the ratio of volumes?
3. Repeat Exercise 2 with a 2-fold increase in C_{A0}. How do the required volumes compare. Repeat for n=1.01. Explain the results.
4. Use a dynamic model for a tank reactor to verify the values calculated by this program. Do the same for a steady state and dynamic tubular reactor model.

Results

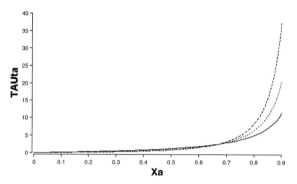

Fig. 2 The required volume for a tank reactor for reaction order n=1, 1.5, and 2.0 is plotted versus fraction conversion.

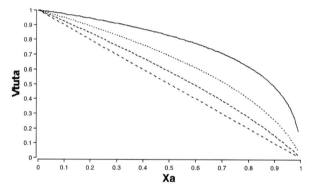

Fig. 3 The ratio of required residence times, tube to tank, is plotted versus conversion for reaction orders n=0.5, 1, 1.5, and 2.0:

Reference

Levenspiel, O. (1972) Chemical Reaction Engineering, 3rd edition, Wiley.

5.4.3
BENZHYD – Dehydrogenation of Benzene

System

The gas-phase dehydrogenation of benzene to diphenyl (D) and further to triphenyl (T) is conducted in an ideal isothermal tubular reactor. The aim is to maximize the production of D and to minimize the formation of T. Two parallel gas-phase reactions occur at atmospheric pressure.

Check these equations

$$2C_6H_6(B) \xrightarrow{k_1} C_{12}H_{10}(D) + H_2(H)$$

$$C_6H_6(B) + C_{12}H_{10}(D) \xrightarrow{k_2} C_{18}H_{14}(T) + H_2(H)$$

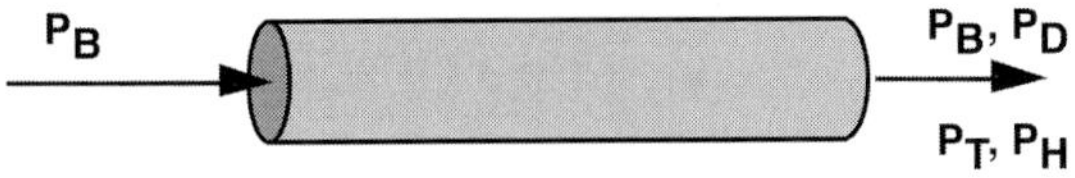

Fig. 1 The partial-pressure variables for the tubular reactor are shown.

Model

The kinetic model is given in terms of Arrhenius relationships and partial-pressure relationships as

$$r_1 = -k_1\left(P_B^2 - \frac{P_D P_H}{K_{eq1}}\right)$$

where

$$k_1 = 1.496E+7 * \exp\left(\frac{8440}{T_r}\right)$$

$$r_2 = -k_2\left(P_B P_H - \frac{P_T P_H}{K_{eq2}}\right)$$

where

$$k_2 = 8.67E+6 * \exp\left(\frac{8440}{T_r}\right)$$

At steady-state conditions, the mass balance design equations for the ideal tubular reactor apply. These equations may be expressed as

$$\frac{V}{F} = -\int \frac{dX_1}{r_1} = \tau$$

$$\frac{V}{F} = -\int \frac{dX_2}{r_2} = \tau$$

where X_1 is the fractional conversion of benzene by reaction 1, X_2 is the fractional conversion of benzene by reaction 2 and τ is the reactor space time.

In differential form

$$\frac{dX_1}{d\tau} = -r_1$$

$$\frac{dX_2}{d\tau} = -r_2$$

Expressing P_B, P_D, P_H and P_T in terms of X_1 and X_2.

$$P_B = (1 - X_1 - X_2)$$

$$P_D = \frac{X_1}{2} - X_2$$

$$P_H = \frac{X_1}{2} + X_2$$

$$P_T = X_2$$

Substituting into the mass balance

$$\frac{dX_1}{d\tau} = k_1 \left((1 - X_1 - X_2)^2 - \frac{\left(\frac{X_1}{2} - X_2\right)\left(\frac{X_1}{2} + X_2\right)}{K_{eq1}} \right)$$

$$\frac{dX_2}{d\tau} = k_2 \left((1 - X_1 - X_2)\left(\frac{X_1}{2} - X_2\right) - \frac{X_2\left(\frac{X_1}{2} + X_2\right)}{K_{eq2}} \right)$$

Nomenclature

Symbols

k_{eq1}, k_{eq2}	Equilibrium constants	–
k_1, k_2	Reaction rate constants	$1/\text{atm}^2$ h
P_B	Partial pressure of benzene	atm
P_D	Partial pressure of diphenyl	atm
P_H	Partial pressure of hydrogen	atm
P_T	Partial pressure of triphenyl	atm
r_1 and r_2	Reaction rates	1/h
Tr	Temperature	K
X_1 and X_2	Fractional conversion	–
τ	Space time	h

Exercises

1. Study the effect of varying space time on the fractional conversions X1 and X2 and evaluate the compositions of benzene, hydrogen, diphenyl and triphenyl.
2. Show that the diphenyl composition passes through a maximum.
3. Study the effect of varying pressure on the reaction by modifying the partial-pressure relationships.
4. Plot the rates r_1 and r_2 as functions of space time.
5. Study the effects of temperature.

Results

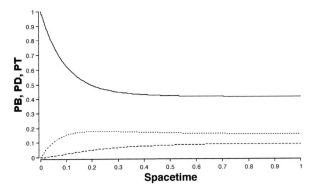

Fig. 2 A maximum in the diphenyl (PD) composition is obtained.

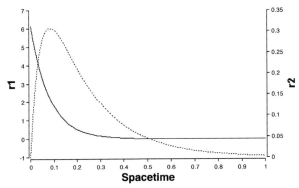

Fig. 3 The reaction rate of the second reaction step exhibits a maximum as a function of reactor space-time.

Reference

Smith, J.M. (1970) Chemical Engineering Kinetics, 2nd edition, p. 158, McGraw-Hill.

5.4.4
ANHYD – Oxidation of O-Xylene to Phthalic Anhydride

System

Ortho-xylene (A) is oxidised to phthalic anhydride (B) in an ideal, continuous flow tubular reactor. The reaction proceeds via the complex consecutive parallel reaction sequence, shown below. The aim of the reaction is to produce the maximum yield of phthalic anhydride and the minimum production of waste gaseous products (C), which are CO_2 and CO.

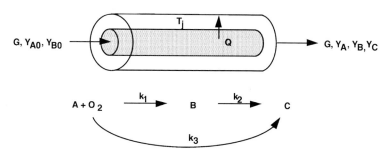

Fig. 1 The fixed-bed reactor and the reaction scheme.
Symbols: A = o-xylene, B = phthalic anhydride, C = waste gaseous products (CO_2 and CO).

The reaction is carried out with a large excess of air, and the total gas mass flow rate can be taken as constant.

Froment and Bischoff (1990) have discussed the modelling of the fixed bed production of phthalic anhydride from o-xylene, including both axial and radial temperature profile effects.

Model

Kinetics

The reaction kinetics is as follows

$$r_A = -(k_1 + k_3) Y_A Y_{O_2}$$

$$r_B = k_1 Y_A Y_{O_2} - k_2 Y_B Y_{O_2}$$

$$r_C = k_3 Y_A Y_{O_2} + k_2 Y_B Y_{O_2}$$

It is assumed that the mole fraction of oxygen does not change, owing to the large excess of air.

The temperature dependencies are according to Arrhenius

$$\ln k_1 = -\frac{27000}{RT} + 19.837$$

$$\ln k_2 = -\frac{31400}{RT} + 20.86$$

$$\ln k_3 = -\frac{28600}{RT} + 18.97$$

Component Mass Balance Equations

Considering Fig. 2, the steady-state balance around any segment ΔV, for component A, is given by

$$0 = n_{AF} - (n_{AF} + \Delta n_{AF}) + r_A \rho \Delta V$$

where n_{AF} is the molar flow rate of A in kmol/h and F_m is the mass flow rate in kg/h ($= G A_c$).

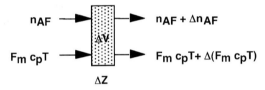

Fig. 2 The balance region ΔV.

Thus the changes in the molar flow rates (kmol/hm) at steady state can be set equal to the rates of production of each component.

$$\frac{dn_{AF}}{dZ} = A_c \rho r_A$$

$$\frac{dn_{BF}}{dZ} = A_c \rho r_B$$

$$\frac{dn_{CF}}{dZ} = A_c \rho r_C$$

The cross-sectional area of the reactor is given by

$$A_c = \pi \frac{d_t^2}{4}$$

The total molar flow (mostly air) does not change. Thus

$$n_{tF} = \frac{G}{M_m} A_c$$

The mole fractions can be calculated by

$$Y_A = \frac{n_{AF}}{n_{tF}}$$

$$Y_B = \frac{n_{BF}}{n_{tF}}$$

$$Y_C = \frac{n_{CF}}{n_{tF}}$$

The fractional conversions of A to B and C are given by

$$X_A = \frac{n_{AF0} - n_{AF}}{n_{AF0}}$$

$$X_B = \frac{n_{FB}}{n_{AF0} - n_{AF}}$$

$$X_C = \frac{n_{FC}}{n_{AF0} - n_{AF}}$$

Energy Balance

At steady state the energy balance for element ΔV is

$$0 = -\Delta(F_m c_p T) + \rho \Delta V \Sigma(r_i \Delta H_i) - UA_t(T - T_j)$$

where the heat transfer area $A_t = \pi d_t \Delta Z$ and the element $\Delta V = A_c \Delta Z$. Thus

$$\frac{dT}{dZ} = \frac{\rho A_c}{F_m c_p}((\Delta H_1 k_1 + \Delta H_3 k_3) Y_A Y_{O2} + \Delta H_2 k_2 2 Y_A Y_{O2}) - \frac{U \pi d_t}{F_m c_p}(T - T_j)$$

The above equations are solved with the initial conditions at the reactor entrance given by

$$n_{AF} = Y_{A0} n_{tF}$$

$$n_{BF} = n_{CF} = 0$$

$$T = T_{in}$$

Program

The particle diameter (not used in the program) can be used to normalize the length Z. The factors 1000 on k_1, k_2 and k_3 in the program convert gmol to kmol.

Nomenclature

Symbols

A_c	Area cross-section	m^2
c_p	Specific heat	kJ/kg K
ΔH_1	Heat of reaction A → B	kJ/kmol
ΔH_3	Heat of reaction A → C	kJ/kmol
d_t	Tube diameter	m
E	Activation energy of reaction	kJ/kmol
F_m	Mass flow rate	kg/h
G	Superficial mass velocity	kg/m² h
k	Rate constant	kmol/(kg cat.) h
M_m	Mean molecular weight	kg/gmol
n	Molar flow rate	kmol/h
nAF	Molar flow rate of A	kmol/h
nBF	Molar flow rate of B	kmol/h
nCF	Molar flow rate of C	kmol/h
ntF	Total molar flow rate	kmol/h
R	Universal gas constant	kJ/kmol K
ρ	Catalyst bulk density	kg/m³
r	Reaction rate	kmol/(kg cat.) h
T and Temp	Temperature	K
U	Heat transfer coefficient	kJ/m² h K
V	Volume	m³
Y	Mole fraction	–
Z	Distance along tube	m

Indices

A, B, C, O_2	Refer to components
F	Refers to flow
j	Refers to jacket
t	Refers to total
1, 2, 3	Refer to reactions
0	Refers to inlet

Exercises

1. Study the effect of tube diameter on the axial profiles.
2. Study the effect of inlet temperature with constant jacket temperature. Note the hot spot effect in the reactor temperature profile.
3. Study the effect of jacket temperature on the temperature profile, by keeping T_0 constant and varying T_j.

4. Repeat Exercise 1 and 2 setting $k_3 = 0$. Observe that the reaction rate A to C is relatively small but has a large influence on the temperature of the reactor.

Results

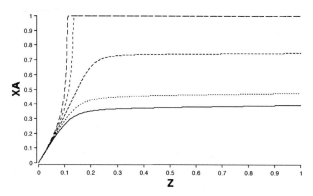

Fig. 3 The influence of jacket temperature ($T_j = 580$ to 590 K) is seen in these steady-state conversion profiles.

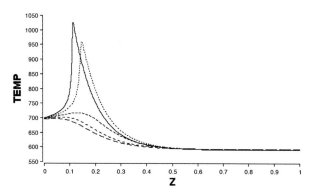

Fig. 4 The temperature profiles are strongly influenced by the feed temperature 695 to 700 K.

Reference

Froment, G. E., Bischoff, K. B. (1990) Chemical Reactor Analysis and Design, 2nd edition, Wiley.

5.4.5
NITRO – Conversion of Nitrobenzene to Aniline

System

Nitrobenzene and hydrogen are fed at the rate of n_{tF} to a tubular reactor operating at atmospheric pressure and 450 K and containing catalyst with a voidage ε.

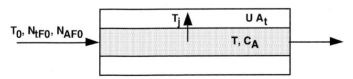

Fig. 1 Tubular reactor with heat transfer.

The heat transfer coefficient, U, to the reactor wall and the coolant temperature is T_j. The nitrobenzene feed rate is N_{AF0}, with hydrogen in very large excess. The heat of reaction is ΔH. The heat capacity of hydrogen is c_p.

The reaction rate is given by

$$r_A = -5.79 \times 10^4 C_A^{0.578} e^{-2958/T}$$

where r_A has the units of kmoles (of nitrobenzene reacting)/(m³ of void space h) and C_A is the concentration of nitrobenzene (kmol/m³). This problem is considered in many chemical engineering texts, but here the solution by digital simulation is shown to be very convenient.

Model

Mass Balance

Consider the steady state balances around the segment ΔV, as shown in Fig. 2

$$0 = N_{AF} - (N_{AF} + \Delta N_{AF}) + r_A \rho \Delta V$$

where N_{AF} is the molar flow rate of A (kmol/h).

$$0 = (N_{tF} c_p T) - (N_{tF} c_p T + \Delta(N_{tF} c_p T)) + r_A \Delta V \Delta H - U A_t (T - T_j)$$

where N_{tF} is the total molar flow rate (kmol/h), which can be assumed constant owing to the large excess of hydrogen in the feed.

The changes in the molar flowrate of A with distance Z at steady-state conditions are thus

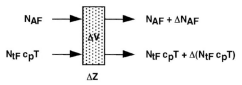

Fig. 2 Balance region for a segment.

$$\frac{dN_{AF}}{dZ} = A_c r_A$$

where

$$A_c = \pi \frac{d_t^2}{4} \varepsilon$$

The fractional conversion of A is

$$X_A = \frac{N_{AF0} - N_{AF}}{N_{AF0}}$$

and

$$dN_{AF} = -N_{AF0} dX_A$$

The balance becomes

$$\frac{dX_A}{dZ} = -\frac{A_c}{N_{AF0}} r_A$$

Energy Balance

At steady-state for constant N_{tF} and c_p

$$0 = -N_{tF} c_p \Delta T + \Delta V r_A \Delta H - U A_t (T - T_j)$$

where the transfer area $A_t = \pi d_t \Delta Z$ and the volume element $\Delta V = A_c \Delta Z$.

Thus

$$\frac{dT}{dZ} = \frac{A_c}{N_{tF} c_p} \Delta H r_A - \frac{U \pi d_t}{N_{tF} c_p}(T - T_j)$$

Kinetics

The reaction rate is

$$r_A = -5.79 \times 10^4 C_A^{0.578} e^{-2958/T}$$

where

$$C_A = C_{A0}(1 - X_A)\frac{T_0}{T} = (1 - X_A)\frac{N_{AF0}}{N_{tF}}\frac{P_{tot}}{RT}$$

The above equations are solved with the initial conditions at the reactor entrance

$$X_A = 0 \quad \text{and} \quad T = T_{in}$$

Nomenclature

Symbols

A_c	Cross-sectional area for flow	m^2
A_t	Transfer area of tube	m^2
C_A	Concentration of nitrobenzene	$kmol/m^3$
c_p	Heat capacity of hydrogen	$kJ/kmol\ K$
ΔH	Heat of reaction	$kJ/kmol$
d_t	Diameter of tube	m
ε	Catalyst with voidage	–
N_{AF}	Nitrobenzene molar flow rate	$kmol/h$
N_{tF}	Total molar flow rate	$kmol/h$
P_{tot}	Total pressure	atm
R	Gas constant	$m^3\ atm/kmol\ K$
r_A	Reaction rate	$kmol/m^3$ voids h
T and Temp	Reactor temperature	K
T_j	Coolant temperature	K
U	Heat transfer coefficient	$kJ/h\ m^2\ K$
X	Fraction conversion	–
Z	Reactor length	m

Indices

0	Refers to entering conditions

Exercises

1. Run the program for varying inlet temperatures 350 to 500 K.
2. Study also the effect of coolant temperature.
3. Comment on the monotonously decreasing temperature profiles in the last part of the reactor.
4. Modify the program such that it models adiabatic behaviour and study the influence of varying feed temperature.

Results

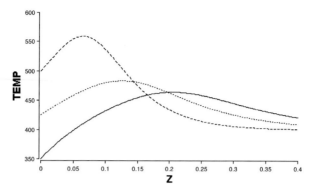

Fig. 3 The inlet temperatures were set at 350, 425 and 500 for these axial temperature profiles.

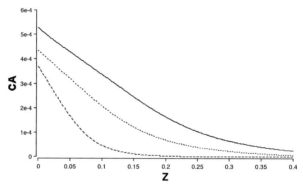

Fig. 4 The reactor concentrations for the conditions of Fig. 3. Note that the inlet concentrations depend on inlet temperature.

Reference

Smith, J. M. (1972) Chemical Engineering Kinetics, 2nd edition, McGraw-Hill.

5.4.6
TUBDYN – Dynamic Tubular Reactor

System

Tanks-in-series reactor configurations provide a means of approaching the conversion of a tubular reactor. In modelling, systems of tanks in series are employed for describing axial mixing in non-ideal tubular reactors. Residence time distributions, as measured by tracers, can be used to characterise reactors, to establish models and to calculate conversions for first-order reactions. The reactor in this example has a recycle loop to provide additional flexibility in modelling the mixing characteristics.

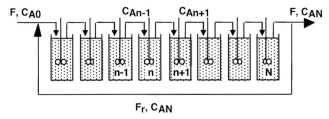

Fig. 1 Tanks-in-series approximation of a tubular reactor with recycle.

Model

For the nth tank the balance for a single component is

$$V \frac{dC_{An}}{dt} = (F + F_r)(C_{An-1} - C_{An}) + r_{An} V$$

If applied to a tracer, r_{An} is zero. Otherwise the reaction rate is taken as nth-order

$$r_{An} = k C^{n,An}$$

Residence time distributions can be measured by applying tracer pulses and step changes and, as described in Levenspiel (1999), the pulse tracer response can be used to obtain the E-curve and to calculate the steady-state conversion for a first-order reaction according to the relationship

$$\frac{C_{Ass}}{C_{A0}} = \int_0^\infty E(t) e^{-kt} dt$$

Program

This program is designed to simulate the resulting residence time distributions based on a cascade of 1 to N tanks-in-series. Also, simulations with nth-order reaction can be run and the steady-state conversion obtained. A pulse input disturbance of tracer is programmed here, as in example CSTRPULSE, to obtain the residence time distribution E curve and from this the conversion for first order reaction.

Nomenclature

Symbols

C_{A0}	Feed reactant concentration	kg/m³
C_{AN}	Concentration in tank N	kg/m³
F	Flow rate	m³/s
k	Reaction constant	various
n	Reaction order	–
N	Number of tanks	–
R	Reaction rate	kg/m³ s
V	Total volume of all tanks	m³

Indices

n	Refers to nth tank
r	Refers to recycle
t	Refers to time
ss	Refers to steady-state

Exercises

1. Set k=0 and simulate tracer experiments. Vary the recycle flow rate and the number of tanks.
2. Use the E-curve to calculate the conversion for n=1 and compare with the results obtained by simulation. This requires changing the program.
3. Vary the operating variables of feed flow rate and C_{A0} to see the influence with n=1 and n=2.
4. Calculate the steady-state conversion for 1, 3 and 8 tanks using the same value of total residence time.
5. Experiment with the influence of recycle flow rate and show by simulation that the model has the limits of a well-stirred tank and a plug flow reactor.

Results

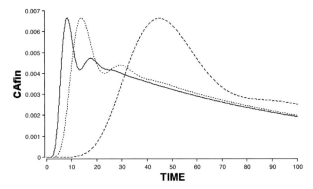

Fig. 2 Pulse response curves for tracer at 3 recycle rates F_r=1, 5.5 and 10.

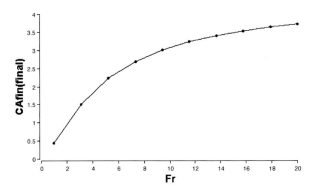

Fig. 3 Steady state outlet concentrations for first order reaction as a function of recycle rate.

Reference

Levenspiel O. (1999) Chemical Reaction Engineering, 3rd edition, Wiley.

5.4.7
DISRE – Isothermal Reactor with Axial Dispersion

System

This example models the dynamic behaviour of an non-ideal isothermal tubular reactor in order to predict the variation of concentration, with respect to both axial distance along the reactor and flow time. Non-ideal flow in the reactor is represented by the axial dispersion flow model. The analysis is based on a simple, isothermal first-order reaction.

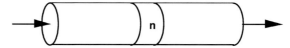

Fig. 1 Tubular reactor with finite-difference element n.

Model

The reaction involves the conversion: A → products, with the kinetics

$$-r_A = kC_A$$

As discussed in Section 4.3.6, the axial dispersion flow model is given by

$$\frac{\partial C_A}{\partial t} = -u\frac{\partial C_A}{\partial Z} + D\frac{\partial^2 C_A}{\partial Z^2} - kC_A$$

where Z is the distance along the reactor, D is the eddy diffusivity and u is the superficial flow velocity.

Thus the problem involves the two independent variables, time t and length Z. The distance variable can be eliminated by finite-differencing the reactor length into N equal-sized segments of length ΔZ, such that $N\Delta Z$ equals L, where L is the total reactor length.

Thus for a segment n omitting subscripts A

$$\frac{dC_n}{dt} = u\left(\frac{C_{n-1} - C_n}{\Delta Z}\right) + D\left(\frac{C_{n-1} - 2C_n + C_{n+1}}{\Delta Z^2}\right) - kC_n$$

Or defining $C'_n = C_n/C_0$, $\Delta z = \Delta Z/L$, $\Theta = tu/L$ in dimensionless terms and omitting the primes for simplicity

$$\frac{dC_n}{d\Theta} = \frac{C_{n-1} - C_n}{\Delta z} + \frac{D}{Lu}\left(\frac{C_{n-1} - 2C_n + C_{n+1}}{\Delta z^2}\right) - \frac{kL}{u}C_n$$

The term Lu/D is known as the Peclet number, Pe, and its inverse as the dispersion number. The magnitude of the Peclet number defines the degree of axial mixing in the reactor.

Thus

$$\frac{dC_n}{d\Theta} = \frac{C_{n-1} - C_n}{\Delta z} + \frac{1}{Pe}\left(\frac{C_{n-1} - 2C_n + C_{n+1}}{\Delta z^2}\right) - \frac{kL}{u}C_n$$

The dimensionless boundary conditions are satisfied by $C_0 = 1$ for a step change in feed concentration at the inlet and by the condition that at the outlet

$C_{N+1}=C_N$, which sets the concentration gradient to zero. The reactor is divided into equal-sized segments.

Program

Array programming is used here which allows the graphing of the axial profile. Closed-end boundary conditions are used for the first and last segments. Also included is a PULSE function for simulating tracer experiments. Thus it should be possible to calculate the E curve and from that the reaction conversion obtained on the basis of tracer experiment. The example CSTRPULSE should be consulted for this.

The full listing of the MADONNA program is shown here both to illustrate the use of array programming and also the use of the PULSE function.

```
{Example DISRE}

{UNSTEADY-STATE FINITE-DIFFERENCED MODEL FOR AXIAL DISPER-
SION AND REACTION IN A TUBULAR REACTOR}
{The reactor length is divided into equal-sized sections,
whose number, Nelem, can be varied. The model equations are
written in dimensionless form.
{The variable Time is dimensionless and equal to tu/L.}
{The Peclet number is equal to Lu/D, where D is the axial dif-
fusivity.}
{All concentrations C[i] are dimensionless and equal to C/C0}

{From HELP: PULSE(v, f, r) Impulses of volume v, first time f,
repeat interval r}
{For the nonreacting system with K=0, making a pulse of tracer
at the inlet with C0=0 and measuring C[Nelem] at the outlet
gives information on the degree of axial mixing. This pulse is
programmed below and can be activated by removing the brack-
ets. This simulated measurement can be used to calculate the
steady state conversion for a first order reaction.}
Method Stiff
DT=0.001
STOPTIME=4

L=10              {Total length, m }
U=1               {Velocity, m/h}
C0=1              {Entering dimensionless conc., -}
Pe=0.5            {Peclet number=Lu/D, -}
K=0.01            {Rate constant, 1/hr}
CI=0              {Initial concentration conditions, -}
```

```
delZ=1.0/Nelem          {Dimensionless length increment, -
                         Dimensionless length has a value of 1.0}

K1=K*L/U                {Dimensionless ratio of reaction rate
                         to convective transport rate}

Nelem=10                {Number of elements}
NIT C[1..Nelem-1]=CI    {Initial concentrations, -}
```

{Dimensionless reactant balances for Nelem sections, with "closed-end" boundary conditions}
d/dt(C[1])=(C0-C[1])/delZ-(C[1]
C[2])/(delZ*delZ*Pe)-K1*C[1]

d/dt(C[2..Nelem-1])=(C[i-1]-C[i])/delZ+(C[i-1]-
2*C[i]+C[i+1])/(delZ*delZ*Pe)-K1*C[i]

d/dt(C[Nelem])=(C[Nelem-1]-C[Nelem])/delZ+(C[Nelem-1]-
C[Nelem])/(delZ*delZ*Pe)-K1*C[Nelem]

LIMIT C>=0

{Brackets can be set or removed for the pulse of tracer. Set K=0 to make a tracer experiment with no reaction.}
C0=Pulse(0.01, starttime, 1000)
Cpulse=C[Nelem] {This is the outlet conc. for the pulse}

Nomenclature

Symbols

C	Concentration	kmol/m^3 and –
D	Effective axial diffusivity	m^2/h
k	Rate constant	1/h
L	Total length	m
Pe	Peclet number (Lu/D)	–
t	Time	h
u	Velocity	m/h
Z	Length variable	m
z	Dimensionless length	–
Θ	Dimensionless time	–

Indices

A	Refers to reactant
n	Refers to increment number
'	Refers to a dimensionless variable

Exercises

1. Obtain the tracer response curve to a step input disturbance of tracer solution by setting k=0.
2. Study the effect of varying Peclet number, Pe, on the resulting tracer response.
3. Note that as Pe becomes large, the conditions approach plug flow, and as Pe approaches zero conditions approach perfect mixing.
4. Study the effect of varying Pe on the performance of the reactor, and compare the resulting performance with perfect plug flow.
5. Use the E curve from a pulse tracer experiment to calculate the first-order conversion.

Results

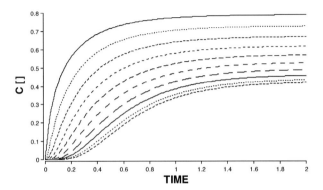

Fig. 2 Dynamic concentration profiles at each axial position in the reactor for a first order reaction.

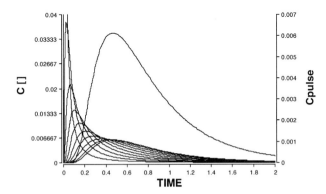

Fig. 3 Response of reactor to a pulse input of tracer. The outlet concentration from the last element is graphed separately.

5.4.8
DISRET – Non-Isothermal Tubular Reactor with Axial Dispersion

System

The dispersion model of example DISRE is extended for the case of non-isothermal reactions and to include the axial dispersion of heat from a first-order reaction.

$$A \rightarrow B + \text{heat}(-\Delta H)$$

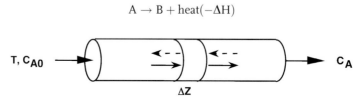

Fig. 1 Tubular reactor showing a finite difference segment.

Model

From the axial dispersion flow model the component balance equation is

$$\frac{\partial C_A}{\partial t} = -u\frac{\partial C_A}{\partial z} + D\frac{\partial^2 C_A}{\partial^2 z} - kC_A$$

with the energy balance equation given by

$$\rho c_p \frac{\partial T}{\partial t} = -c_p u \frac{\partial T}{\partial z} + \lambda \frac{\partial^2 T}{\partial z^2} - kC_A(-\Delta H)$$

where λ is the eddy diffusivity for heat transfer.

The reaction kinetics are

$$r_A = kC_A$$

and

$$k = k_0 e^{-E/RT}$$

Defining $Z=z/L$, $t=T/T_0$, $C=C_A/C_{A0}$, $\Theta=Dt/L^2$, the model equations can be recast into dimensionless terms to give the following dimensionless model equations:

Mass Balance

$$\frac{\partial C}{\partial \Theta} = \frac{\partial^2 C}{\partial Z^2} - P_1 \frac{\partial C}{\partial Z} - B_1 C e^{-\varepsilon/t}$$

Energy Balance

$$\frac{\partial t}{\partial \Theta} = \frac{a}{D}\frac{\partial^2 t}{\partial Z^2} - P_2 \frac{\partial t}{\partial Z} - B_1 B_2 C e^{-\varepsilon/t}$$

where

$$P_1 = \frac{uL}{D}, \quad P_2 = \frac{uL}{a}, \quad a = \frac{\lambda}{\rho c_p}, \quad B_1 = \frac{k_0 L^2}{D}, \quad B_2 = (-\Delta H)\frac{C_{A0}}{\rho c_p T_0}$$

and $\varepsilon = \dfrac{E}{RT_0}$

A full treatment of this problem is given by Ramirez (1976, 1989) and by Clough and Ramirez (1971).

Finite-differencing the dimensionless length of the reactor (Z=1) into N equal-sized segments of length ΔZ, such that $N\,\Delta Z = L$, gives for segment n

$$\frac{dC_n}{d\Theta} = \frac{C_{n-1} - 2C_n + C_{n+1}}{\Delta Z^2} + P_1 \frac{C_{n-1} - C_n}{\Delta Z} - B_1 C_n e^{\varepsilon/t_n}$$

and

$$\frac{dt_n}{d\Theta} = \frac{a}{D}\frac{t_{n-1} - 2t_n + t_{n+1}}{\Delta Z^2} + P_2 \frac{t_{n-1} - t_n}{\Delta Z} - B_1 B_2 C_n e^{-\varepsilon/t_n}$$

The boundary conditions used here are $C_0 = t_0 = 1$ for a step change of inlet concentration or temperature, and at the outlet by $t_{N+1} = t_N$ and $C_{N+1} = C_N$.

Program

Array programming is used, which allows the graphing of the axial profiles. Closed-end boundary conditions are used for the first and last segments.

PULSE functions could be used for studying the axial dispersion of heat or, alternatively, the axial dispersion of mass (see DISRE).

Nomenclature

Symbols

C_A	Concentration	kmol/m^3
C_{A0}	Inlet concentration	kmol/m^3
c_p	Specific heat	kJ/kg K
D	Mass transfer eddy diffusiviy	m^2/s
ΔZ	Dimensionless length increment	–

E	Activation energy	kJ/kmol
$(-\Delta H)$	Heat of reaction	kJ/kmol
L	Reactor length	m
λ	Heat transfer eddy diffusivity	kJ/m s K
R	Universal gas constant	kJ/kmol K
r_A	Reaction rate	kmol/m^3 s
ρ	Density	kg/m^3
t	Time	s
T_0	Inlet temperature	K
u	Flow velocity	m/s
Z	Frequency factor	1/s
a	Diffusivity of heat	m^2/s

Dimensionless Factors

B_1	Dimensionless group	–
B_2	Dimensionless group	–
C	Dimensionless concentration	–
P_1	Peclet number for mass transfer	–
P_2	Peclet number for heat transfer	–
t	Dimensionless temperature	–
Z	Dimensionless length	–
ε	Activation energy group	–
Θ	Dimensionless time	–

Exercises

1. Setting B_1, and hence $k_0=0$, models the reactor response curve to a step input disturbance $C_{A0}=1$.
2. Vary reactor inlet temperature up to 700 K.
3. Study the effect of varying mass transfer and heat transfer diffusivities (D and λ, respectively) and hence Peclet numbers (P_1 and P_2) on the resulting dimensionless concentration and temperature reactor profiles.
4. Based on a linearisation approach as applied to the nonlinear model equations, Clough and Ramirez (1971) predict multiple, steady-state solutions for the reactor. Is this confirmed and if not, why not?
5. Verify that the unusual mixed units used here actually give the dimenionless groups as stated and that the dimensionless balances are consistent.
6. Revise the model to allow for open-end boundary conditions. This would be the case if the reactor were a catalytic packed bed with non-catalytic packing beyond the reactor ends (Fig. 2).

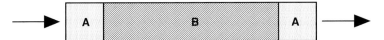

Fig. 2 Catalytic packed bed reactor with non-reactive end sections.

Results

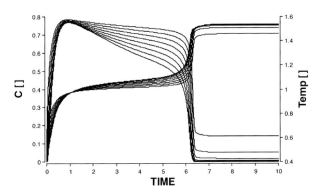

Fig. 3 The reactor responding in C and T for each segment before reaching steady state.

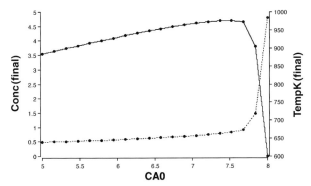

Fig. 4 Steady states for C and T as a parametric run varying inlet concentration.

References

Clough, D. E. and Ramirez, W. F. (1971) Simulation, 207.
Ramirez, W. F. (1976) Process Simulation, Lexington.

5.4.9
VARMOL – Gas-Phase Reaction with Molar Change

System

A tubular reactor is considered in which the gas-phase reaction leads to a change in the molar flow rate and thus in the linear gas velocity. The reaction stoichiometry is represented by

$$A \rightarrow mB$$

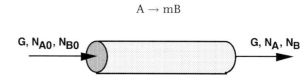

Fig. 1 Tubular reactor with variable molar flow.

Model

The model is described in Section 4.3.3. The steady-state balances are written in terms of moles. From the molar flow at each point in the reactor, the Ideal Gas Law is used to calculate the volumetric flow rate. This gives also the possibility of considering the influence of temperature and pressure profiles along the tube:

$$\frac{d(y_A G)}{dZ} = -k y_A A$$

$$\frac{d(y_B G)}{dZ} = +m k y_A A$$

$$G = \frac{(N_A + N_B + N_{inerts})}{P}$$

$$N_A = \frac{y_A G P}{RT}$$

$$N_B = \frac{y_B G P}{RT}$$

Nomenclature

Symbols

A	Cross-sectional area	m²
G	Volumetric flow rate	m³/s
k	Reaction rate	1/s

m	Stoichiometric constant	–
N	Molar flow rate	kmol/s
P	Pressure	N/m²
R	Gas constant	kJ/kmol K
Temp	Temperature	K
X_A	Fraction conversion	–
y	Mole fraction	–
Z	Length	m

Indices

0	Refers to inlet
inerts	Refers to inerts
A	Refers to component A
B	Refers to component B

Exercises

1. Vary the stoichiometry to see the influence of m. Note that if m=1, G must be constant through the reactor.
2. Set the molar feed rate of inerts to a high value, and note that G does not change much with position.
3. Change the model and program to account for a linear temperature profile.
4. Change the model and program to account for a linear pressure profile, allowing for pressure drop through the reactor.

Results

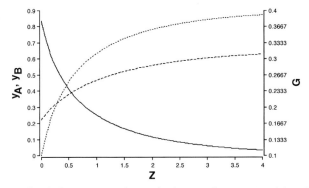

Fig. 2 Variation of mole fractions y_A and y_B and volumetric flow rate G with length.

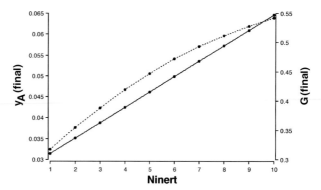

Fig. 3 Outlet mole fraction y_A as a function of molar flow rate of inerts.

5.5
Semi-Continuous Reactor Examples

5.5.1
SEMIPAR – Parallel Reactions in a Semi-Continuous Reactor

System

A semi-continuous reactor is used to carry out the following parallel reaction, where P is valuable product and Q is unwanted waste. The problem is to optimise the feed strategy to the reactor such that the maximum favourable reaction selectivity is obtained, for similar systems but of differing kinetic rate characteristics.

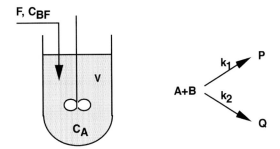

Fig. 1 Parallel reactions in a semi-continuous reactor.

Model

The total mass balance for constant density is

$$\frac{dV}{dt} = F$$

The mass balances for components A, B, P and Q are

$$\frac{d(VC_A)}{dt} = -(r_1 V + r_2 V)$$

$$\frac{d(VC_B)}{dt} = F\,C_{BF} - (r_1 V + r_2 V)$$

$$\frac{d(VC_P)}{dt} = r_1 V$$

$$\frac{d(VC_Q)}{dt} = r_2 V$$

The kinetics for the parallel reactions are

$$r_1 = k_1 C_A^{n_{A1}} C_B^{n_{B1}}$$

$$r_2 = k_2 C_A^{n_{A2}} C_B^{n_{B2}}$$

The relative magnitudes of the orders of reaction, n_{B1} and n_{B2}, determine whether the concentration C_B in the reactor should be maintained at a high or low level. For optimum selectivity, the total reaction time must be adjusted such that P is obtained at its maximum concentration as affected by the rate at which B is fed to the reactor.

Program

Note that, at high rates of feeding, the maximum reactor volume may be reached before the simulation time is complete. Under these conditions, the feed of B to the reactor will stop. The reaction then continues batchwise until the simulation is complete.

Nomenclature

Symbols

C	Concentration	kmol/m³
V	Volume	m³
F	Flow rate	m³/min
r	Reaction rate	kmol/m³min

Indices

A, B, P, Q	Refer to components
F	Refers to inlet feed
n	Refers to reaction orders
1, 2	Refer to reactions

Exercises

1. Study the system for the kinetic case $n_{A1} = n_{A2}$ and $n_{B1} < n_{B2}$. Run the reactor (i) semi-continuously with slow feeding of B and (ii) as a batch reactor with B charged initially to the reactor and zero feed of B. Compare the results obtained for the two differing modes of reactor operation.
2. Repeat Exercise 1 but for the kinetic case $n_{B1} > n_{B2}$.

3. Study the semi-continuous mode of operation further, using different values of feed rate and feed concentration to the reactor. Determine the feeding conditions that are most beneficial in terms of selectivity.
4. Assign arbitrary cost values ($/kmol) for the reactants A and B, a profit value for the product P and waste disposal and handling charges for the waste Q. Modify the program to incorporate these cost values and define a net cost function for the reaction and hence determine the best "environmental" operating condition. Study the effect of differing cost values on the above optimum.

Results

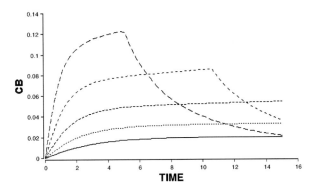

Fig. 2 Concentration of B for five feeding rates from F=0.005 to 0.1 m^3/min.

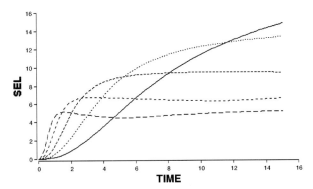

Fig. 3 Selectivity for P for the five feeding rates of Fig. 2.

5.5.2
SEMISEQ – Sequential-Parallel Reactions in a Semi-Continuous Reactor

System

A complex reaction is run in a semi-batch reactor with the purpose of improving the selectivity for the desired product P, compared to that of the waste Q, which is costly to treat and dispose. The kinetics are sequential with respect to components A, P and Q but are parallel with respect to B. The relative magnitudes of the orders of the two reactions determine the optimal feeding policy.

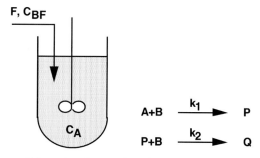

Fig. 1 Complex sequential reactions in a semi-continuous reactor.

Model

The total material balance under constant density conditions is

$$\frac{dV}{dt} = F_B$$

and the component balances are

$$\frac{d(VC_A)}{dt} = -r_1 V$$

$$\frac{d(VC_B)}{dt} = F\,C_{BF} - r_1 V - r_2 V$$

$$\frac{d(VC_P)}{dt} = r_1 V - r_2 V$$

$$\frac{d(VC_Q)}{dt} = r_2 V$$

The kinetic model equations are

$$r_1 = k_1 C_A^{n_A} C_B^{n_{B1}}$$

$$r_2 = k_2 C_P^{n_P} C_B^{n_{B2}}$$

The relative magnitudes of the orders of reaction, n_{B1} and n_{B2}, determine whether the concentration C_B in the reactor should be maintained high or low. The total reaction time must be adjusted such that product P is obtained at its maximum concentration, and this will also influence how B should be fed to the reactor.

Program

If the feeding rate is high the maximum reactor volume may be reached, which will cause the feeding rate of B to stop. The reaction then continues batchwise until the simulation is stopped. In the program the reaction orders are set as $n_A = n_P = 1$ and $n_{B1} < n_{B2}$. The selectivity for the desired product will be highest at the lowest feeding rates, since the desired reaction is then promoted by low concentrations of B.

The objective function used here, OBJ, assumes that product concentration, CP, product purity, Pfrac, and productivity, PROD, are important factors.

Nomenclature

Symbols

C	Concentration	kmol/m^3
V	Volume	m^3
F	Flow rate	m^3/min
r	Reaction rate	kmol/m^3 min

Indices

A, B, P, Q	Refer to components
F	Refers to inlet feed
n	Refers to reaction orders
1, 2	Refer to reactions

Exercises

1. Determine the optimal time to stop the reaction for a range of input flow rates to the reactor, such that the amount of product is greatest. How do the values of k_1 and k_2 influence this?
2. Study the influence of flow rate from the standpoint of maximum production of desired product P and for the minimisation of waste Q. Discuss the meaning and significance of the function OBJ. Study the influence of flowrate on its maximum value. How do the selectivity SEL and the maximum product concentration CP change with flowrate?
3. Study the system for the kinetic case $n_A=n_P=1$ and $n_{B1}<n_{B2}$. Run the reactor (i) semi-continuously with slow feeding of B and (ii) as a batch reactor with B charged initially to the reactor and zero feed of B. Compare the results obtained for the two differing modes of reactor operation.
4. Plot the ratio of the reaction rates for a range of conditions as an aid in understanding the effects of Exercise 2.
5. Modify the objective function to simulate a larger importance of investment costs by introducing an exponent for PROD larger than 1. How does the value of the exponent influence the value of the objective function OBJ? How is the time of the maximum of the objective function changed?
6. Assign the following cost values ($/kmol) for the reactants A (300) and B (100), a profit value for the product P (500) and waste disposal and handling charges for the waste Q (50). Modify the program to incorporate these cost values, define a net cost function for the reaction only considering these raw material and waste treatment costs and determine the best operating conditions. How strongly does the cost of A influence the total economics. How much does a 10% difference make?

Results

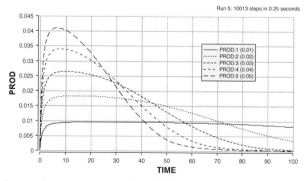

Fig. 2 Multiple runs showing the productivity PROD as a function of time for five feed rates of B between 0.05 and 0.01 m³/min. The highest feed rate gives the lowest selectivity.

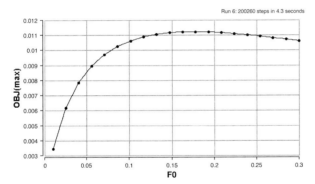

Fig. 3 Parametric run plotting the maximum in the objective function as a function of feed rate of B 0.01 to 0.3 m³/min.

5.5.3
HMT – Semi-Batch Manufacture of Hexamethylenetetramine

System

Aqueous ammonia is added continuously to an initial batch charge of formaldehyde solution. The reaction proceeds instantaneously and is highly exothermic.

$$4\,NH_3 + 6\,HCHO \rightarrow N_4(CH_2)_6 + 6\,H_2O$$

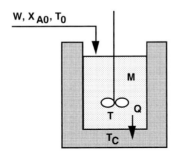

Fig. 1 Variables for the HMT example.

Model

For the semi-batch operation the total mass balance is

$$\frac{dM}{dt} = W$$

For the instantaneous reaction, the rate is equal to the reactant feed rate. The energy balance then becomes

$$M c_p \frac{dT}{dt} = W c_p(T_0 - T) + W X_{A0}(-\Delta H) - UA(T - T_c)$$

where X_{A0} is the mass fraction of ammonia in the feed stream.

Program

Since the reaction rate is instantaneous, this is set equal to the feed rate of ammonia.

Nomenclature

Symbols

A	Heat exchange area	m²
c_p	Specific heat	kJ/kg C
M	Mass in reactor	kg
T and Temp	Temperature	C
U	Overall heat transfer coefficient	kJ/m² C
W	Mass feeding rate	kg/h
X_{A0}	Ammonia feed mass fraction	kg/kg
X_{F0}	Formaldehyde charge fraction	kg/kg
ΔH	Heat of reaction	kJ/kg

Indices

c	Refers to cooling
F	Refers to formaldehyde
0	Refers to feed values

Exercises

1. Find the size of cooling area required such that the reactor temperature just rises to 100 °C, by the end of the feeding at 90 min time.
2. Carry out simulations with differing inlet temperatures, initial charges and feeding rates.

Results

The results indicate that the mass flow rate has a strong influence on the reactor temperature.

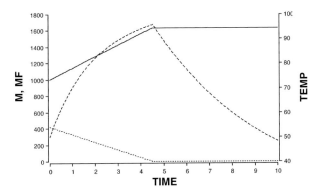

Fig. 2 Time profiles for total mass M, mass of formaldehyde MF and temperature. A slider for cooling area A was used to find the value necessary to keep T<100 °C.

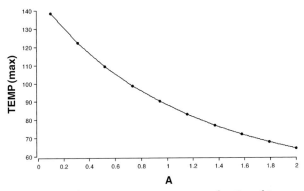

Fig. 3 Parametric run giving the maximum temperature as a function of A.

Reference

Smith, J. M. (1970) Chemical Engineering Kinetics, 2nd edition, McGraw-Hill.

5.5.4
RUN – Relief of a Runaway Polymerisation Reaction

System

The manufacture of a polyol lubricant by the condensation of an alcohol with propylene oxide in a semi-batch reactor proceeds according to the following reaction

$$C_4H_9OH + (n+1)C_3H_6O \rightarrow C_4H_9(OC_3H_6)_nOCH_2CHOHCH_3 + \text{heat}$$

The reaction is highly exothermic and the reactor contains large quantities of volatile oxide. Careful control of temperature is therefore required to avoid a runaway reaction and excessive pressure generation.

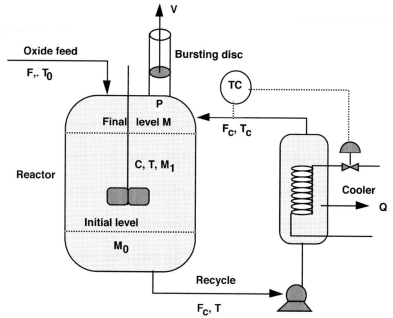

Fig. 1 Polymerisation reactor with its cooling unit and bursting disc.

The alcohol is charged initially to the reactor and the oxide is then fed to the reactor at constant rate. Heat generated by reaction is removed via an external heat transfer loop. In this, the batch contents are recycled through an external cooler with sufficient capacity to maintain a constant return temperature to the reactor. With high rates of cooling the reactor operation is stable, but a product of low molecular weight is produced. As the cooling is decreased, higher molecular weight product is obtained, but at the risk of a runaway reaction and reactor failure. In the event of this or the case of a cooling failure, a bursting disc in the reactor will rupture and the feed of oxide would usually be stopped. This will discharge oxide vapour to the atmosphere.

The simulation emphasises that when reactions start to run away, they do so extremely fast. In this example the rate of pressure generation is compounded by the double exponential terms in the Arrhenius and vapour pressure equations.

Model

The total mass balance is

$$\frac{dM}{dt} = F - V$$

where M is the reactor mass at time t (kg) and V is the vapour venting rate (kg/min) and F is the feed rate (kg/min).

The oxide component mass balance is

$$\frac{d(MC)}{dt} = F - V - R_1$$

where C is the oxide concentration (kg/kg) and R_1 is the oxide reaction rate (kg/min).

The mass of oxide reacted is given by

$$\frac{dX}{dt} = +R_1$$

where X is the mass of oxide reacted at time t.

The energy balance is

$$Mc_p \frac{dT_R}{dt} = Fc_p(T_0 - T_R) - V\lambda - R_1(-\Delta H) - Q$$

where T_R is the temperature of reaction (°C), Q is the rate of cooling (kJ/min) and λ is the heat of vaporisation of the oxide.

The heat removed via the recirculation through the external cooler is given by

$$Q = F_c c_p (T_R - T_c)$$

where F_c is the recycle mass flow rate (kg/min).

The reaction kinetics are given by

$$R_1 = k\,MC \text{ (kg oxide/min)}$$

where

$$k = Z e^{-\frac{E}{R(T_R+273)}}$$

The polymer molecular weight is

$$MW = \frac{M_0 + X}{N}$$

where N is the kmol of initial alcohol charge.

Assuming instantaneous equilibrium, the reactor pressure and oxide vapour pressure is given by:

$$P = \left[e^{\left(\frac{3430}{T_R+273}+11.7\right)} + 1.45 \times 10^3 \text{ MW} \right] C$$

The vapour discharge rate is given by the following.

(1) For normal conditions and up to disc rupture

$$V = 0$$

(2) For sonic discharge to atmospheric pressure

$$V = \frac{k_v \, 0.85 \, P}{\sqrt{T_R + 273}} \text{ kg/min for } P > 1.9 \text{ bar}$$

(3) For subsonic discharge for $1 < P < 1.9$ bar

$$V = \frac{k_v \, P}{\sqrt{T_R + 273}} \sqrt{1 + (1/P)^2} \text{ kg/min}$$

where k_v is the valve discharge coefficient.

Program

The MADONNA program RUN with KV=0 represents operation of the reactor under zero venting (V=0) conditions, i.e., the bursting disc remains closed. With KV>0 (100–1000) the program simulates the performance under emergency venting conditions. The program will continue to run unless coolant failure occurs, corresponding to FC=0.

The program ensures that the bursting disc remains integral up to the bursting pressure, that the vapour discharge is sonic for reactor pressures greater than critical, that the discharge is subsonic for pressures less than critical and that the discharge stops when the pressure is atmospheric. The feed rate is programmed to either stop or continue following disc failure.

For convenience, the exit temperature from the cooler, T_c, is assumed equal to the feed temperature, T_0, i.e., the heat transfer capacity of the cooler is in large excess.

This program contains an example of conditional statements. A conditional variable DISK is defined such that if DISK = 1 then the bursting disk is intact, and if DISK = 0 the disk has burst. The discharge velocity through the disk varies with pressure as given by the complex conditional statement in the partial listing below:

{Example RUN}

{Vapor pressure in reactor}
P = (EXP(-3430/(TR+273)+11.7)+1.45E-3*MW)*C
Limit P>=1

{Bursting disc control: If Disk=1 then bursting disk OK; if Disk=0 disk has burst}
F = IF (Disk<1) THEN 0 ELSE F0 {Feed flow rate of oxide, kg/min. Oxide feed shuts off on bursting. This can be deactivated to maintain feeding}

{Venting velocity control}
V = IF(Disk<1 AND P>1.9) {Velocity through disk varies
                          with P}
   THEN 0.85*KV*P/SQRT(TR+273)
     ELSE IF (Disk<1 AND P<=1.9 AND P>1.1)
       THEN KV*P/SQRT(TR+273)*SQRT(1+(1/P)*(1/P))
         ELSE IF P<=1.1
           THEN 0.0
           ELSE 0

MW = (M0+X)/N   {Molecular weight of polymer at any time,
                 kg/kmol}
R1 = K*MC   {Rate of polymerization reaction, kg oxide/min}
K = Z*EXP(-E/(R*(TR+273))) {Arrhenius relation}

{Total mass balance taking into account the initial alcohol M0 and the oxide fed and vaporized}
d/dt(M) = F-V   {Total mass balance}
d/dt(MC) = F-V-R   {Component mass balance for oxide}
LIMIT MC >= 0   {Prevents negative values}
C = MC/M   {Mass fraction conc, kg/kg}
LIMIT C >= 0   {Prevents negative values}

{Heat balance, considering flow, vaporization, reaction and cooling}
d/dt(MTR) = (F*Cp*(T0-TR)-V*LAMBDA-R1*HR-FC*Cp*(TR-T0))/Cp
LIMIT MTR >= 0
TR = MTR/M   {Calculation of TR}
d/dt(X) = R1   {Oxide reacted}

{State of the disk, starts at 1 and decreases quickly after
bursting; the value of Disk is used to detect if the disk has
burst. After bursting Disk = 0}
d/dt(Disk) = -Fdisk
Limit Disk >= 0
INIT Disk = 1
Fdisk = IF (P >= PBURST) THEN 10 ELSE 0.0

{Cooling failure}
FC = IF TIME >= TIMEfail THEN 0.0 ELSE FC1   {Conditional re-
                                              cycle cooling flow}
TIMEfail = 1500   {This is the time of failure, min}

Nomenclature

Symbols

C	Oxide concentration	kg/kg
c_p	Specific heat	kJ/kg C
E	Activation energy	kJ/mol
F	Oxide mass feed rate	kg/min
FC and FC1	Coolant flow rate	kg/min
λ	Heat of vaporisation of oxide	kJ/kg
M	Charge mass of alcohol	kg
MC	Mass of oxide	kg
MW	Molecular weight	kg/mol
Mol	Molecular weight of charge	kg/kmol
N	Initial alcohol charge	kmol
P	Reactor pressure	bar
P1	Bursting disc relief pressure	bar
P2	Maximum safe working pressure	bar
R	Gas constant	kJ/mol C
R1	Reaction rate	kg/min
T_R	Reaction temperature	C
t	Batch time	min
T_0	Charge, feed and recycle temperature	C
V	Venting rate	kg/min
X	Mass of oxide	kg
Z	Frequency factor	1/min
ΔH	Heat of reaction	kJ/kg

Indices

0	Refers to initial and inlet values
c	Refers to cooler

Exercises

1. Study the MADONNA program so that the conditional logic based on the IF_THEN_ELSE statements is fully understood.
2. With the bursting disc remaining closed (KV=0 and V=0), establish the minimum external recycle cooling flow (F_{c1}) needed to ensure stable operating conditions and satisfactory polymer product (Mol. wt=2500). Determine the minimum cooling capacity required of the cooler. Study the effects of "runaway" following coolant failure by interactive computation. By carefully decreasing F_c from an initially "safe" value, the absolute limit of safe operation can be established. In this the reaction starts to run away, but then the reaction corrects itself by having consumed all the oxide and then continues with further stable operation.
3. Using the value of F_c found in Exercise 2, study the effect of differing valve discharge capacities (K_V) following cooling failure at different stages of reaction time (F_c=0).
4. Run the simulation under disc-bursting conditions for both with and without shutting off the feed.
5. As an alternative to venting the vapour to atmosphere, the vapour may be condensed and returned to the reactor. Modify the program accordingly and study the resultant behaviour. See simulation examples REFRIDGE1 and REFRIDGE2.
6. Alter the program to include a complete model for the heat exchanger. This will allow a more realistic calculation of the heat removed.

Result

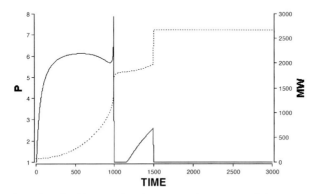

Fig. 2 Set at the point of stable operation the pressure increases until the oxide MC is consumed (Fig. 3) and then at lower temperature the process continues.

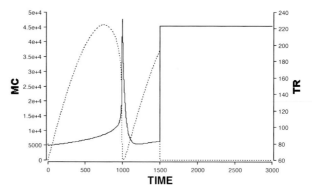

Fig. 3 The run of Fig. 2, showing the oxide MC falling to zero and the process continuing at lower temperature.

References

Kneale, M. and Forster, G. M. (1968) I. Chem. E. Symp. Series No. 25, 98.
Kneale, M. and Binns, J. S. (1977) I. Chem. E. Symp. Series No. 49.

5.5.5
SELCONT – Optimized Selectivity in a Semi-Continuous Reactor

System

Keller (1998) describes the semi-continuous reaction process of a vinyl ketone K with lithium acetylide LA to yield lithium ethinolate LE an intermediate in the vitamin production. In an undesired side reaction an oligomer byproduct BP is produced. During the process, reactant K is fed to the semi-batch reactor at a rate to maximize the selectivity for LE.

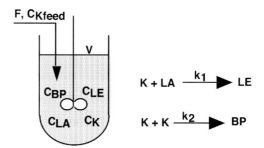

Fig. 1 Semi-continuous reactor with feeding of ketone reactant.

Model

The total balance for the semi-batch reactor assuming constant density is

$$\frac{dV}{dt} = F$$

Written in terms of masses (C_iV) the component balances are

$$\frac{d(C_{LA}V)}{dt} = r_1 V$$

$$\frac{d(C_{LE}V)}{dt} = -r_1 V$$

$$\frac{d(C_K V)}{dt} = F\, C_{Kfeed} + r_1 V + r_2 V$$

$$\frac{d(C_{BP}V)}{dt} = -r_2 V$$

and the concentrations for each component i are found from relations of the form $C_i = C_i V/V$.

The reaction rates are known to be given by

$$r_1 = -k_1 C_K C_{LA}$$

$$r_2 = -k_2 C_K^2$$

An objective function can be defined with respect to the positive impact by the productivity of LE and the negative contribution by that of BP weighted by a factor a.

$$OF = \frac{C_{LE}V - C_{BP}Va}{t_B}$$

The final time t_B is reached when the reactor is full and when the residual concentration of K is 0.01.

Exercises

1. Find the optimal constant feeding rate of ketone to the reactor for $a = 50$.
2. Investigate the influence of the value of a on the optimum.
3. Try to find a higher value for the objective function by using two time periods with two different constant feeding rates. Find the optimal values for the time periods and the flow rates.

4. Assume a ramp function for the flow rate ($F = F_0 + kt$) and find the values of F_0 and k which give the highest value for the objective function. Remember that F must remain positive.
5. Try to improve the value of the objective function by using a PI-controller to maintain constant selectivity throughout the batch, where

$$SEL = \frac{r_1}{r_2 + r_1}$$

Find the value of the selectivity setpoint which gives the highest value of the objective function, as defined above with $a = 50$. Compare the feeding profile found with those found in the previous exercises.

Nomenclature

C	Concentration	kmol/m^3
k	Rate constants	m^3/kmol h
F	Flow rate of feed	m^3
OF	Objective function	kmol/h
r	Reaction rate	kmol/m^3 h
V	Volume of reactor	m^3
a	Penalty constant	–
SEL	Selectivity	–
T_B	Total time of reaction	h

Subscripts

BP	Refers to by-products
K	Refers to vinylketone
LA	Refers to lithium acetylide
LE	Refers to lithium ethinolate
1	Refers to desired reaction
2	Refers to undesired reaction

Results

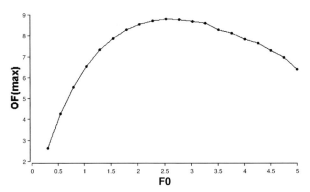

Fig. 2 The objective function varying with feed rate, with a maximum at 2.5 m³/h.

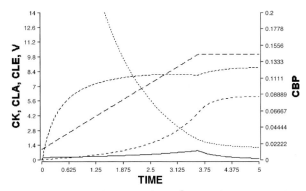

Fig. 3 Concentration profiles for a feed rate of 2.5 m³/h. The feed is stopped at time = 3.6 h.

Reference

Keller, A. (1998) ETH PhD Dissertation No. 12607, Zürich

5.5.6
SULFONATION – Space-Time-Yield and Safety in a Semi-Continuous Reactor

System

In this case history, the control of the TMR_{ad} (adiabatic Time-to-Maximum-Rate) is to be achieved in a semi-continuous reactor process by the dynamic optimization of the feed rate. Here it is desired to have the highest possible space-time-yield STY and it is necessary to achieve a thermally safe process (Keller, 1998). The reaction involves the addition of a sulfur trioxide on a nitro-aromatic compound

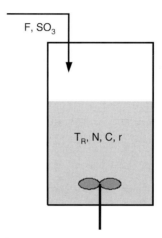

The reaction takes place in the liquid phase of a semi-continuous process, which allows the feed rate manipulation to control the rate of heat release, as shown below.

Fig. 1 Semi-continuous reactor process.

Model

Although the kinetics is not completely known, the following second order rate equation is assumed

$$r_1 = k_{1\infty} e^{\left(-\frac{E_{a1}}{RT_{RM}}\right)} [\text{Nitroaromatic}] [SO_3]$$

where T_{RM} is the temperature of the reacting mass, k_1 is the preexponential factor and E_{a1} is the activation energy.

Decomposition reactions will occur at high temperatures and their rates are assumed to be first order as follows

$$r_2 = k_{2\infty} e^{\left(-\frac{E_{a2}}{RT_{RM}}\right)} [\text{Sulfonic acid}]$$

$$r_3 = k_{3\infty} e^{\left(-\frac{E_{a3}}{RT_{RM}}\right)} [\text{Nitroaromatic}]$$

The rate of heat production by the three reactions is written as

$$\frac{dQ_R}{dt} = -r_1(-\Delta H_{R1}) + r_2(-\Delta H_{R2}) + r_3(-\Delta H_{R3})$$

Calculating by dynamic mass balance the moles of trioxide at any time

$$\frac{dN_{SO_3}}{dt} = -r_1 V + F_D \frac{\Delta N_{SO_3}}{\Delta V}$$

where the experimental method of feeding allowed determining the concentration from the moles added for each volume. The initial volume of the reaction fluid is 3.7 m³ having a total mass of 5586 kg. This consists of 18 kmol nitroaromatic compound dissolved in sulfuric acid. After the addition of 27.7 kmol SO_3 in the form of 65% oleum, the mass is increased by 4496 kg and the volume by 2.3 m³. This determines $\Delta N_{SO_3}/\Delta V$.

Since the moles of trioxide will be less than the amount of nitroaromatic in the initial period of the process, trioxide will usually limit the extent of reaction. Later, the amount of trioxide will be larger. Therefore the maximum temperature that could be reached by complete conversion of the trioxide, MTSR, is

$$MTSR = T_R + \frac{-\Delta H_{R1} N_{Acc}}{M\, c_{p,RM}}$$

N_{Acc} is the molar amount of the limiting reactant. From the MTSR the time to the maximum rate TMR_{ad} can be determined as

$$TMR_{ad} = \frac{c_{p,RM}\, R\, MTSR^2}{q_{ZR}\, E_{a2}}$$

Here q_{ZR} is the rate of heat produced by the decomposition reactions. The derivation of this equation is not shown here, but assumes a zero-order, worst-case reaction (Townsend and Tou, 1980).

The feed rate of SO_3 can be adjusted in three ways:

- Constant feed rate F_{p1} if CONTROLSET = 0 and STEPF = 0;
- Stepwise feed with feed rates F_1 to F_{20} with time interval PUMPTIME if CONTROLSET = 0 and STEPF > 0; or
- Controlled feeding depending on the calculated time-to-maximum rate TMR_{ad} if CONTROLSET > 0. The controller is of PI type and the set-point TMRset is slightly above the minimum tolerated time-to-maximum rate, TMR_{lim}, which is fixed at 24 hours. This is necessary because of possible slight undershooting of TMR_{ad} below TMR_{set}. In this case the flow rate is bound between 0 and F_{max}.

Objective Function

In this example only the main reaction and the two decomposition reactions are considered. For reasons of product quality, a conversion of 99% must be obtained. Also the production rate per unit volume, or space time yield, should be high. This can be expressed as

$$\text{STY} = \frac{M_{prod}}{V\, t_{Cycl}}$$

For this reaction, that is first order for each of the reactants, it follows that the larger the concentrations the larger is the STY. However the safety of the process would be in jeopardy because TMR_{ad} also decreases with higher trioxide concentrations. For safety the lowest acceptable allowable value of TMR_{ad} is often taken as 24 h (Stoessel, 1993).

For calculation purposes an objective function is defined which penalizes the process if the value of TMR_{ad} takes a value lower than this value. It has the following form

$$J = \frac{M_{prod}}{V\, t_{Cycl}} - \theta \int_0^\infty \Delta TMR_{ad}^2\, dt$$

such that

$\Delta TMR_{ad}^2 = 0$ for $\Delta TMR_{ad}(\lim) - \Delta TMR_{ad} \leq 0$

$\Delta TMR_{ad} = TMR_{ad}(\lim) - TMR_{ad}$ for $\Delta TMR_{ad}(\lim) - \Delta TMR_{ad} \geq 0$

Nomenclature

The definition of the symbols and their values are listed in the program.

Program

The program SULFONATION is on the CD. The line
Fp2 = IF CONTROLSET > 0 THEN Fpreg2 ELSE if STEPF > 0 THEN Fstep ELSE Fp1

Serves to switch between the three selectable feed modes as described above.
The step feeding is programmed using nested if statements. 19 time points with a time interval of PUMPTIME (time1 to time time 19) are defined at which flow rate is switched to the next predefined value F2 to F20.

```
Fstep =
IF TIME < time1 THEN F1 ELSE
  IF TIME < time2 THEN F2 ELSE
    IF TIME < time3 THEN F3 ELSE
      IF TIME < time4 THEN F4 ELSE
        IF TIME < time5 THEN F5 ELSE
          IF TIME < time6 THEN F6 ELSE
            IF TIME < time7 THEN F7 ELSE
              IF TIME < time8 THEN F8 ELSE
                IF TIME < time9 THEN F9 ELSE
                  IF TIME < time10 THEN F10 ELSE
                    IF TIME < time11 THEN F11 ELSE
                      IF TIME < time12 THEN F12 ELSE
                        IF TIME < time13 THEN F13 ELSE
                          IF TIME < time14 THEN F14 ELSE
                            IF TIME < time15 THEN F15 ELSE
                              IF TIME < time16 THEN F16 ELSE
                                IF TIME < time17 THEN F17 ELSE
                                  IF TIME < time18 THEN F18 ELSE
                                    IF TIME < time19 THEN F19 ELSE
F20
```

In this way each individual flow rate can be freely adjusted or automatically optimized.

Exercises

1. Study the operation with constant feed rate. Set the variable CONTROLSET = 0 for constant feeding. Here it is of interest to understand the operation by following the volume, flow rate, moles of A and B, accumulated moles of A and B, the amount of limiting reactant N_{acc}, concentrations of A, B and P, TMR_{ad}, STY and J. Check whether TMR_{lim} has been under-run.
2. Investigate the influence of temperature TR by running at slightly higher and lower temperatures. Watch changes in all variables listed in Exercise 1. See if safe conditions result, such that TMR_{ad} is always greater than 24 h.
3. Repeat the study of Exercise 2 for changes in feed rate.
4. Using the Parameter-Optimize feature in MADONNA, determine the maximum value of J in both programs (controlled and uncontrolled). Here the function to maximize is –J, since a minimum is sought.
5. Check to see that J was really a maximum using Parameter Plots. Plot also the STY values here.
6. Turn the controller on, repeat Exercise 1 and note the differences in results compared with the constant feed case of Exercise 1.

7. For the controlled case determine the optimal temperature using the Parameter-Optimize feature. How does this J value compare to that from Exercise 4?
8. Find out an optimal profile when using the step feeding alternative. After setting CONTROLSET = 0 and STEPF = 1 first try to find an optimal profile by manually adjusting F1 to F20. Make sure that TMR_{ad} is not dropping below TMR_{lim}.
9. Using the Parameter-Optimize feature in MADONNA, determine the maximum value of J by letting the program find the best values for F1 to F20. Define minimum flow rates 0. Compare now all three results using the overlay feature of MADONNA and adjusting CONTROLSET and STEPF.

References

Keller, A. H. (1998) „Stufengerechte Beurteilung und Optimierung der thermischen Prozesssicherheit mittels dynamischer Modellierung". Diss. ETH #12607.

Stoessel, F. (1993) What is your thermal risk? Chem. Eng. Prog. October: 68–75.

Townsend, D. I. and Tou, J. C. (1989) Thermal hazard evaluation by an accelerating rate calorimeter. Thermochim Acta 37:1–30.

Results

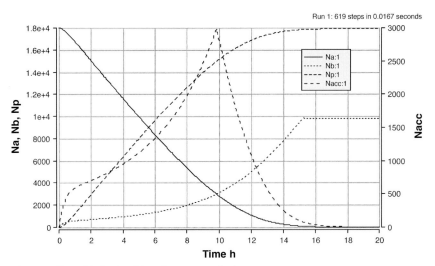

Fig. 1 Profiles of A, B, P and N_{acc} for constant feed rate.

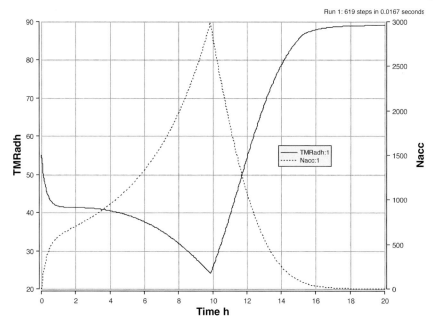

Fig. 2 Profiles of TMR$_{ad}$ and N$_{acc}$ for the run in Fig. 1.

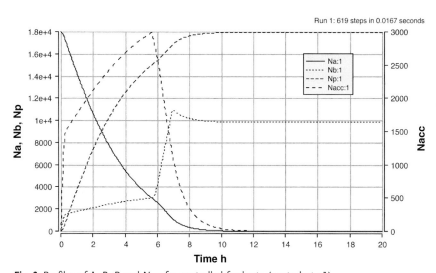

Fig. 3 Profiles of A, B, P and N$_{acc}$ for controlled feed rate (controlset=1).

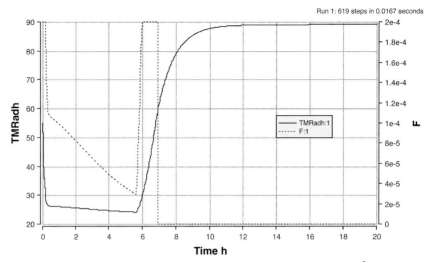

Fig. 4 Profiles of TMR_{adh} and F for controlled feed rate (stepfeed=1, $KP=1\times10^{-8}$).

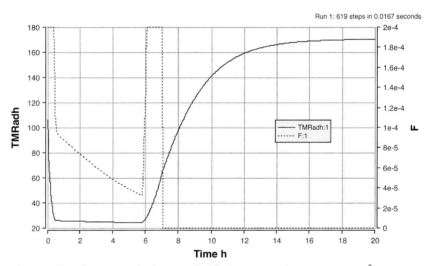

Fig. 5 Profiles of TMR_{adh} and F for controlled feed rate (controlset=1, $KP=1\times10^{-8}$) at lower temperature, 388 K.

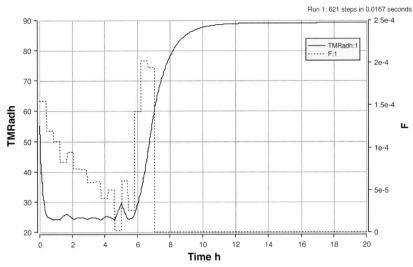

Fig. 6 Profiles of TMR_{adh} and F for a step feeding run.

5.6
Mixing-Model Examples

5.6.1
NOCSTR – Non-Ideal Stirred-Tank Reactor

System

In this model of non-ideal reactor mixing, a fraction, f_1, of the volumetric feed rate, F, completely bypasses the mixing in the reactor. In addition, a fraction, f_2, of the reactor volume, V, exists as dead space. F_3 is the volumetric rate of exchange between the perfectly mixed volume V_1 and the dead zone volume V_2 of the reactor.

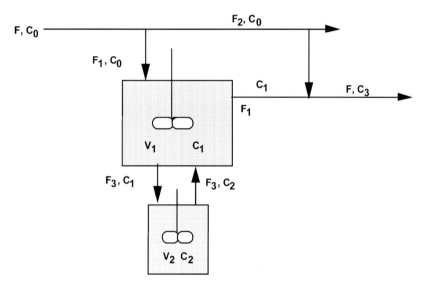

Fig. 1 The non-ideal flow reactor model.

Model

The reaction is assumed to be first order, where

$$r = kC$$

The component mass balances are as follows.

For the well-mixed volume

$$V_1 \frac{dC_1}{dt} = F_1(C_0 - C_1) + F_3(C_2 - C_1) - V_1 k_1 C_1$$

where $V_1 = (1-f_2)V$ and $F_1 = (1-f_1)F$.

For the dead space volume

$$V_2 \frac{dC_2}{dt} = F_3(C_1 - C_2) - V_2 k C_2$$

where

$$V_2 = f_2 V$$

and

$$F_3 = aF$$

Note that the dead space is treated as having a region of uniform well-mixed composition.

For the reactor outlet

$$FC_3 = F_1 C_1 + F_2 C_0$$

where

$$F_2 = f_1 F$$

For a perfectly mixed tank with outlet concentration C_4 and total volume V

$$V \frac{dC_4}{dt} = F(C_0 - C_4) - V k C_4$$

Program

Note that in the program that while the by-passing flow fraction f_1 (FRAC1) can be set to zero, the fractional dead space volume f_2 (FRAC2) cannot be set to zero without modifying the program. This would cause the derivative value for C_2 to be divided by a zero value for V_2.

Nomenclature

Symbols

C_0	Feed concentration	kmol/m³
C_1	Active zone concentration	kmol/m³
C_2	Deadzone concentration	kmol/m³
C_3	Effluent concentration	kmol/m³
C_4	Ideal tank concentration	kmol/m³
F	Total flow rate	m³/min
f_1	Fractional by-pass flow	–

F_2	By-pass flow rate	m^3/min
f_2	Dead volume fraction	–
F_3	Deadzone exchange flow	m^3/min
R	Rate constant	1/min
X	Fractional conversion	–
a	Fractional deadzone flow	–

Indices

0	Refers to feed
1	Refers to interchange flow and outlet concentration
2	Refers to dead volume region
3	Refers to deadzone flow

Exercises

1. Study the effects of fractional bypassing and fractional dead space volume on the steady-state conversion, X3, compared to that of the ideal tank, X4.
2. Note if K=0, the program generates a tracer step-response curve for the non-ideal reactor.
3. Study the effect of variations in the rate of exchange F3 between the two reactor volumes, by varying the value of ALPHA.
4. Modify the program to allow for zero dead space volume and study the effect of fractional bypassing only. Compare the results to that for a perfectly stirred tank.
5. Modify the system to that shown below by Froment and Bischoff (1990), in which the volumetric flow, F, having passed through a partial well-mixed volume, V_1, is then diverted such that a fraction, f, then passes through the residual volume, V_2, before rejoining the main flow stream. Formulate the model equations and record the F-curve response for the system for varying values of V_1 and V_2 keeping the total volume ($V = V_1 + V_2$) constant and varying the values of f.

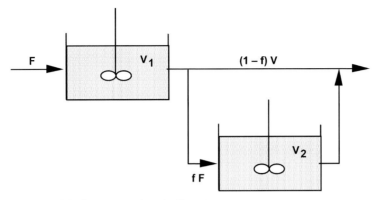

Fig. 2 Mixing model of Froment and Bischoff (1990).

Results

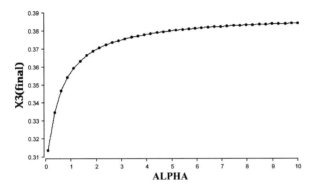

Fig. 3 A parametric run showing the influence of ALPHA from 0.1 to 10.

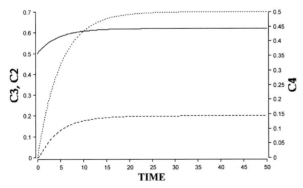

Fig. 4 A response of the reactor model to a step change in concentration. The well-mixed case C4 is shown for comparison.

Reference

Froment, G. F., Bischoff, K. B. (1990) Chemical Reactor Analysis and Design, 2nd ed., Wiley Interscience.

5.6.2
TUBEMIX – Non-Ideal Tube-Tank Mixing Model

System

Non-ideal mixing conditions in a reactor can often be modelled as combinations of tanks and tubes. Here a series of stirred tanks are used to simulate a tubular, by-passing condition.

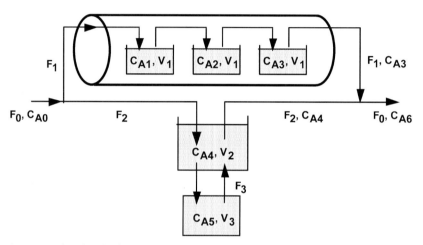

Fig. 1 A combined tank-tube mixing model.

Model

For the tubular by-pass flow representation, the balances for reactant A have the form

$$\frac{V_1 dC_{An}}{dt} = F_1(C_{An-1} - C_{An}) + r_{An}V$$

for each of the sections n = 1, 2, ... N.

The main tank reactor region has the balance

$$\frac{V_2 dC_{A4}}{dt} = F_2 C_{A0} + F_3 C_{A5} - F_2 C_{A4} - F_3 C_{A4} + r_{A4} V_2$$

The dead-zone region is balanced by

$$\frac{V_3 dC_{A5}}{dt} = F_3 C_{A4} - F_3 C_{A5} + r_{A5} V_3$$

The final outlet concentration is given by the balance at the mixing point

$$F_0 C_{A6} = F_1 C_{A3} + F_2 C_{A4}$$

The flow rates are given in terms of the flow fractions as

$$F_1 = F_0 R_1$$

and

$$F_3 = F_2 R_2$$

and the volumes are given by

$$V_2 = \frac{V_{tub}}{R_3}$$

and

$$V_3 = \frac{V_2}{R_4}$$

The reaction rates are

$$r_{An} = k C_{An}^m$$

where n refers to the section number and m refers to the reaction order.

Program

The program is written in array form so that the number of tanks within the tubular section can be varied.

Nomenclature

Symbols

C_{A0}	Feed concentration	kg/m^3
C_{Ainit}	Initial concentrations	kg/m^3
F_0	Feed to reactor system	m^3/min

F_1	Flow rate to tube	m³/min
F_2	Flow rate to tank	m³/min
F_3	Interchange flow rate	m³/min
k	Reaction constant	various
m	Reaction order	
R_1	Feed fraction tube/tank	–
R_1	Flow fraction F_1/F_0	–
R_2	Feed fraction tank1/tank2	–
R_2	Flow fraction F_3/F_2	–
R_3	Volume fraction tube/tank1	–
R_4	Volume fraction tank1/tank2	–
V_1	Volume of tank sections in tube	m³
V_2	Volume of main reactor section	m³
V_3	Volume of deadzone section	m³
V_{tub}	Volume of tube reactor	m³

Results

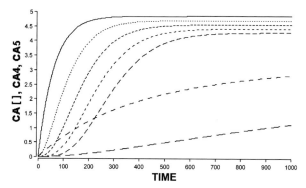

Fig. 2 Shown here for first-order reaction is a step response of the 5 tanks, as well as the slower responses of CA4 and CA5.

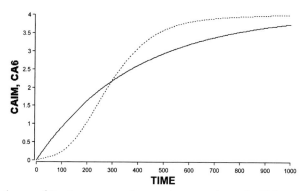

Fig. 3 From the run of Fig. 2, as an overall comparison, are shown the CA6 outlet response and the response for a completely mixed system, given by CAiM.

5.6.3
MIXFLO1 and MIXFLO2 – Mixed–Flow Residence Time Distribution Studies

System

The following systems represent differing combinations of ideal plug-flow, mixing, dead space, flow recycle and flow by-pass.

Case 1

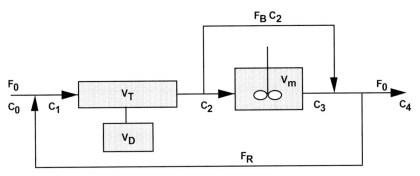

Fig. 1 Mixing model for residence time distribution Case 1.

Case 2

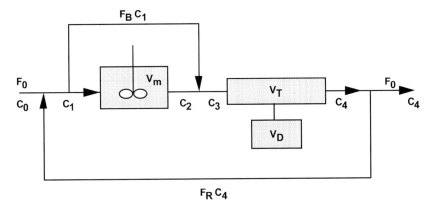

Fig. 2 Mixing model for residence time distribution Case 2.

It is desired to compare the residence time distributions (RTD) for these cases.

Model

Ideal plug-flow regions, such as V_T, are represented by standard Madonna time delay functions.

Case 1

The mixing junction is given by the steady-state balance,

$$F_0 C_0 + F_R C_4 = (F_0 + F_R) C_1$$

Thus the concentration leaving the junction is

$$C_1 = \frac{F_0 C_0 + F_R C_4}{F_0 + F_R}$$

For the plug-flow region V_T, the outlet concentration C_2 is equal to the input concentration C_1, but delayed in time by

$$\frac{V_T - V_D}{F_0 + F_R}$$

The well-mixed region is given by

$$V_m \frac{dC_3}{dt} = (F_0 + F_R - F_B)(C_2 - C_3)$$

The final outlet concentration is given by a steady-state junction relation as,

$$C_4 = \frac{F_B C_2 + (F_0 + F_R F_B) C_3}{F_0 + F_R}$$

Case 2

The relations used here are analogous to Case 1.

For the junction

$$C_1 = \frac{F_0 C_0 + F_R C_4}{F_0 + F_R}$$

For the mixing tank

$$V_m \frac{dC_2}{dt} = (F_0 + F_R - F_B)(C_1 - C_2)$$

For the by-pass junction

$$C_3 = \frac{F_B C_1 + (F_0 + F_R F_B) C_2}{F_0 + F_R}$$

For the plug-flow time delay

$$C_4 \text{ is equal to } C_3 \text{ delayed in time by } \frac{V_T - V_D}{F_0 + F_R}$$

Program

The Madonna program MIXFLO1 models the situation of Case 1, and program MIXFLO2 models that of the Case 2. The plug-flow section of the plant is modelled with the time delay function, DELAYT. This is suitable for residence time distribution studies, but is not suitable to calculate the conversion for non-first order reactions. The Madonna logic is such that a dynamic concentration mixing point of finite volume is necessary within the program to provide an initial condition, owing to the time delay and the recycle stream.

Nomenclature

Symbols

V	Volume	m^3
F	Volumetric flow	m^3/s
C	Tracer concentration	kg/m^3

Indices

T	Refers to plug flow
M	Refers to mixing
D	Refers to dead space
B	Refers to by-pass
R	Refers to recycle
1, 2, 3, 4	Refer to positions according to figure

Exercises

1. Vary the relative system parameters VT, VD and VM for the two differing flow cases, keeping the total volume of the system constant. Compare the resulting residence time distributions.
2. Keeping the total volume of the system constant, vary the relative flowrates, F0, FB and FR and note the influence on the RTD's.

3. Evaluate the conversion for first-order reaction from a tracer pulse response curve using the method in example CSTRPULSE. Show that although the residence time distributions may be the same in the two cases, the overall chemical conversion is not, excepting for the case of first-order reaction.

Results

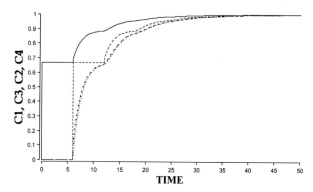

Fig. 3 For MIXFLO1 a step in C0 creates this response in C1, C2, C3 and C4. The time delay function gives a delay between C1 and C2.

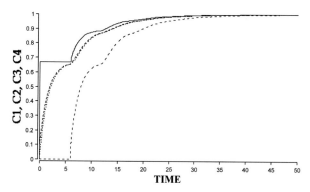

Fig. 4 For MIXFLO2 a step in C0 creates this response in C1, C2, C3 and C4. The time delay function gives a delay between C3 and C4.

Reference

Szekely, J., Themelis, N. J. (1971) Rate Phenomena in Process Metallurgy, Wiley Interscience.

5.6.4
GASLIQ1 and GASLIQ2 – Gas–Liquid Mixing and Mass Transfer in a Stirred Tank

System

The gas–liquid flow characteristics of stirred vessels depend both on the level of agitation and the rate of gas flow and can vary from the case of bubble column type operation to that of a full circulating tank, as shown in Fig. 1. The mixing characteristics and gas distribution obtained obviously exert a considerable influence on the rate of mass transfer obtained (Harnby et al., 1985).

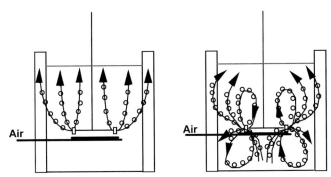

Fig. 1 Gas–liquid mixing modes: (a) bubble column behaviour (high gas flow, low rotor speed); (b) circulating tank behaviour (high rotor speed, reduced gas flow).

In this example it is required to model the behaviour of the mixing vessel according to the conceptual models indicated below.

Liquid Mixing

The basic liquid mixing pattern is assumed to remain constant, as shown in Fig. 2.
In this, the liquid acts as three well-stirred regions, the smaller one representing the region close to the impeller. The two larger ones represent the volumes of liquid above and below the impeller, with the flow from the impeller discharging equally into the two main tank regions.

Gas Mixing

The gas phase mixing is represented by three equally-sized tanks in series.
In the case of the full bubble column, the bottom gas phase mixed region is coincident with the impeller, and no significant gas mixing occurs below the impeller, in the lower region of the tank, as shown in Fig. 3. The dashed arrows in Figs. 3 and 4 represent mass transfer interchange between the gas space and liquid volumes of the tank. The liquid circulation, which is not shown, is identical to Fig. 2.

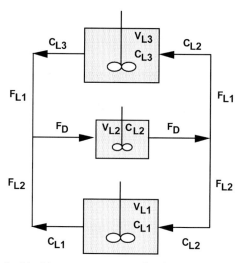

Fig. 2 Model for the liquid mixing pattern in a stirred tank.

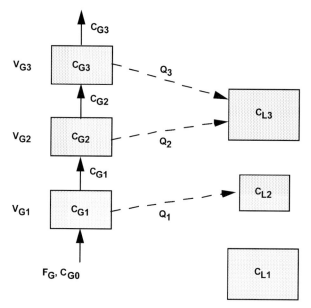

Fig. 3 Gas phase mixing pattern and mass transfer for full bubble column.

With full circulation, substantial gas phase mixing occurs in all tank regions, as shown in Fig. 4.

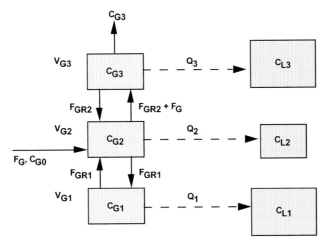

Fig. 4 Gas phase mixing and mass transfer for full circulation case.

Model

Bubble Column Operation

Liquid phase balances

$$V_{L1}\frac{dC_{L1}}{dt} = F_{L2}(C_{L2} - C_{L1})$$

$$V_{L2}\frac{dC_{L2}}{dt} = F_{L2}C_{L1} + F_{L1}C_{L3} - F_D C_{L2} + Q_1$$

$$= F_{L2}(C_{L1} - C_{L2}) + F_{L1}(C_{L3} - C_{L2}) + Q_1$$

$$V_{L3}\frac{dC_{L3}}{dt} = F_{L1}(C_{L2} - C_{L3}) + Q_2 + Q_3$$

Pump discharge rate

$$F_D = F_{L1} + F_{L2}$$

Gas phase balances

$$V_{G1}\frac{dC_{G1}}{dt} = F_G(C_{G0} - C_{G1}) - Q_1$$

$$V_{G2}\frac{dC_{G2}}{dt} = F_G(C_{G1} - C_{G2}) - Q_2$$

$$V_{G3}\frac{dC_{G3}}{dt} = F_G(G_{G2} - C_{G3}) - Q_3$$

Rates of mass transfer

$$Q_1 = K_L a (C_{L1}^* - C_{L2}) V_{G1}$$

$$Q_2 = K_L a (C_{L2}^* - C_{L3}) V_{G2}$$

$$Q_3 = K_L a (C_{L3}^* - C_{L3}) V_{G3}$$

The equilibrium relationships are represented by means of Henry's law where

$$C_{L1}^* 0 \frac{RT C_{G1}}{H}$$

$$C_{L2}^* = \frac{RT C_{G2}}{H}$$

$$C_{L3}^* = \frac{RT C_{G3}}{H}$$

Full Circulation

Liquid phase balances

$$V_{L1} \frac{dC_{L1}}{dt} = F_{L2}(C_{L2} - C_{L1}) + Q_1$$

$$V_{L2} \frac{dC_{L2}}{dt} = F_{L2}(C_{L1} - C_{L2}) + F_{L1}(C_{L3} - C_{L2}) + Q_2$$

$$V_{L3} \frac{dC_{L3}}{dt} = F_{L1}(C_{L2} - C_{L1}) - Q_1$$

Gas phase balances

$$V_{G1} \frac{dC_{G1}}{dt} = F_{GR1}(C_{G2} - C_{G1}) - Q_1$$

$$V_{G2} \frac{dC_{G2}}{dt} = F_G C_{G0} + F_{GR1}(C_{G1} - C_{G2}) + F_{GR2} C_{G3} - (F_{GR2} + F_G) C_{G2} - Q_2$$

$$V_{G3} \frac{dC_{G3}}{dt} = (F_{GR2} + F_G)(C_{G2} - C_{G3}) - Q_3$$

Rates of mass transfer

$$Q_1 = K_L a (C_{L1}^* - C_{L1}) V_{G1}$$

$$Q_2 = K_L a (C_{L2}^* - C_{L2}) V_{G2}$$

$$Q_3 = K_L a (C_{L3}^* - C_{L3}) V_{G3}$$

The equilibrium relationships are represented by Henry's law as before.

Program

Program GASLIQ1 simulates the case of the full bubble column and program GASLIQ2 that of the full circulation case. For simplicity a constant value of overall mass transfer capacity coefficient, $K_L a$, is assumed to apply for all the liquid regions, but the value obtained in the case of full circulation is taken as five times that for the bubble column.

Nomenclature

Symbols

V	Volume	m^3
F	Volumetric flow rate	m^3/s
C	Oxygen concentration	kmol/m^3
Q	Rates of oxygen transfer	kmol/s
$K_L a$	Mass transfer capacity coefficient	1/s
R	Ideal gas constant	m^{3}atm/kmol K
T	Temperature	K
H	Henry's law constant	m^{3}atm/kmol

Indices

0, 1, 2, 3	Refer to streams and zones (see figures)
L	Refers to liquid
G	Refers to gas
R	Refers to recycle
*	Refers to equilibrium

Exercises

1. Simulated tracer experiments can be performed on either program by setting $K_L a = 0$ and starting with suitable initial conditions and inlet concentrations. The methods given in example CSTRPULSE can be used for this.
   For example:
   a) In GASLIQ1 set initial values as follows: CL2=1, CL1 and CL3=0. Observe the tracer response curve for different values of FL1 and FL2.
   b) In GASLIQ2 set CGI=0 and the inflow CG0=1. Observe the response of the outflow CG3 for various values of the circulation flow FGR1 and FGR2.
2. Set $K_L a$ to a suitable value and experiment with the influence of the mixing parameters FL1 and FL2 in GASLIQ1 and FGR1 and FGR2 in GASLIQ2.

Results

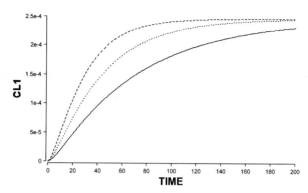

Fig. 5 GASLIQ1 with transfer showing the response of CL1 for three values of FL1 = FL2 from 0.005 to 0.02 m^3/s.

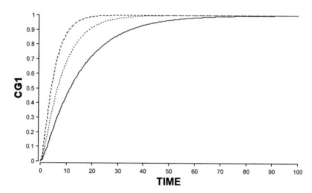

Fig. 6 Step tracer response of GASLIQ2 without transfer showing the response of CG1 for three values of FGR1 = FGR2 from 0.0005 to 0.002 m^3/s.

Reference

Harnby, N., Edwards, M. F. and Nienow, A. W. (Eds.) (1985) Mixing in the Process Industries, Butterworths.

5.6.5
SPBEDRTD – Spouted Bed Reactor Mixing Model

System

In this example, a stirred-tank model is employed to model the mixing behavior of an air–solid, spouted, fluidised-bed reactor. The central spout is modelled as two tanks in series, the top fountain as a further tank and the down flowing annular region of the bed as six equal-sized tanks in series. It is assumed that a constant fraction of the total solids returns from each stage of the annular region into the central two-tank region, as depicted below.

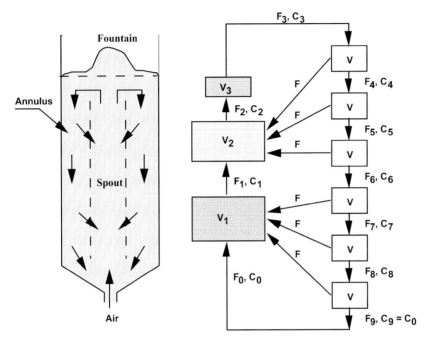

Fig. 1 The spouted bed and its mixing model.

Experimental tracer studies are to be carried out in which the solid in the bottom half of the bed is initially mixed with tracer, the bed started and timed samples taken from various locations in the bed annular region. It is desired to model the resulting tracer distribution in the bed in order to find the fractional slippage rate between the annular tanks and the central bed regions.

Model

The initial expansion of the bed consequent to the start of the air flow is neglected from the analysis. Component balance equations for each region of the reactor are given.

For the bed central flow region

$$V_1 \frac{dC_1}{dt} = F_0 C_0 + F[C_7 + C_8 + C_9] - F_1 C_1$$

$$V_2 \frac{dC_2}{dt} = F_1 C_1 + F[C_4 + C_5 + C_6] - F_2 C_2$$

$$V_3 \frac{dC_3}{dt} = F_2 C_2 - F_3 C_3$$

For the annular flow region

$$V \frac{dC_4}{dt} = F_3 C_3 - [F_4 + F]C_4$$

$$V \frac{dC_5}{dt} = F_4 C_4 - [F_5 + F]C_5$$

$$V \frac{dC_6}{dt} = F_5 C_5 - [F_6 + F]C_6$$

$$V \frac{dC_7}{dt} = F_6 C_6 - [F_7 + F]C_7$$

$$V \frac{dC_8}{dt} = F_7 C_7 - [F_8 + F]C_8$$

$$V \frac{dC_9}{dt} = F_8 C_8 - [F_9 + F]C_9$$

where $F_1 = F_0 + 3\,F$
$F_2 = F_1 + 3\,F$
$F_3 = F_2$
$F_4 = F_3 - F$
$F_5 = F_4 - F$
$F_6 = F_5 - F$
$F_7 = F_6 - F$
$F_8 = F_7 - F$
$F_9 = F_8 - F$

and $C_0 = C_9$

Total bed volume

$$V_{Total} = V_1 + V_2 + V_3 + 6V$$

The initial distribution of tracer within the bed is represented by

At $t = 0$, $C_0 = C_1 = C_7 = C_8 = C_9 = C_{Start}$

$C_2 = C_3 = C_4 = C_5 = C_6 = 0$

Program

If desired, the model and program can be extended by adding a continuous inlet and outlet stream and by incorporating the E-curve methods found in Example CSTRPULSE.

Nomenclature

Symbols

V	Volume	m³
F	Volumetric flow rate	m³/h
C	Tracer concentration	mg/m³

Indices

0, 1,…9   Refer to designated regions and flow rates (see Fig. 1).

Exercises

1. Determine the time required for the tracer to be nearly completely mixed throughout the vessel. Which criterion did you use for uniform mixing?
2. Change the initial conditions to place the tracer at a different region of the vessel. Does this effect the time for complete mixing?
3. Double the flowrates in the vessel, and note the influence on the time for mixing.

Results

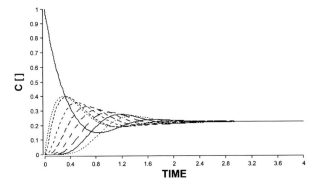

Fig. 2 A pulse in compartment 1 mixes with all the compartments in about 3 h.

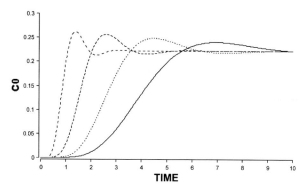

Fig. 3 As expected, a 10 fold change in F0 from 0.1 to 1.0 m³/s causes a similar difference in the mixing times.

5.6.6
BATSEG, SEMISEG and COMPSEG – Mixing and Segregation in Chemical Reactors

System

Chemical reactions will take place only when the reactant molecules are in intimate contact. In some cases, especially with very fast reactions or viscous liquids, segregation of the reactants can exist, which make the reaction rates and selectivities dependent on the mixing intensity. In chemical reactor engineering, the assumption is usually made that only mean concentrations need be considered. In reality, concentration values fluctuate about a mean, and in some cases these fluctuations must be considered in detail. This field is very complex and is still the subject of much research. This example serves only to introduce these concepts and to show how simulations can be made for certain simple situations.

The concept of segregation and its meaning to chemical reactors was first described by Danckwerts (1953). The intensity or degree of segregation is given the symbol I, which varies between one and zero. Shown in Fig. 1 is a tank with two components, A and B, which are separated into volume fractions, q_A and $1-q_A$; this condition represents complete initial segregation (I=1). Stirring or

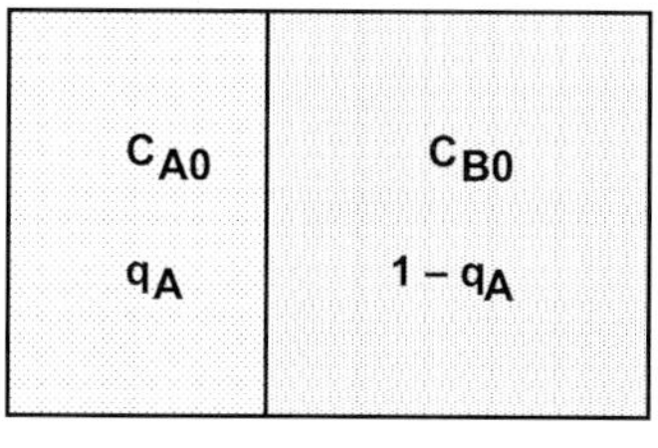

Fig. 1 Complete segregation (I=1) of reactants A and B in two volume fractions.

mixing this tank would cause the segregation to decay until complete homogeneity would be achieved (I=0). In this situation, the rate of any reaction between the reactants could obviously be influenced by the rate of mixing, as measured by the change in I.

Segregation may also be important if the reactants are fed to a reactor in an unmixed condition. This could be the case in any continuously fed reactor, either tubular (Fig. 2) or tank (Fig. 3).

The intensity of segregation, I, can be described in terms of concentration fluctuations, as shown in Fig. 4. Here the time-varying fluctuations above and below an average value are shown.

Fig. 2 Feeding of unmixed reactants to a tubular reactor.

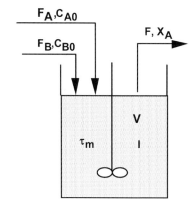

Fig. 3 Unmixed reactants fed into a continuous stirred tank.

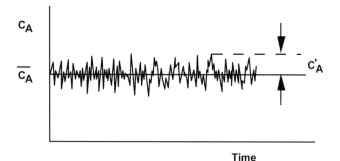

Fig. 4 Time-variant turbulent fluctuations of concentration about a mean value ($\overline{C_A}$), showing the fluctuation value C'_A.

The unmixedness can be characterised by the mean of the square of the fluctuations or concentration variance, according to

$$\overline{C'^2_A} = \overline{(C_A - \overline{C_A})^2}$$

To compare the "state of unmixedness", Danckwerts introduced the intensity of segregation, which is calculated in terms of the mean square of the fluctuations, as

$$I = \frac{\overline{C'^2_A}}{[\overline{C'^2_A}]_0}$$

where the subscript 0 denotes the initial or feed value.

In the case shown in Fig. 1, the initial mean square fluctuations would be

$$\left[\overline{C'^2_A}\right]_0 = C^2_{A0}\overline{q_A}(1 - \overline{q_A})$$

For unpremixed feed streams, the mean volume fraction is defined in terms of the mean entering flow rates as

$$\overline{q_A} = \frac{\overline{F_A}}{\overline{F_A} + \overline{F_B}}$$

Segregation or unmixedness of non-reactive systems, as measured by I or $\left[\overline{C'^2_A}\right]$, can be balanced using the general macroscopic population balance equation as follows

$$\begin{pmatrix} \text{Rate of} \\ \text{accumulation} \\ \text{of unmixedness} \\ \text{in the system} \end{pmatrix} = \begin{pmatrix} \text{Flow of} \\ \text{unmixedness} \\ \text{into} \\ \text{the system} \end{pmatrix} - \begin{pmatrix} \text{Flow of} \\ \text{unmixedness} \\ \text{out of} \\ \text{the system} \end{pmatrix}$$

$$\begin{pmatrix} \text{Dissipation rate} \\ \text{of unmixedness} \\ \text{by mixing} \\ \text{in the system} \end{pmatrix} + \begin{pmatrix} \text{Rate of} \\ \text{production} \\ \text{of unmiedness} \\ \text{in the system} \end{pmatrix}$$

The dissipation rate term, r_{dm}, can be taken as a first-order decay of concentration variance whose rate constant is the inverse mixing time

$$r_{dm} = -\frac{1}{\tau_m}\overline{C'^2_A}$$

5.6 Mixing-Model Examples

The mixing time can be related to power per unit volume and the geometry. Turbulent dispersion can produce unmixedness within the system, if there are gradients of mean concentration. This is not considered here. A useful discussion of these equations was given by Brodkey (1975) and Baldyga (1989). This area is treated in a recent book by Baldyga and Bourne (1999).

Model

In what follows, the above balance for unmixedness is applied to individual reactor cases. The relation for reaction rate in terms of I is then considered, and finally this is applied for simple and complex reactions.

Intensities of Segregation in a Batch Reactor

For a batch reactor system only the accumulation and dissipation terms are important. Thus

$$V \frac{d(\overline{C_A'^2})}{dt} = -\frac{1}{\tau_m} \overline{C_A'^2} V$$

With the initial value of $\left[\overline{C_A'^2}\right]_0$, integration gives

$$\frac{\overline{C_A'^2}}{\left[\overline{C_A'^2}\right]_0} = I = e^{-t\tau_m}$$

This equation predicts the intensity of segregation to decay with time in the batch reactor. It is also applicable to a steady-state plug flow system, where t is the residence time.

Intensities of Segregation in a Continuous Reactor

For a steady-state continuous tank reactor only the flow terms and the dissipation terms are important. Thus

$$0 = F\left[\overline{C_A'^2}\right]_0 - F\overline{C_A'^2} - \frac{1}{\tau_m}\overline{C_A'^2}V$$

With the mean residence time $\tau = V/F$, this becomes

$$\frac{\overline{C_A'^2}}{\left[\overline{C_A'^2}\right]_0} = I = \frac{1}{1+\tau/\tau_m} = \frac{\tau_m/\tau}{1+\tau_m/\tau} = \frac{\Theta_m}{\Theta_m + 1}$$

where $\Theta_m = \tau_m/\tau$.

Intensities of Segregation in a Semi-Continuous Reactor

In a semi-batch system the volume varies from zero to V_{max} with total flow rate F as

$$\frac{dV}{dt} = F_A + F_B = F$$

or for constant F

$$V = Ft$$

During filling, the accumulation, inflow and dissipation terms are important. Thus the balance gives

$$\frac{dI}{dt} = \frac{1-I}{t} - \frac{I}{\tau_m}$$

where feed and the initial tank contents are completely segregated, i.e., $I(0) = 1$.

Introducing dimensionless time, $\Theta = t/\tau$, and dimensionless mixing time, $\Theta_m = \tau_m/\tau$, during the filling period ($\Theta < 1$) gives

$$\frac{dI}{d\Theta} = \frac{1-I}{\Theta} - \frac{I}{\Theta_m}$$

After the reactor is full and overflowing $V = V_{max}$, and therefore $\tau = V_{max}/F$. Then $\Theta \geq 1$ and the outflow term is also important. The balance becomes

$$\frac{dI}{d\Theta} = 1 - I - \frac{I}{\Theta_m}$$

Reaction Rate in Terms of Intensity of Segregation

The rate of a simple $A + B \rightarrow C$ reaction, which is given by the instantaneous concentrations is expressed in terms of the mean concentrations and the extra variable $\overline{C'_A \, C'_B}$, becoming

$$\overline{r_A} = -kC_A C_B = -k\left(\overline{C_{A0}} \, \overline{C_{B0}} + \overline{C'_A \, C'_B}\right)$$

The derivation of this rate expression depends on the theory developed by Toor (1969) and assumes that with and without reaction the covariances of the reactant fluctuations are the same, as given by

$$\overline{C'_A \, C'_B} = -I\left(\overline{C_{A0}} \, \overline{C_{B0}}\right)$$

Mixing effects are particularly important for complex reactions, since selectivity is changed. The following reaction is considered

$$A + B \xrightarrow{k_1} R \quad \text{(desired reaction)}$$

$$R + B \xrightarrow{k_2} S \quad \text{(undesired reaction)}$$

The kinetics in terms of the intensities of segregation and the mean concentrations can be shown to be

$$\overline{r_A} = -k(\overline{C_A}\,\overline{C_B} - I\overline{C_{A0}}\,\overline{C_{B0}})$$

$$\overline{r_B} = -k_1(\overline{C_A}\overline{C_B} - I\overline{C_{A0}}\,\overline{C_{B0}}) - k_2\overline{C_B}\,\overline{C_R}$$

$$\overline{r_R} = +k_1(\overline{C_A}\overline{C_B} - I\overline{C_{A0}}\,\overline{C_{B0}}) - k_2\overline{C_B}\,\overline{C_B}$$

$$\overline{r_S} = +k_2\overline{C_B}\,\overline{C_B}$$

This result was obtained by Toor and Bourne (1977) by neglecting the influence of the covariance, $C'_B C'_R$ and by assuming that $C'_A C'_B$ could be estimated by Toor's hypothesis, as explained above. This simplification is restricted to a relatively slow second reaction and to a volume ratio q_A/q_B near one.

Batch Reactor

The batch reactor balance for a simple $A + B \rightarrow C$ reaction becomes

$$\frac{d(\overline{C_A})}{dt} = \overline{r_A} = -k(\overline{C_A}\,\overline{C_B} - I\overline{C_{A0}}\,\overline{C_{B0}})$$

Dimensionless variables can be defined as follows: $\Theta = t/\tau_m$, $\eta = \overline{C_{B0}}/\overline{C_{A0}}$, $Da = k\overline{C_{A0}}\tau$, and $y = \overline{C_A}/\overline{C_{A0}}$. Here $\overline{C_{B0}} = C_{B0}q_B/(q_A+q_B)$ and $\overline{C_{A0}} = C_{A0}q_A/(q_A+q_B)$. The dimensionless balance becomes

$$\frac{dy}{d\Theta} = -Da[y(y+\eta-1) - I\eta] = -Da[y(y+\eta-1) - e^{-\Theta}\eta]$$

where the initial value $y(0) = 1$.

For finite y, when $Da \rightarrow \infty$ this gives

$$[y(y+\eta-1) - I\eta] \rightarrow 0$$

and thus y_∞ is given by the quadratic formula

$$y_\infty = \frac{-(\eta-1) \pm \sqrt{(\eta-1)^2 + 4\eta I}}{2}$$

In this case only the + sign has a physical meaning.

The conversion X in the batch tank is given by $X = 1 - y$, which becomes

$$\frac{dX}{d\Theta} = Da[(1-X)(\eta-X) - I\eta] = Da[(1-X)(\eta-X) - e^{-\Theta}\eta]$$

The above equations also apply to a plug flow reactor, where Θ is the dimensionless residence time, which varies with distance.

Continuous Stirred-Tank Reactor

For a simple $A + B \rightarrow C$ reaction in a continuous stirred-tank reactor, as shown in Fig. 3, in terms of fraction conversion of the reactant A, the balance becomes

$$X = Da[(1-X)(\eta-X) - I\eta]$$

with the fractional conversion defined as

$$X = 1 - \frac{C_A}{\overline{C_{A0}}}$$

where $\overline{C_{A0}} = C_{A0} F_A/(F_A + F_B)$

If X is to be finite when $Da \rightarrow \infty$, the term $[(1-X)(\eta-X) - I\eta] \rightarrow 0$, and then as before

$$X_\infty = \frac{(1+\eta) \pm \sqrt{(1+\eta)^2 - 4\eta(1-I)}}{2}$$

since $0 < X_\infty < 1$, only the minus sign has meaning here.

Generally

$$X = \frac{\left(1+\eta+\frac{1}{Da}\right) \pm \sqrt{\left(1+\eta+\frac{1}{Da}\right)^2 - 4\eta(1-I)}}{2}$$

Semi-Batch Reactor

For a simple $A + B \rightarrow C$ reaction the semi-batch reactor balance is

$$\frac{d(\overline{C_A})}{dt} = \frac{F}{V}(\overline{C_{A0}} - \overline{C_A}) - k(\overline{C_A\,C_B} - I\overline{C_{A0}\,C_{B0}})$$

using the same dimensionless quantities, this becomes

$$\frac{dy}{d\Theta} = \frac{1}{A}(1 - y) - \text{Da}[y(y + \eta - 1) - I\eta]$$

with $y(0) = 1$. Here the constant $A = \Theta$ during the filling period, and $A = 1$ after the filling period, when $\Theta > 1$.

The requirement to have y finite when $\text{Da} \rightarrow \infty$ gives

$$y_\infty = \frac{-(\eta - 1) \pm \sqrt{(\eta - 1)^2 + 4\eta I}}{2}$$

The component balances for the complex reaction

$$A + B \xrightarrow{k_1} R \quad \text{(desired reaction)}$$

$$R + B \xrightarrow{k_2} S \quad \text{(undesired reaction)}$$

become as follows:

A-balance

$$\frac{d(\overline{C_A})}{dt} = \frac{F}{V}(\overline{C_{A0}} - \overline{C_A}) - k_1(\overline{C_A\,C_B} - I\overline{C_{A0}\,C_{B0}})$$

B-balance

$$\frac{d(\overline{C_B})}{dt} = \frac{F}{V}(\overline{C_{B0}} - \overline{C_B}) - k_1(\overline{C_A\,C_B} - I\overline{C_{A0}\,C_{B0}}) - k_2\overline{C_B\,C_R}$$

R-balance

$$\frac{d(\overline{C_R})}{dt} = -\frac{F}{V}\overline{C_R} + k_1(\overline{C_A\,C_B} - I\overline{C_{A0}\,C_{B0}}) - k_2\overline{C_B\,C_R}$$

S-balance

$$\frac{d(\bar{C}_S)}{dt} = -\frac{F}{V}\bar{C}_S + k_2\bar{C}_B\bar{C}_R$$

During the filling period $V < V_{max}$.

When combined with the total balance with constant F, the dimensionless form of these equations apply both to the filling and full periods. These are

$$\frac{dy_A}{d\Theta} = \frac{1}{A}(1 - y_A) - \beta\,Da(y_Ay_B - I\eta)$$

$$\frac{dy_B}{d\Theta} = \frac{1}{A}(\eta - y_B) - \beta\,Da(y_Ay_B - I\eta) - Da\,y_By_R$$

$$\frac{dy_R}{d\Theta} = -\frac{1}{A}y_R + \beta\,Da(y_Ay_B - I\eta) - Da\,y_By_R$$

$$\frac{dy_S}{d\Theta} = -\frac{1}{A}y_S + Da\,y_By_R$$

where the dimensionless terms are $\Theta = t/\tau$, $\eta = \bar{C}_{B0}/\bar{C}_{A0}$, $Da = k_2\bar{C}_{A0}\tau$, $y = \bar{C}_i/\bar{C}_{A0}$ and $\beta = k_1/k_2$. Here $\tau = V/F$, and the feed concentrations are $y_{A0}=1$, $y_{B0}=h$, $y_{R0}=0$ and $y_{S0}=0$. During the filling period $A=\Theta$, and $A=1$ when $\Theta > 1$. For the desired selectivity, $\beta = k_1/k_2 > 1$, and $\eta < 1$.

Selectivity at any time Q can be defined as:

$$X_S = \frac{2y_S}{y_R + 2y_S}$$

Here a value of $X_S = 1$ would mean that no desired product was obtained.

Program

The programs are as follows: BATSEG for a simple batch reactor, SEMISEG for a semi-continuous reactor that fills up and becomes continuous and COMPSEG for the complex reaction in a semi-continuous reactor. This example was developed by J. Baldyga., Dept. of Chem. and Proc. Eng., Warsaw University of Technology, Poland.

Nomenclature

Symbols

C	Concentration	kmol/m^3
$\overline{C'^2_A}$	Concentration fluctuation variance	kmol^2/m^6
Da	Damköhler number (= $k\overline{C_{A0}}\tau$ and $k_2\overline{C_{A0}}\tau$)	–
F	Volumetric flow rate	m^3/s
I	Degree of segregation	–
q	Fractional flow rate	–
t	Time	s
V	Volume	m^3
X	Fractional conversion	–
y	Dimensionless concentration	–
β	Ratio of reaction rate constants (= k_1/k_2)	–
η	Initial concentration ratio (= $\overline{C_{B0}}/\overline{C_{A0}}$)	–
Θ	Dimensionless time (= t/τ) or (= τ/τ_m)	–
τ	Time constant or mean residence time	s

Indices

'	Refers to fluctuations around mean value
0	Refers to initial or input value
dm	Refers to dissipation of unmixedness
m	Refers to mixing
max	Refers to maximum volume
T	Refers to tank
∞	Refers to Da = infinity
–	Refers to mean value

Exercises

1. Very high values of Da correspond to instantaneous reactions. Verify that simulations with BATSEG and SEMISEG give results corresponding to the quadratic formulae presented.
2. Show by simulation that the corresponding conversion of B for instantaneous reaction is given by solving the equation $(1-X)(\eta-X) - I\eta = 0$
3. Investigate the influence of concentration ratio, η, from 0.05 to 0.8 on the selectivity and concentration of desired product using COMPSEG.
4. With all three programs, vary the Damköhler number, Da, over a range from 1 to 500 and explain the results. Note that when Da no longer has any influence on the rate, then mixing must be controlling.

Results

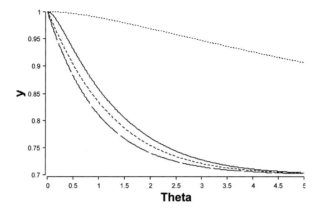

Fig. 5 Program BATSEG: Dimensionless concentration y versus dimensionless time Θ for Da=0.1, 5.0, and 100. Shown also is the decay of segregation I with time.

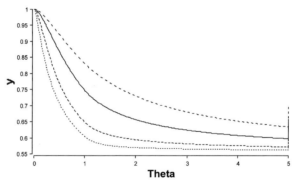

Fig. 6 Program SEMISEG: Varying THETAMIX (0.03, 0.1, 0.3 and 0) gave these results for the semi-batch reactor. Note that when THETA reaches 1.0 the reactor is full and becomes continuous.

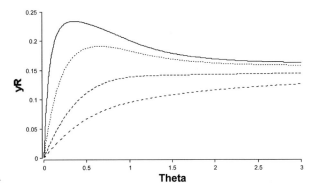

Fig. 7 Program COMPSEG: Dimensionless concentration of desired product YR versus dimensionless time THETA obtained by varying THETAMIX (0.03, 0.1, 0.3, and 0.6. Parameter values used: ETA = 0.3, BETA = 2000, DA = 1.0.

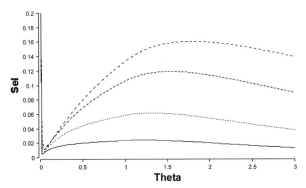

Fig. 8 Selectivity SEL profiles corresponding to the simulations in Fig. 7. Low values of SEL are desired. Mixing is seen to greatly influence the selectivity, even though the second reaction is much slower.

References

Baldyga, J. and Bourne J. R. (1999) Turbulent Mixing and Chemical Reactions, Wiley.
Baldyga, J. (1989) Chem. Eng. Sci. 44, 1175.
Bourne, J. R. and Toor, H. L (1977) AIChE J. 23, 602.
Brodkey, R. S. (1975) Chap. 2 in "Turbulence in Mixing Operations", R. S. Brodkey ed., Academic Press.
Danckwerts, P. V (1953) Appl. Sci. Res., A3, 209.
Toor, H. L. (1969) Ind. Eng. Chem. Fundam. 8, 655.

5.7
Tank Flow Examples

5.7.1
CONFLO 1, CONFLO 2 and CONFLO 3 – Continuous Flow Tank

System

Liquid flows from an upstream source, at pressure P_1, via a fixed position valve into the tank, at pressure P_2. The liquid in the tank discharges via a second fixed position valve to a downstream pressure P_3. The tank can be open to atmosphere or closed off to the atmosphere and can work either under isothermal or adiabatic temperature conditions. The detailed derivation of this problem is discussed both by Franks (1972) and Ramirez (1976, 1989). Data values in the problem are similar to those of Ramirez, who also presented simulation studies.

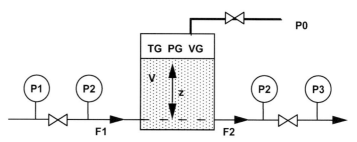

Fig. 1 Schematic of the tank flow problem.

Model

A mass balance for the liquid gives

$$\frac{dM}{dt} = F_1\rho - F_3\rho$$

For constant density ρ

$$\frac{dV}{dt} = A\frac{dz}{dt} = F_1 - F_3$$

where $V = Az$.

The flow through each fixed-position valve is given by

$$F_1 = K_{V1}\sqrt{P_1 - P_2}$$

$$F_3 = K_{V2}\sqrt{P_2 - P_3}$$

where K_{V1} and K_{V2} are the effective valve constants.

Tank pressure and temperature

a) For the tank open to atmosphere, the outlet pressure from the tank P_2 is given by

$$P_2 = P_0 + \rho g z$$

b) For a closed tank under isothermal conditions, the air space above the liquid obeys the Ideal Gas Law

$$PV = nRT$$

so that

$$\frac{P_0 V_G}{T_G} = \frac{P_{G0} V_{G0}}{T_{G0}}$$

where P_{G0} and T_{G0} are the defined initial conditions.

c) Under well insulated conditions, the air in the tank will behave adiabatically and its temperature will change isotropically, where

$$\frac{T_G}{T_{G0}} = \left(\frac{V_{G0}}{V_G}\right)^{\gamma - 1}$$

where γ is the adiabatic expansion coefficient for air, equal to 1.4.

Both the upstream pressure P and final downstream pressure P_3 are fixed, together with the atmospheric pressure P_0.

The above model equations are also presented in dimensionless terms by Ramirez (1989).

Program

The cases (a), (b) and (c) are represented respectively by the simulation programs CONFLO1, CONFLO2 and CONFLO3.

Nomenclature

Symbols

A	Area	m^2
K_V	Valve proportionality factor	$m^4/kN^{0.5}$ s

g	Acceleration of gravity	m/s^2
M	Mass	kg
n	Molar mass	kmol
P	Pressure	N/m^2
R	Ideal gas constant	kNm/kmol K
ρ	Density	kg/m^3
T	Temperature	K
V	Volume	m^3
z	Depth of liquid	m
γ	Adiabatic expansion coeff., c_p/c_v	–

Indices

0, 1, 2, 3   Refer to positions (see figures)

Exercises

1. For the open valve case (CONFLO1), make a step change in P_1 and observe the transients in flow rate and liquid level. Try this for a sinusoidally varying inlet pressure. Experiment with a sudden change in the outlet valve setting.
2. For the closed valve case (CONFLO2), repeat the changes of Exercise 1, observing also the variation in the gas pressure and volume. Change the gas volume and note the influence of this variable.
3. Using CONFLO3, determine the importance of including the adiabatic work effects in the model, over a range of flow rate, pressure and valve conditions by following the changes in the gas space temperature and pressure.
4. Study the dimensionless forms of the model equations given by Ramirez (1989); program these and compare the results.

Results

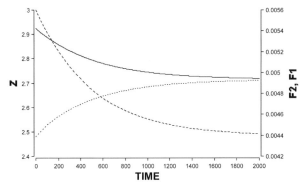

Fig. 2 In CONFLO1, as the liquid height drops the inlet and outlet flows eventually become equal.

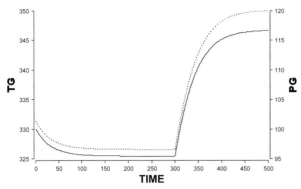

Fig. 3 In CONFLO3, the inlet pressure was changed at time=300 s from 150 to 200 kN/m². Seen here are the temperature and pressure effects of the adiabatic expansion and compression.

References

Franks, R. G. E. (1972) Modelling and Simulation in Chemical Engineering, Wiley-Interscience.
Ramirez, W. F. (1976) Process Simulation, Lexington Books.
Ramirez, W. F. (1989) Computational Method for Process Simulation, Butterworth.

5.7.2
TANKBLD – Liquid Stream Blending

System

Two aqueous streams containing salt at differing concentrations flow continuously into a well-mixed tank, and a mixed-product stream is removed. In this case, the densities of the streams are not assumed constant but vary as a function of concentration. All three streams can have time varying flow rates, and hence the concentration and volume of liquid in the tank will also vary with time.

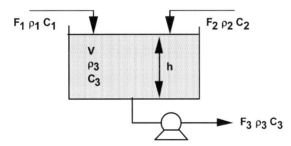

Fig. 1 Tank with density variation.

Model

Total mass balance

$$\frac{d(V\rho_3)}{dt} = F_1\rho_1 + F_2\rho_2 - F_3\rho_3$$

Component mass balance

$$\frac{d(VC_3)}{dt} = F_1C_1 + F_2C_2 - F_3C_3$$

The density of salt solution varies linearly with the concentration of each stream, according to

$$\rho = \rho_0 + bC$$

where b is a constant of proportionality.

The above system was treated in detail by Russell and Denn (1972), and an analytical solution was obtained.

Program

The tank is filled with streams 1 and 2, and when the tank is approximately full, the outlet volumetric flow rate is set to the approximate steady state value, as would be controlled by a pump. This leads to some further volume change as the final product density is reached.

Nomenclature

Symbols

b	Coefficient in density relation	–
C	Concentration	kg/m^3
F	Volumetric flow rate	m^3/min
V	Volume	m^3
ρ	Density	kg/m^3

Indices

1, 2, 3	Refer to stream values
0	Refers to water

Exercises

1. Carry out simulations for differing tank volumes, flow rates and feed concentrations, in which the inlet and outlet flow rates are set equal ($F_1 + F_2 = F_3$), and observe the approach to steady state. Relate the time taken to approach steady state to the mean tank residence time ($\tau = V/F_3$).
2. Study the case of tank washout, with the tank initially full, at high salt concentration and the liquid input streams free of salt, $C_1 = C_2 = 0$. Show that the tank concentration, C_3, decreases exponentially with respect to time.
3. Russell and Denn, assuming constant liquid density, showed analytically that

$$C_3 = Z_1 + (C_{30} - Z_1)\left(\frac{1}{1 + Z_2 t}\right)^{Z_3}$$

where

$$Z_1 = \frac{F_1 C_1 + F_2 C_2}{F_1 + F_2}$$

$$Z_2 = \frac{F_1 + F_2 - F_3}{V_0}$$

$$Z_3 = \frac{F_1 + F_2}{F_1 + F_2 - F_3}$$

Here C_{30} is the starting tank concentration at time t=0 and V_0 is the starting tank volume at t=0. Compare the results of the simulation to the analytical solution which can also be calculated with MADONNA. For the constant density assumption, simply set b=0.
The analytical solution shows that the approach to steady state is very rapid when V_0 is small and that the concentration in the tank is always constant, when starting with a relatively empty tank. It also indicates that the rate of change of volume in the tank is equal to the net volumetric flow rate, but only for a linear density concentration relationship. Check the above analytical conclusions numerically and test the case of a non-linear density–concentration relationship by simulation.
4. Modify the program to enable the outlet flow to vary as a function of liquid depth and density, e.g.,

$$F_3 = k\sqrt{\rho_3 V}$$

and test the system with time-varying inlet flows F_1 and F_2.

Results

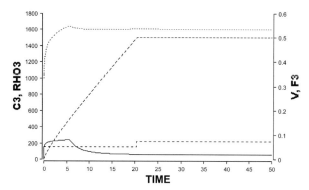

Fig. 2 Variation of C_3 and V during the approach to steady state. Note that V and RHO3 continue to change after the final value of F_3 is set.

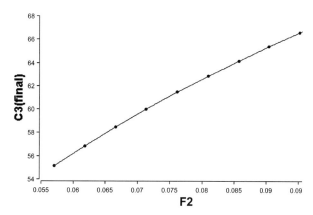

Fig. 3 The steady state values of C_3 as a function of F_2.

Reference

Russell, T. W. F. and Denn, M. M. (1972) Introduction to Chemical Engineering Analysis, Wiley.

5.7.3
TANKDIS – Ladle Discharge Problem

System

The discharge of molten steel from a ladle into a vacuum degassing chamber has been treated by Szekely and Themelis (1971). The transfer is effected via a discharge nozzle, in which the effects of both friction and wear can be important.

5.7 Tank Flow Examples

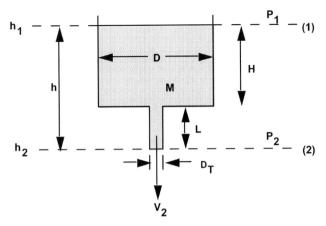

Fig. 1 Schematic drawing of the tank discharge problem.

Model

Total mass balance

$$\frac{dM}{dt} = -W$$

where for the nozzle mass flow

$$W = \frac{\pi D_N^2}{4} \rho v_2$$

For the cylindrical tank geometry the mass as a function of head is

$$M = AH\rho = A\rho(h - L)$$

where $A = (\pi D^2)/4$ and gives the result that

$$h = \frac{M}{A\rho} + L$$

Neglecting acceleration effects in the metal, the energy balance equation is given by

$$\frac{v_1^2}{2g} + h_1 + \frac{P_1}{\rho g} = \frac{v_2^2}{2g} + h_2 + \frac{P_2}{\rho g} + \Delta H_f$$

where ΔH_f represents the losses due to friction and

$$\Delta H_f = \frac{2fLv_2^2}{D_N g}$$

The friction factor, f, depends on the relative surface roughness of the nozzle (ε) and the Reynolds number (Re) where

$$\text{Re} = \frac{r v_2 D_N}{\mu}$$

A constant value of the friction factor $f=0.009$ is assumed, for fully developed turbulent flow and a relative pipe roughness $\varepsilon = 0.01$. The assumed constancy of f, however, depends upon the magnitude of the discharge Reynolds number, which is checked during the program. The program also uses the data values given by Szekely and Themelis (1971), but converted to SI units.

Since the velocity of the metal at the ladle surface is very low ($v_1 = 0$), the metal discharge velocity is then given by

$$v = v_2 = \sqrt{\frac{gh + \frac{P_1 - P_2}{\rho}}{\frac{1}{2} + 2f \frac{L}{D_N}}}$$

Note that a full treatment of this problem would also involve an energy balance and a momentum balance, together with relations for possible physical property changes during discharge and cooling. Acceleration effects are ignored in this analysis, which is solved analytically by Szekely and Themelis.

Nomenclature

Symbols

D	Tank diameter	m
D_N	Nozzle diameter	m
f	Friction factor	–
g	Acceleration of gravity	m/s^2
H	Tank height	m
h	Effective head	m
L	Nozzle length	m
M	Mass	kg
P	Pressure	kg/m s^2
ρ	Metal density	kg/m^3
Re	Reynolds number	–
t	time	s
v	Metal velocity	m/s
W	Mass flow rate	kg/s
μ	Metal viscosity	kg m s
ΔH_f	Pressure head lost by friction	m

Indices

1	Metal surface
2	Tank outlet

Exercises

1. Compare the rates of discharge with and without allowance for friction.
2. Modify the program to allow for nozzle wear.
3. How would you expect cooling to affect the discharge rate? How could you make allowance for this in your model?

Results

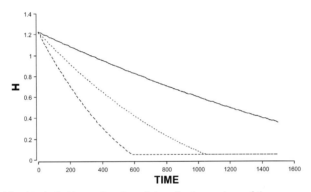

Fig. 2 Liquid level in ladle H as a function of time for three values of the nozzle diameter 0.3 to 0.6 m.

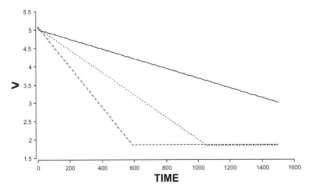

Fig. 3 Discharge velocity as a function of time for three values of the nozzle diameter 0.3 to 0.6 m.

Reference

Szekely, T. and Themelis, N. T. (1971) Rate Phenomena in Process Metallurgy, Wiley Interscience.

5.7.4
TANKHYD – Interacting Tank Reservoirs

System

Two open reservoirs with cross-sectional areas A_1, A_2 and liquid depths h_1 and h_2 are connected by a pipeline of length L and diameter D. The flow of liquid back and forth creates a decaying oscillatory response to a final steady state. Ramirez (1976) provides a detailed model derivation for this problem and shows how the problem is solved by analogue computation. The parameters are the same as those used by Ramirez.

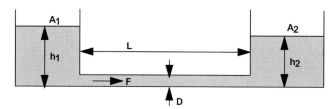

Fig. 1 Dynamics of interacting reservoir levels.

Model

The mass balance for tank 1 at constant liquid density is

$$\frac{d(A_1 h_1)}{dt} = -F$$

and for tank 2

$$\frac{d(A_2 h_2)}{dt} = F$$

where F is the volumetric exchange flow rate between tank 1 and tank 2.

Momentum balance for the pipeline:

$$\begin{pmatrix} \text{The rate of} \\ \text{accumulation} \\ \text{of momentum} \end{pmatrix} = \begin{pmatrix} \text{Rate of} \\ \text{momentum} \\ \text{in} \end{pmatrix} - \begin{pmatrix} \text{Rate of} \\ \text{momentum} \\ \text{out} \end{pmatrix} + \begin{pmatrix} \text{Sum of} \\ \text{forces} \\ \text{acting} \end{pmatrix}$$

Since the pipeline is of uniform diameter, the rate of momentum at the inlet is equal to that at the outlet and

$$\begin{pmatrix} \text{Rate of accumulation} \\ \text{of momentum} \end{pmatrix} = \begin{pmatrix} \text{Sum of the forces} \\ \text{acting on the pipeline} \end{pmatrix}$$

Thus

$$\frac{d(F\rho L)}{dt} = P_1 - P_2 + (-F_x)$$

where P_1 and P_2 are the pressure forces acting on the system

$$P_1 = \frac{\pi D^2}{4} \rho g h_1$$

$$P_2 = \frac{\pi D^2}{4} \rho g h_2$$

and F_x is the frictional force, given by the Fanning frictional equation as

$$F_x = \frac{8F^2}{\pi D^3} f \rho L$$

where f is the Fanning friction factor, assumed constant in this analysis.

Program

Note that the frictional force must always act in the opposite direction to that of the direction of flow and that this must be allowed for in the program.

Nomenclature

Symbols

A	Cross-sectional area of tanks	m²
D	Diameter of pipe	m
F	Flow rate	m³/s
f	Overall loss factor	–
F_x	Frictional force	N
g	Acceleration of gravity	m/s²
h	Liquid depth	m
L	Length of pipe	m
P	Hydrostatic pressure force	N/m²
t	Time	s
ρ	Liquid density	kg/m³

Indices

1, 2   Refer to tanks 1 and 2

Exercises

1. Observe the oscillatory approach to steady-state for different initial liquid depths. Plot the depths and the flow rate versus time.
2. Investigate the influence of pipe length and diameter on the time to achieve steady state.
3. Suppose tank 1 represents a body of tidal water with tidal fluctuations, and tank 2 is a tide meter used to measure the height of the tide. The connection is made by 100 m of pipeline and tank 2 is 1 m in diameter. Assume that the tide varies in a sine wave fashion with an amplitude of 2 m with a wave length of 12 h. Change the program to investigate the minimum pipe diameter required to ensure less than a 1 cm difference between the actual tide level and the level in the tide meter.

Results

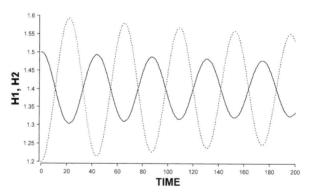

Fig. 2 Oscillating heights of the liquid in the tanks.

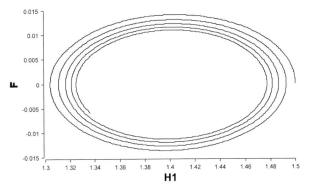

Fig. 3 Phase-plane of liquid height in tank 1 versus flow rate. Note the reversal in the flow direction.

Reference

Ramirez, W. F. (1976) Process Computation, Lexington Books.

5.8
Process Control Examples

5.8.1
TEMPCONT – Control of Temperature in a Water Heater

System

The simple feedback control system below consists of a continuous-flow stirred tank, a temperature measurement device, a controller and a heater.

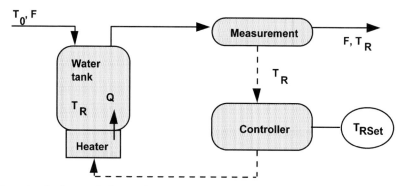

Fig. 1 Feedback control of a simple continuous water heater.

Model

The energy balance for the tank is

$$V \rho c_p \frac{dT_R}{dt} = F \rho c_p (T_0 - T_R) + Q$$

where Q is the delayed heat input from the heater represented by a first order lag

$$\frac{dQ}{dt} = \frac{(Q_c - Q)}{t_Q}$$

The measurement of temperature is also delayed by a sensor lag given by

$$\frac{dT_{sens}}{dt} = \frac{T_R - T_{sens}}{\tau_{sens}}$$

A proportional-integral feedback controller is modelled by

$$Q_c = Q_0 + K_P \varepsilon + \frac{K_P}{\tau_I} \int \varepsilon \, dt$$

where the control error is given by

$$\varepsilon = (T_{Rset} - T_{sens})$$

Nomenclature

Symbols

c_p	Specific heat	kJ/(kg °C)
f	Frequency of oscillations	1/h
F	Flow rate	m³/h
K_P	Proportional control constant	kJ/(h °C)
Q	Heat input	kJ/h
T	Temperature	°C
V	Reactor volume	m³
ε	Error	°C
ρ	Density	kg/m³
τ_D	Differential control constant	h
τ_I	Integral control constant	h
τ_Q	Time constant for heater	h
τ_{sens}	Time constant for measurement	h

Indices

C	Refers to controller
R	Refers to reactor
sens	Refers to sensor
set	Refers to setpoint
0	Refers to inlet or initial

Exercises

1. Disconnect the controller (set controller=0), and measure the temperature response to step changes in water flow (set stepflow=1 and steptemp = 0). Observe how the temperature reaches a steady state. From the response curve measure the time constant (63% response)
2. Repeat Exercise 1 for step changes in inlet temperature (set stepflow=0 and steptemp = 1). From the response curve determine T_L and K from the process reaction curve as described in Chapter 2.

3. Measure the controlled response to a step change in F (set controller=1; step-flow=1 steptemp=0) with proportional control only (set τ_I very high). Notice the offset error ε and its sensitivity to K_P. Increase K_P with a small factor, e.g. 1.1, until you observe sustained oscillations. Here it is useful to use the slider tool with a minimum of 1 and a maximum of 1000. Study the influence of K_P on the error ε and the integral of its square, EINT2, using a parameter plot again with logarithmic scale for K_P. Is there any relation between the onset of oscillations and the value of EINT2?
4. Study the influence of the integral part of the controller by changing τ_I to a low value. Does the offset disappear?
5. Referring to Exercise 1, use Cohen–Coon settings from Table 2.2 to obtain the best controller settings for P and PI control from the process reaction curve parameters. Try these out in a simulation. Observe the suitability of the settings from the error e and the integral of its square, EINT2.
6. Operate with proportional control only and a step change in flowrate (set τ_I very high, controller=1; stepflow=1 steptemp=0). Increase K_P until oscillations in the response occur at K_{P0}. Use this oscillation frequency, f_0, to set the controller according to the Ultimate Gain Method ($K_P=0.45$ K_{P0}, $\tau_I = 1/(1.2\ f_0)$), where f_0 is the frequency of the oscillations at $K_P=K_{P0}$ (see Sec. 2.3.3). How high is EINT2?
7. Using EINT2 as objective function, vary Kp (KP) and τ_I (TAUI) using the Optimize function in Madonna. Set controller=1; stepflow=1 and steptemp=0. Use parameter plots to verify the minimum.

Results

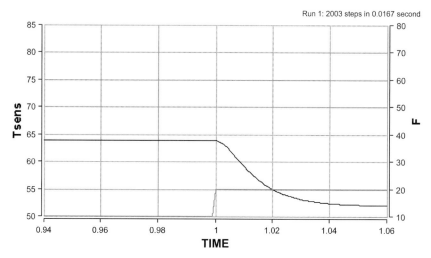

Fig. 2 Response of the system to a step change F as in Exercise 1.

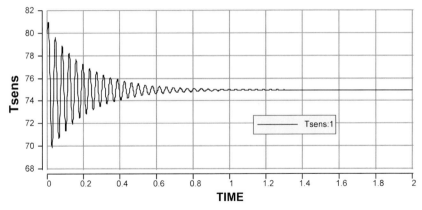

Fig. 3 Response showing damped oscillations at Kp=290, as in Exercise 5.

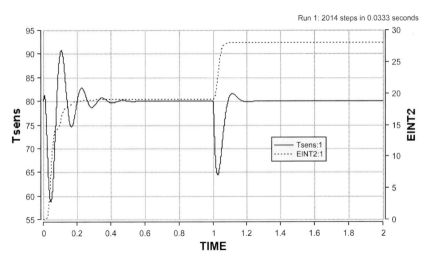

Fig. 4 Response using the control values from the optimisation, as in Exercise 6. Here Kp was limited to a maximum of 10.

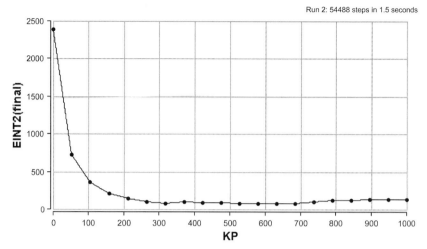

Fig. 5 A parameter plot of Kp showing its influence on EINT2.

5.8.2
TWOTANK – Two Tank Level Control

System

Liquid flows through a system of two tanks arranged in series, as shown below. The level control of tank 2 is based on the regulation of the inlet flow to the tank 1. This tank represents a considerable lag in the system. The aim of the controller is to maintain a constant level in tank 2, despite disturbances that occur in the flow F_3.

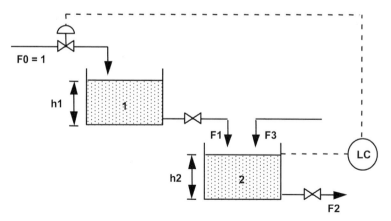

Fig. 1 Level control in a two-tank system.

Model

Mass balance equations with constant density yields for tank 1

$$A_1 \frac{dh_1}{dt} = F_0 - F_1$$

and for tank 2

$$A_2 \frac{dh_2}{dt} = F_1 + F_3 - F_2$$

Flow equations are given by

$$F_1 = K_1 \sqrt{h_1}$$

$$F_2 = K_2 \sqrt{h_2}$$

The disturbance is generated with a sine wave

$$F_3 = F_{30} + F_{3AMP} \sin \frac{t}{3}$$

and the proportional control equation is

$$F_0 = F_0 + K_p(h_2 - h_{2set})$$

Program

Note the change in the value of the proportional gain KP is changed at 90 h by use of an IF-THEN-ELSE statement:

$$KP = IF\ TIME\ <=\ 90\ THEN\ 4\ ELSE\ 20$$

Nomenclature

Symbols

A	Tank area	m^2
F	Flow rate	m^3/h
h	Liquid depth	m
K	Valve constant	$m^3/h\ m^{0.5}$
K_p	Controller constant	m^2/h
V	Volume	m^3

Indices

0	Refers to inlet
1	Refers to tank 1
2	Refers to tank 2
3	Refers to feed of tank 2
amp	Refers to amplitude
set	Refers to setpoint

Exercises

1. With no control $K_p=0$ and constant flow of F_3, study the response of the levels h_1 and h_2 and final steady-state values for differing flow rates F_0 and F_3.
2. Apply sinusoidal variations in the flow rate F_3, $F_{3AMP}=0.05$, and again study the responses in h_2 to changes in F_3.
3. Study the influence of the ratio of A_2/A_1 on the response in h_2.
4. Set $F_{3AMP}=0$, for constant flow F_3 and study the system response to h_{2set}. Repeat the simulations with control added.
5. Repeat Exercise 3 but with a sinusoidal variation in F_3, as in Exercise 2.
6. Change the valve constants and note the influence on the dynamics.
7. Change the controller constant and note its influence on the levels and the flow rates.
8. Many other simple combinations of filling and emptying tanks, which can be envisaged and programmed to form the basis of simulation examples.

Results

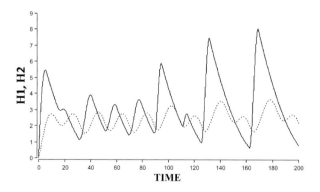

Fig. 2 At T=90 the value of the controller constant was increased (K_p from 4 to 20).

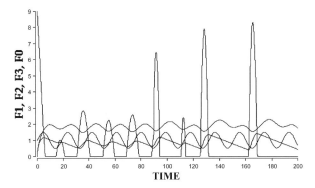

Fig. 3 The increase in K_p at T=90 gives very high peaks in F0. The sine wave disturbance in F3 is shown here.

Reference

Smith, C.L. Pike, R.W. and Murril, P.W. (1970) Formulation and Optimisation of Mathematical Models, Intext

5.8.3
CONTUN – Controller Tuning Problem

System

The temperature of a continuous flow of material through a steam-heated stirred tank is controlled by regulating the flow of steam. The tank temperature is measured by a thermocouple set inside a thermowell, giving a delayed temperature measurement response. This example is based on that of Robinson (1975).

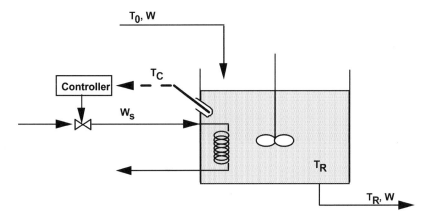

Fig. 1 Tank reactor with control.

Model

Tank heat balance

$$M c_p \frac{dT_R}{dt} = W c_p (T_0 - T_R) + W_S \lambda$$

where λ is the latent heat of the steam entering at the mass flowrate W_S. The temperature of both thermocouple and thermowell are each described by first-order lag equations, so that:

for the thermocouple

$$\frac{dT_C}{dt} = \frac{T_W - T_C}{\tau_C}$$

and for the thermowell

$$\frac{dT_W}{dt} = \frac{T_R - T_W}{\tau_W}$$

where τ_C and τ_W are the lag constants.

Controller Equation

For proportional and integral control

$$W_S = W_{S0} + K_p \varepsilon + \frac{K_p}{\tau_I} \int \varepsilon \, dt$$

$$\varepsilon = T_{Rset} - T_C$$

Nomenclature

Symbols

c_p	Specific heat	kJ/kg °C
K_p	Proportional gain	kg/min °C
τ_I	Integral time constant	min
M	Mass	kg
T_R	Temperature of tank	°C
t	Time	min
W	Mass flow rate	kg/min
λ	Latent heat of vapourisation of steam	kJ/kg
τ	Time constant	min

Indices

C	Refers to thermocouple
S	Refers to steam
W	Refers to thermowell
0	Refers to feed condition
I	refers to integral control mode
set	refers to setpoint

Exercises

1. Allow the tank to achieve steady-state operation in the absence of control ($K_p=0$). Use the resulting process reaction curve to estimate combined proportional and integral control parameters. Then use the obtained steady-state values as the initial values for a following sequence of runs.
2. Using proportional control only (τ_I very high), vary the proportional gain constant, in conjunction with a temperature set point change of 10 C and determine the critical value of K_p giving continuous oscillation. Tune the controller for both proportional and integral actions using the Ziegler-Nichols-Criteria, given in Section 2.3.3.2.
3. Experiment with the influence of different values of the measurement time constants, τ_C and τ_W. Plot all the temperatures to compare the results.

Results

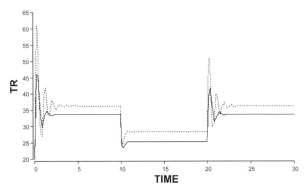

Fig. 2 Tank temperature versus time for two values of KP (1.5 and 2.0), with $\tau_I=10000$. The changes at T=10 and T=20 are programmed step changes in the inlet water flow rate. Oscillations and offset are caused by sub-optimal controller tuning.

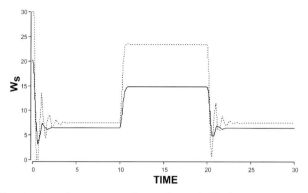

Fig. 3 Variations in steam flow corresponding to the run in Fig. 1.

References

Robinson, E.R. (1975) Time Dependent Processes, Applied Science Publishers.

5.8.4
SEMIEX – Temperature Control for Semi-Batch Reactor

System

An exothermic reaction involving two reactants is run in a semi-continuous reactor. The heat evolution can be controlled by varying the feed rate of one component. This is done via feedback control with reactor temperature measurement used to manipulate the feed rate. The reactor is cooled by a water jacket, for which the heat transfer area varies with volume. Additional control could involve the manipulation of the cooling-water flow rate.

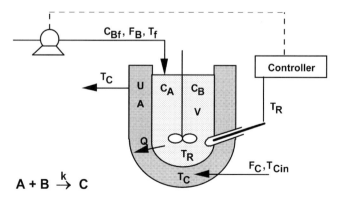

Fig. 1 Exothermic semi-batch reactor with feed control.

Model

The working volume of the reactor is given by the total mass balance, assuming constant density conditions

$$\frac{dV}{dt} = F_B$$

The component balances with second-order kinetics for the variable volume reactor are as follows:

$$\frac{d(VC_A)}{dt} = -k\,C_A\,C_B\,V$$

$$\frac{d(VC_B)}{dt} = F_B C_{Bf} - k\,C_A\,C_B\,V$$

$$\frac{d(VC_C)}{dt} = k\,C_A\,C_B\,V$$

The reaction rate constant varies with absolute temperature as

$$k = k_0 e^{-E/RT_{abs}}$$

The heat balance for the reactor contents assuming constant physical property values, is

$$V\rho\, c_p \frac{dT_R}{dt} = F_B \rho\, c_p (T_f - T_R) - U A(T_R - T_C) + (-\Delta H)(-k\,C_A\,C_B\,V)$$

which simplifies to

$$\frac{dT_R}{dt} = \frac{F}{V}(T_f - T_R) - \frac{U A}{V\rho\, c_p}(T_R - T_C) + \frac{\Delta H}{\rho c_p} k\,C_A\,C_B$$

The cooling jacket is assumed to be well-mixed and is modelled by

$$\rho V_C c_{pc} \frac{dT_C}{dt} = F_C \rho_C c_{pc}(T_{Cin} - T_C) + U A(T_R - T_C)$$

simplifying to

$$\frac{dT_C}{dt} = \frac{F_C}{V_C}(T_{Cin} - T_C) + \frac{UA}{\rho V_C c_{pc}}(T_R - T_C)$$

5 Simulation Tools and Examples of Chemical Engineering Processes

The level in the tank, h, varies with volume and tank diameter, D, according to

$$h = \frac{V}{\pi(D/2)^2}$$

The heat transfer area changes with liquid level as

$$A = \pi D h$$

The reactor temperature is controlled by manipulating the feed rate according to

$$F_B = F_{B0} + K_p \varepsilon + \frac{K_p}{\tau_I} \int \varepsilon \, dt$$

where

$$\varepsilon = T_{Rset} - T_R$$

A similar equation could be applied to the manipulation of the cooling water flow.

Program

A partial listing is given below.
{Example SEMIEX}

```
TRabs=TR+273
d/dt(V)=Fb              {Total balance}
A=A0+(V/C)              {Variable transfer area}
k=k0*EXP(-ER/Trabs)     {Arrhenius eq.}

{Mass balances}
d/dt(VCa)=-k*Ca*Cb*V
d/dt(VCb)=Fb*Cbf-k*V*Ca*Cb
d/dt(VCc)=k*Ca*Cb*V
Ca=VCa/V
Cb=VCb/V
Cc=VCc/V

{Energy balances}
Q=U*A*(TR-TC)/(Rho*Cpr)
d/dt(TR)=(Fb*(Tf-TR)-Q+H*k*Ca*Cb*V/(Rho*Cpr))/V
d/dt(TC)=(Fc/Vc)*(TCin-TC)+U*A*(TR-TC)/(Rho*Cpc*Vc)

{Feed flow rate control}
Err=TR-Tset
d/dt(Interr)=Err
Fb=Fb0-Kp*Err-(Kp/TI)*Interr
```

Nomenclature

Symbols

A_0	Initial heat transfer area	m^2
C_A	Concentration of A in tank	$kmol/m^3$
C_{Bf}	Concentration of B fed into tank	$kmol/m^3$
c_p	Heat capacity of reactor and feed	$kJ/kg\ K$
c_{pC}	Heat capacity of cooling water	$kJ/kg\ K$
D	Diameter of tank	m
ER	Activation energy/gas constant, E/R	K
F_{B0}	Feed rate basis	m^3/h
Fc	Cooling water flow	m^3/h
H	Heat of reaction	$kJ/kmol$
h	Level of reactor liquid	m
k	Reaction rate constant	$m^3/kmol\ h$
k_0	Frequency factor	$m^3/kmol\ h$
K_P	Proportional control constant	$m^3/h\ K$
Pi	value of π	–
ρ	Density of all streams	kg/m^3
T_0	Initial temperature in tank	°C
T_A	Absolute reactor temperature	K
T_{Cin}	Cooling water temperature	°C
T_f	Feed temperature	°C
τ_I	Integral control constant	h
T_R and T_{Rabs}	Reactor temperature	°C and K
U	Heat transfer coefficient	$kW/m^2\ K$
V_0	Initial tank volume	m^3
V_C	Volume of cooling jacket	m^3

Indices

abs	Refers to absolute temperature
C	Refers to cooling water
f	Refers to feed values
I	Refers to initial values
in	Refers to cooling water inlet
out	Refers to cooling jacket
set	Refers to setpoint

Exercises

1. Operate the reactor without control using a constant flow of reactant B. Note the reactor temperature.
2. Choose a suitable temperature setpoint and simulate the reactor with control, first with proportional control only and then including integral control. Adjust the controller constants to obtain adequate control.
3. Using the controller constant found in Exercise 2, make a step decrease in the cooling water flow rate. Did the controller decrease the reactant flowrate to compensate for this change?
4. Modify the program and change the controller to manipulate the cooling water flow rate. Experiment with this program.

Results

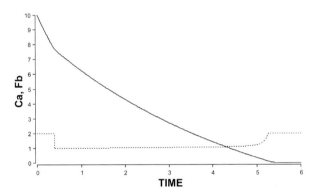

Fig. 2 The feed rate of B, FB, was manipulated by the control equation to control the reactor temperature. At the end of the batch, when the reaction rate decreased due to decreasing CA, FB increased gradually to its maximum.

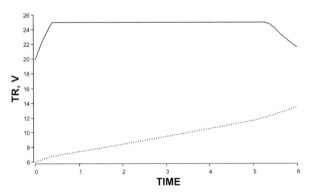

Fig. 3 During the run of Fig. 1, the temperature, TR, at first increased to its setpoint and then decreased at the end of the run. Liquid volume, V, increased throughout the run.

5.8.5
TRANSIM – Transfer Function Simulation

System

This is based on the example of Matko, Korba and Zupancic (1992), who describe general methods for simulating dynamic systems, when represented in the form of process transform functions. Here the transfer function is given by

$$G(s) = \frac{Y(s)}{U(s)} = \frac{1 - 4s}{(1 + 4s)(1 + 10s)}$$

and this is converted into an equivalent differential equation form for solution by Madonna, via the use of intermediate dummy variables.

Model

The above transfer unit may be written as

$$G(s) = G_1(s) G_2(s) G_3(s)$$

where $G_1(s)$, $G_2(s)$ and $G_3(s)$ relate the outlet response $Y(s)$ to the inlet disturbance $U(s)$, via the assumed dummy variables $W_1(s)$ and $W_2(s)$.

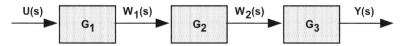

Fig. 1 Transfer function consisting of three transfer functions in series.

Here

$$G_1(s) = \frac{W_1(s)}{U(s)} = \frac{1}{1 + 4s}$$

$$G_2(s) = \frac{W_2(s)}{W_1(s)} = 1 - 4s$$

$$G_3(s) = \frac{Y(s)}{W_2(s)} = \frac{1}{1 + 10s}$$

Hence

$$sW_1 = 0.25U - 0.25W_1$$

$$W_2 = W_1 - 4s\,W_1$$

$$sY = 0.1W_2 - 0.1Y$$

Defining the variables Y, W_1 and W_2 as deviations from an initial steady state, the original transfer function can now be expressed as

$$\frac{dW_1}{dt} = 0.25U - 0.25W_1$$

$$W_2 = W - 1 + 4\frac{dW_1}{dt}$$

$$\frac{dY}{dt} = 0.1W_2 - 0.1Y$$

Exercises

1. Vary the numerical parameters of the transform and study the resulting response characteristics.
2. Modify the form of the transform function, convert this into differential equation format and study the resulting response characteristics.
3. Use Madonna to study the response for transfer functions including time delays.

Results

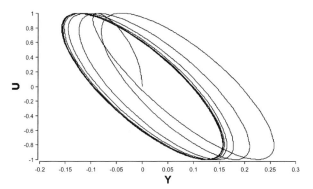

Fig. 2 Phase-plane output for the transfer function with a sine wave disturbance at a frequency of F=0.1.

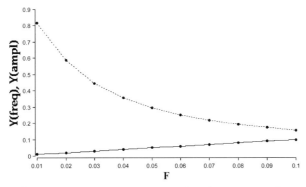

Fig. 3 Varying F from 0.01 to 0.1 in a parametric run that plots automatically the frequency and amplitude of the response.

Reference

Matko, D., Korba, R. and Zupancic, B. (1992) "Simulation and Modelling of Continuous Systems: A Case Study Approach", Prentice-Hall.

5.8.6
THERMFF – Feedforward Control of an Exothermic CSTR

System

A first-order, exothermic reaction occurs within a continuous stirred-tank reactor, via jacket cooling. The kinetics are given by:

$$A \rightarrow \text{products} + \text{heat}$$

$$-r_A = k\, C_A \qquad k = Z\, e^{-E/RT_R}$$

The process model and parameters are the same as in the simulation example THERM.

In this example, the reactor is equipped with a feedforward controller that calculates the flowrate F and the jacket temperature T_J required to maintain the reactor temperature constant for variations in inlet feed concentration C_{A0} and temperature T_0.

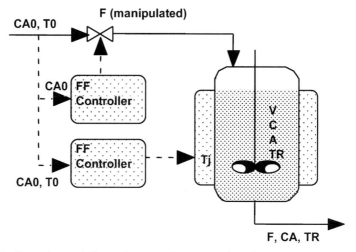

Fig. 1 Feedforward control of a continuous tank reactor with exothermic reaction.

Model

The dynamic process model involves a component balance, energy balance, kinetics and Arrhenius relationship. Hence

$$V T_J = T_R + \frac{C_A k V}{U A}\left(\Delta H + \frac{\rho c_p (T_R - T_0)}{C_{A0} - C_A}\right) = F C_{A0} + F C_A + k V C_A$$

$$V \rho c_p \frac{dT_R}{dt} = F \rho c_p (T_0 - T_R) - k V C_A(-\Delta H) - UA(T_R - T_J)$$

The steady-state forms of these equations allow the manipulation of the feed flowrate F and jacket temperature Tj to maintain the steady state set point under conditions of time-varying C_{A0} and T_0

$$0 = F C_{A0} - F C_A + k V C_A$$

Giving

$$F = \frac{C_A \cdot kV}{C_{A0} - C_A}$$

From this equation the steady state, feed-forward control value of F can be calculated to compensate for variations in C_{A0} with respect to time and based on the setpoint values for k and C_A.

Using the steady state energy balance

$$0 = F\rho c_p (T_0 - T_R) - k V C_A(-\Delta H) - UA(T_R - T_J)$$

C_A may be substituted from the steady state mass balance to give the steady state feed-forward control equation for jacket temperature Tj

$$T_J = T_R + \frac{C_A k V}{UA}\left(\Delta H + \frac{\rho c_p(T_R - T_0)}{C_{A0} - C_A}\right)$$

Note that the values of T_R, C_A, and k are evaluated at the desired setpoint, but that both C_{A0} and T_0 may vary with time as represented here by input sine-wave disturbances.

Programs

The program THERMFF solves the same dynamic process model equations as THERM, where it was shown that all the parameters, including the inlet temperature and concentration will influence the steady state. In the case of multiple steady states the values of the steady state parameters cannot be set, because they are not unique. This example should, therefore, be run under parameter conditions that will guarantee a single steady state for all expected values of the C_{A0} and T_0. These can be selected with the aid of the programs THERMPLOT and THERM.

Nomenclature

Symbols

c_p	Specific heat	kJ/kg K
ρ	Density	kg/m^3
F	Volumetric feed rate	m^3/s
V	Reactor volume	m^3
UA	Heat transfer capacity coefficient	kJ/K s
ΔH	Exothermic heat of reaction	kJ/kmol
T	Feed temperature	K
T_R	Reactor temperature	K
C	Concentration	kmol/m^3
Z	Collision frequency	1/s
E	Activation energy	kJ/kmol
R	Gas constant	kJ/kmol K

Indices

0	Refers to feed condition
A	Refers to reactant
J	Refers to jacket

Exercises

1. Using THERMPLOT, vary F to establish a single steady state at the low temperature, low conversion range. With the same parameters, obtain the steady states values of C_A and T_R using THERM. Enter these values as C_{Aset} and T_{Rset} in THERMFF. Using sliders vary the values of F_{min}, T_{Jmin} and T_{Jmax}.
2. Repeat Exercise 1 for a steady state value of high temperature and high conversion.
3. Devise a feedback control loop to correct for any imperfections in the feedforward controller.
4. Investigate the influence of incorrect parameter values in the feedforward control model. Vary the parameters by 10% effects in the control model only to see the effects of uncertainties. This will involve renaming parameters used in the control algorithm, such that the main process parameters still remain the same.

Results

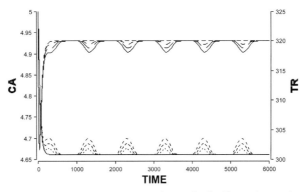

Fig. 2 Reactor response to sine waves in C_{A0} and T_0 under feedforward control, manipulating F and T_J. Tjmin was varied from 180 K (perfect control) to 230 K and 260 K

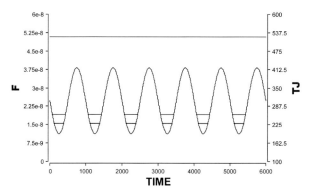

Fig. 3 Manipulated variables showing constant F and T_J variations for the three values of Tjmin as in the runs of Fig. 2.

Reference

Luyben, M. L. and Luyben W. L. (1997) Essentials of Process Control, McGraw-Hill.

5.9
Mass Transfer Process Examples

5.9.1
BATEX – Single Solute Batch Extraction

System

Consider a batch two-phase extraction system, with a single solute transferring from the feed phase into an immiscible solvent phase. The background to the problem is discussed in Section 3.3.1.1

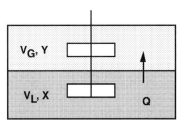

Fig. 1 Batch liquid–liquid extraction.

Solute balances give:

for the feed phase

$$V_L \frac{dX}{dt} = -K_L a(X - X^*)V_L$$

and for the solvent phase

$$V_G \frac{dY}{dt} = +K_L a(X - X^*)V_L$$

where $K_L a$ is the overall mass transfer capacity coefficient (1/s), V_L and V_G are the relative phase volumes (m³) and where for a linear equilibrium relationship:

$$X^* = \frac{Y}{m}$$

The fractional extraction is given by

$$Z = \frac{X_0 - X}{X_0}$$

where X_0 is the initial charge concentration (kg/m³).

Program

The program includes a Pulse function, which has been added to the X phase. This is used to represent the effect of sudden addition of solute to this phase at TIME = Tswitch.

Nomenclature

Symbols

$K_L a$	Mass transfer capacity coefficient	1/h
m	Equilibrium distribution constant	–
V	Volume	m^3
X	Concentration in X-phase	kg/m^3
Y	Concentration in Y-phase	kg/m^3

Indices

*	Refers to equilibrium
L,G	Refers to L and G phases
0	Refers to the feed condition

Exercises

1. Observe the dynamic approach to equilibrium, and show how the driving force for mass transfer changes with time.
2. Change the value of the mass transfer capacity coefficient $K_L a$ and observe how this affects the time to reach equilibrium. Is there a relationship between $K_L a$ and the system time constant τ?
3. Vary the value of the equilibrium constant m and study its effect on system behaviour.
4. Study the effect of varying the solvent to feed charge ratio V_L/V_G.
5. Verify that the total mass of solute is conserved.
6. Modify the program to allow for a non-linear equilibrium relationship.

Results

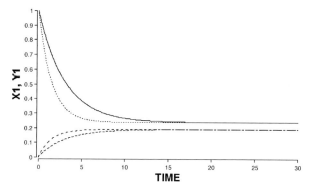

Fig. 2 Dynamic concentration profiles: Effect of $K_L a = 2.5$ and $5.0 \ 1/h$

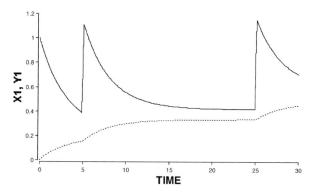

Fig. 3 The X phase is pulsed to suddenly change the concentration.

5.9.2
TWOEX – Two-Solute Batch Extraction with Interacting Equilibria

System

This example considers the interactions involved in multicomponent extraction and takes the particular case of a single batch extractor with two interacting solutes, as shown in Section 3.3.1.2.

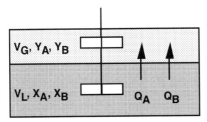

Fig. 1 Batchwise two-solute batch extraction.

Model

Two solutes, A and B, distribute between the feed and solvent phases as concentrations X_A and Y_A, and X_B and Y_B, respectively.

Balances for the two solutes give:

for the feed phase

$$\frac{dX_A}{dt} = -K_{LA}a(X_A - X_A^*)$$

$$\frac{dX_B}{dt} = -K_{LB}a(X_B - X_B^*)$$

and for the solvent phase

$$\frac{dY_A}{dt} = +K_{LA}a(X_A - X_A^*)\frac{V_L}{V_G}$$

$$\frac{dY_B}{dt} = +K_{LB}a(X_B - X_B^*)\frac{V_L}{V_G}$$

The respective equilibrium concentrations, X_A^* and X_B^*, are functions of both solute concentrations, where

$$X_A^* = 1.25Y_A - 0.2Y_B$$

$$X_B^* = Y_B + 0.5Y_A^2$$

Nomenclature

Symbols

V	Phase volume	m³
X	Solute concentration in the feed phase	kg/m³
Y	Solute concentration in the solvent phase	kg/m³
$K_L a$	Mass transfer capacity coefficient	1/h

Indices

A,B   Solutes A and B
L,G   Feed and solvent phases
*   Equilibrium values

Exercises

1. Observe the influence of the mass transfer capacity coefficients $K_{LA}a$ and $K_{LB}a$ on the system dynamics and on the final approach to equilibrium.
2. Confirm that the total balances for both solutes are correct.
3. Overall mass transfer coefficients are only constant when both liquid film coefficients are constant and also when the slope of the equilibrium line is constant. Thus, for a non-linear equilibrium relationship, the overall mass transfer coefficient will vary with concentration. How would you implement this effect into the program?

Results

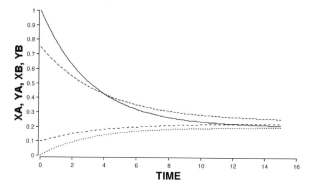

Fig. 2 Time-dependent concentrations for the two-solute extraction.

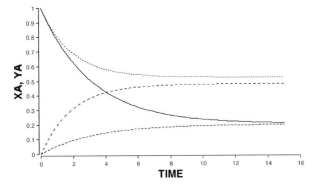

Fig. 3 Time-dependent concentration profiles for solute "A": Effect of phase ratio, with $V_G/V_L = 1.0$ and 4.0.

5.9.3
EQEX – Simple Equilibrium Stage Extractor

System

This is represented by a single well mixed, constant volume, continuous flow extraction stage.

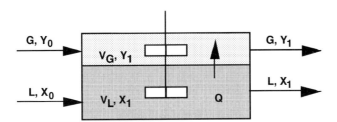

Fig. 1 Continuous equilibrium stage extraction.

Solute balances for each phase give

$$V_L \frac{dX_1}{dt} = LX_0 - LX_1 - K_L a(X_1 - X_1^*)V_L$$

and

$$V_G \frac{dY_1}{dt} = GY_0 - GY_1 + K_L a(X_1 - X_1^*)V_L$$

and where for a linear equilibrium relationship

$$X_1^* = Y_1/m$$

As shown in Section 3.3.1.3, near equilibrium conditions can be obtained by choosing an arbitrarily high value for the mass transfer capacity constant K_La.

Nomenclature

Symbols

G	Flow rate	m³/h
K_La	Mass transfer capacity constant	1/h
L	Flow rate	m³/h
m	Equilibrium constant	–
V	Volume	m³
X	Concentration in X-phase	kg/m³
Y	Concentration in Y-phase	kg/m³

Indices

*	Refers to equilibrium
0	Refers to inlet
1	Refers to outlet
L,G	Refer to L and G phases

Exercises

1. Vary the system model parameters and see how these affect the time to steady state
2. Compare the residence time for the L phase, V_L/L, with the time constant for mass transfer, $1/K_La$. Maintain a constant ratio of these two time constants but vary the individual parameters. How is the approach to steady state influenced by these changes?
3. What minimum value of K_La is required to obtain 99% equilibrium conditions? (Hint – plot X_1 versus X_1^*).

Results

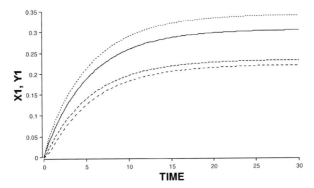

Fig. 2 Approach to steady state [$K_La=1$ and 5 $1/h$]. Note that equilibrium steady state is apparently not attained.

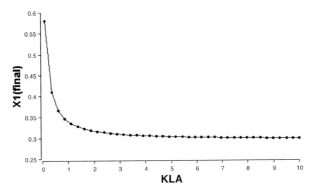

Fig. 3 A parametric run showing that equilibrium conditions are not obtained at low K_La.

5.9.4
EQMULTI – Continuous Equilibrium Multistage Extraction

System

A countercurrent multistage extraction system is to be modelled as a cascade of equilibrium stages.

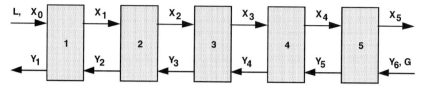

Fig. 1 Countercurrent multistage equilibrium extraction unit.

Model

The solute balances for any stage n are given by

$$V_L \frac{dX_n}{dt} = L(X_{n-1} - X_n) - Q_n$$

$$V_G \frac{dY_n}{dt} = G(Y_{n+1} - Y_n) + Q_n$$

where

$$Q_n = K_L a(X_n - X_n^*)V_L$$

and for a linear equilibrium

$$X_n^* = Y_n/m$$

As in example EQEX, the mass transfer capacity coefficient $K_L a$ may be set high to obtain near equilibrium stage conditions.

Program

Note that the program is written such that the number of stages in the cascade, N_{stage}, can be varied as an additional parameter in the simulation. The stage numbers can be plotted versus the final values by choosing [i] as the X-axis. In this way the steady state concentration profiles can be graphed. A part of the program is shown below.

```
{Example EQMULTI}
INIT X[1..Nstage]=0
INIT Y[1..Nstage]=0

{Mass transfer rates kg/h}
Q[1..Nstage]=KLa*(X[i]-Y[i]/M)*VL
```

{Component balances for stage 1}
d/dt(X[1])=(L*X0-L*X[1]-Q[1])/VL
d/dt(Y[1])=(G*(Y[2]-Y[1])+Q[1])/VG

{Extract outlet concentration}
Y1= Y[1]

{Component balances for stages 2 to Nstage-1}
d/dt(X[2..Nstage-1])=(L*(X[i-1]-X[i])-Q[i])/VL
d/dt(Y[2..Nstage-1])=(G*(Y[i+1]-Y[i])+Q[i])/VG

{Component balances for stage Nstage}
d/dt(X[Nstage])=(L*(X[i-1]-X[i])-Q[i])/VL
d/dt(Y[Nstage])=(G*YNstage-G*Y[i]+Q[i])/VG

{Raffinate outlet concentration}
YN= Y[Nstage]

Nomenclature

Symbols

G	Flow rate	m^3/h
K_La	Mass transfer capacity constant	$1/h$
L	Flow rate	m^3/h
m	Equilibrium constant	–
Q	Solute transfer rate	kg/h
V	Volume	m^3
X_1	Concentration in X-phase	kg/m^3
Y_1	Concentration in Y-phase	kg/m^3

Indices

*	Refers to equilibrium
0	Refers to inlet
L,G	Refer to L and G phases
n	Refers to stage

Exercises

1. Run the model for the 5-stage model and ascertain if equilibrium is attained for each stage. Reduce K_La by a factor of 10 and check again. Reduce L and G by the same factor. Explain the effects in terms of the process time constants.
2. Confirm that the overall solute balance is satisfied at steady state.
3. Carry out further simulations on the model and investigate the number of stages required for say 99% separation recovery as a function of the other system parameters.

4. Modify the above program to model the case of a discontinuous five-stage extraction cascade, in which a continuous flow of aqueous phase L is passed through the cascade. The solvent is continuously recycled through the cascade and also through a solvent holding tank, of volume V_S, as shown below. This problem of a stagewise discontinuous extraction process has been solved analytically by Lelli (1966).
5. Try to predict the form of the L and G phase responses before carrying out simulations for the above problem.
6. Vary L and G and observe the influence on the dynamics. Explain the results in terms of the system time constants.

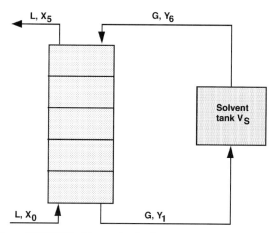

Fig. 2 Modification of EQMULTI with recycle of solvent.

Results

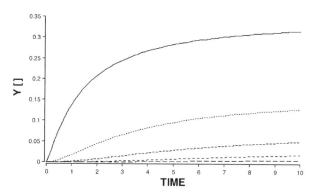

Fig. 3 Approach to near steady state for the L phase solute concentrations.

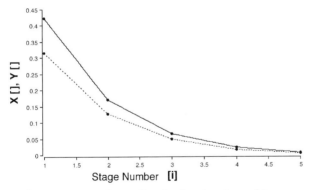

Fig. 4 Near steady state concentration profiles showing almost complete extraction to have been obtained.

Reference

Lelli, U. (1966) Annali di Chimica 56, 113.

5.9.5
EQBACK – Multistage Extractor with Backmixing

System

The full model is described in Section 3.3.1.5 and consists of a countercurrent stagewise extraction cascade with backmixing in both phases.

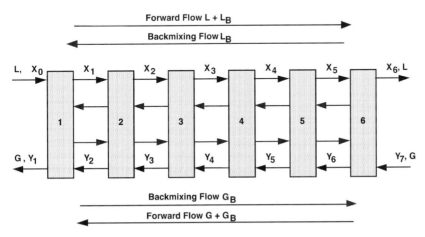

Fig. 1 Multistage extractor with backmixing.

Model

Allowing for the backmixing flow contributions, the continuity equations for any stage n are

$$V_L \frac{dX_n}{dt} = (L + L_B)(X_{n-1} - X_n) + L_B(X_{n+1} - X_n) - Q_n$$

$$V_G \frac{dY_n}{dt} = (G + G_B)(Y_{n+1} - Y_n) + G_B(Y_{n-1} - Y_n) + Q_n$$

where $\quad Q_n = K_L a(X_n - X_n^*)V_L \quad$ and $\quad X_n^* = Y_n/m$

The model equations are also often written in terms of backmixing factors, where

$$a_L = L_B/L \quad \text{and} \quad a_G = G_B/G$$

Special balances also apply to the end sections owing to the absence of backflow, exterior to the column.

Nomenclature

Symbols

a	Backmixing factor	–
G	Flow rate	m³/h
$K_L a$	Mass transfer capacity coefficient	1/h
L	Flow rate	m³/h
m	Equilibrium constant	–
Q	Transfer rate	kg/h
V	Volume	m³
X	Concentration in X-phase	kg/m³
Y	Concentration in Y-phase	kg/m³

Indices

*	Refers to equilibrium
0	Refers to inlet
L,G	Refer to L and G phases
B	Refers to backmixing
n	Refers to stage number

Program

Note that the program is written such that the number of stages in the cascade may be used an additional parameter in the simulation.

Exercises

1. Vary the backmixing flows for each phase and notice the effect of changes in conditions between plug flow and full backmixing on the performance of the extractor.
2. For which conditions of backmixing does the extraction process operate best?
3. Check the steady-state overall solute balances.

Results

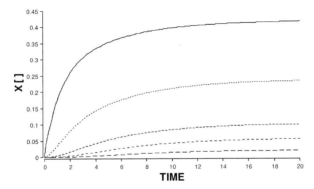

Fig. 2 Time-dependent concentrations for the feed phase in stages 1 to 5 [$a_L=0.15$ and $a_G=0.15$].

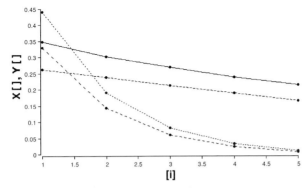

Fig. 3 High backmixing [$a_L=5$, $a_G=5$] flattens the steady state concentration profiles and therefore reduces extraction efficiency, as compared to zero backmixing conditions [$a_L=0$, $a_G=0$], as seen in this axial profile plot.

5.9.6
EXTRACTCON – Extraction Cascade with Backmixing and Control

System

In this example of a five stage extraction column with backmixing, proportional plus integral control of the exit raffinate concentration is to be achieved by regulating solvent flowrate.

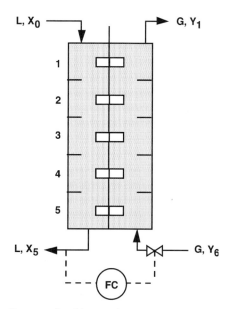

Fig. 1 Five stage extraction cascade with control.

Model

The solute balance equations for each phase are

$$V_L \frac{dX_n}{dt} = (L + L_B)(X_{n-1} - X_n) + L_B(X_{n+1} - X_n) - Q_n$$

$$V_G \frac{dY_n}{dt} = (G + G_B)(Y_{n+1} - Y_n) + G_B(Y_{n-1} - Y_n) + Q_n$$

where

$$Q_n = K_L a(X_n - X_n^*)V_L \quad \text{and} \quad X_n^* = Y_n/m$$

The PI controller equation is

$$G = G_0 + K_C \varepsilon + \frac{K_C}{\tau_I} \int \varepsilon \, dt$$

where

$$\varepsilon = X_n - X_{n \text{ set}}$$

and $X_{n \text{ set}}$ is the controller set point.

Program

The simulation starts from initially zero concentration conditions along the cascade. Control is implemented from time t=0, in order to obtain an outlet raffinate concentration equal to the controller set point. The program is essentially the same as that used in EQBACK, but with added controller equations.

Nomenclature

Symbols

G	Flow rate	m³/h
G_0	Base solvent flow rate	m³/h
K_C	Proportional control constant	(m³/hr)/(kg/m³)
$K_L a$	Mass transfer capacity coefficient	1/h
L	Flow rate	m³/h
m	Equilibrium constant	–
Q	Transfer rate	kg/h
V	Phase volume	m³
X	Concentration in X-phase	kg/m³
X_{SET}	Raffinate concentration setpoint	kg/m³
Y	Concentration in Y-phase	kg/m³
ε	Controller error	kg/m³
τ_I	Integral control time constant	h

Indices

*	Refers to equilibrium
0	Refers to inlet
L,G	Refer to L and G phases
B	Refers to backmixing
n	Refers to stage number

Exercises

1. Study the response of the system with only proportional control (τ_I very large) and determine the resultant steady state offset. Add an increasing degree of integral control and study its effect.
2. Try tuning the controller to obtain the best response to a change in feed concentration and feed rate.
3. Set the initial concentration profiles in the two phases equal to the steady state open loop response ($K_C = 0$), and study the response of the system to another set point change or flow rate change.

Results

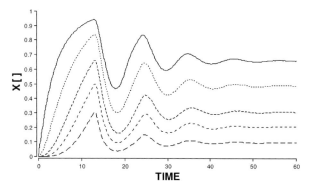

Fig. 2 Concentration variations in the feed phase for stages 1 to 5 and for start up from initially zero concentration conditions with full control implementation ($K_C = 5$).

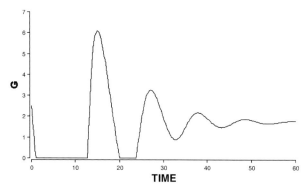

Fig. 3 The corresponding solvent flow variations.

5.9.7
HOLDUP – Transient Holdup Profiles in an Agitated Extractor

System

The fractional holdup of the dispersed phase in agitated extraction columns varies as a function of flowrate. Under some circumstances it may be important to model the corresponding hydrodynamic effect. The system is represented below as a column containing seven agitated compartments or stages.

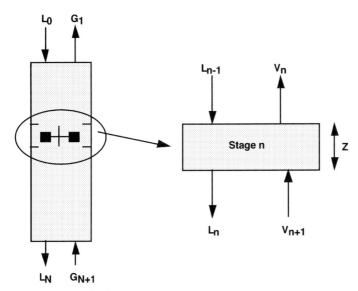

Fig. 1 Seven-stage agitated column contactor.

Model

The variation in the holdup of solvent dispersed phase is given by

$$V_n \frac{dh_n}{dt} = G_{n+1} - G_n$$

Since the two liquid phases are also incompressible, then

$$L_{n-1} + G_{n+1} = L_n + G_n$$

and for the overall column

$$L_0 + G_{N+1} = L_N + G_1$$

In this example, it is assumed that the holdup relationship is given as a direct empirical function of the continuous and dispersed phase flow rates. This avoids any difficulties due to a possible algebraic implicit loop in the solution (see Section 3.3.1.11).

For a given range of feed flow rate, L, it is assumed that the fractional holdup, h, and the solvent phase flow rate, G, can be correlated in the form

$$G = h^{a/b}$$

where "a" and "b" are empirical constants and the flow rates are expressed as superficial phase velocities.

Program

The program is written in array form, with N_{cmpts} the parameter representing the desired number of compartments in the column. The simulation starts with a uniform holdup distribution throughout the column and with the column operating at steady-state. At time, T_{switch}, the organic solvent flow is suddenly increased, as effected by the use of a MADONNA IF-THEN-ELSE command.

Nomenclature

Symbols

a,b	Holdup relation parameters	
G	Organic, dispersed phase flow velocity	$cm^3/cm^2\ h$
h	Fractional holdup of dispersed phase	–
L	Aqueous, continuous phase flow velocity	$cm^3/cm^2\ h$
N_{cmpts}	Compartment number	–
t	Time	h
T_{switch}	Switching time	h
V	Volume	cm^3

Index

n	Compartment number

Exercises

1. Vary the holdup parameters a and b to study their effect on the resulting holdup variations.
2. The holdup changes are the result of interdroplet interactions, which may also have their own dynamics. Assume that this can be represented by a first order lag equation, and show how this effect may be implemented into the simulation.
3. An important limit to extraction column behaviour is of course that of column flooding. How could this effect also be incorporated into the simulation?

Results

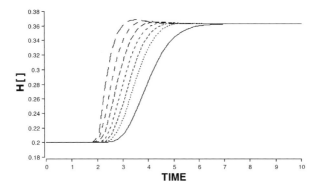

Fig. 2 Holdup response to an increase in organic (dispersed) phase flow.

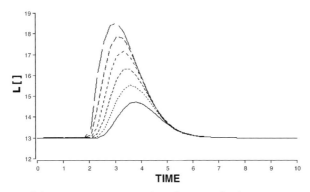

Fig. 3 Response of the continuous aqueous phase flow rates for the Fig. 2 run.

5.9.8
KLADYN, KLAFIT and ELECTFIT – Dynamic Oxygen Electrode Method for K_La

System

A simple and effective means of measuring the oxygen transfer coefficient (K_La) in an air-water tank contacting system involves first degassing the batch water phase with nitrogen (Ruchti et al., 1981). Then the air flow is started and the increasing dissolved oxygen concentration is measured by means of an oxygen electrode.

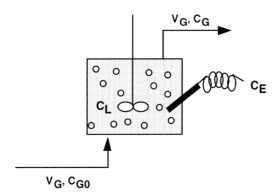

Fig. 1 Aerated tank with oxygen electrode.

As shown below, the influence of three quite distinct dynamic processes plays a role in the overall measured oxygen concentration response curve. These are the processes of the dilution of nitrogen gas with air, the gas-liquid transfer and the electrode response characteristic, respectively. Whether all of these processes need to be taken into account when calculating K_La can be determined by examining the mathematical model and carrying out simulations.

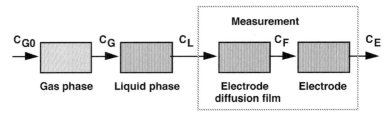

Fig. 2 Representation of the process dynamics.

Model

The model relationships include the mass balance equations for the gas and liquid phases and equations representing the measurement dynamics.

Oxygen Balances

The oxygen balance for the well-mixed flowing gas phase is described by

$$V_G \frac{dC_G}{dt} = G(C_{G0} - C_G) - K_L a (C_L^* - C_L) V_L$$

where $V_G/V = \tau_G$ and $K_L a$ is based on the liquid volume.

The oxygen balance for the well-mixed batch liquid phase is

$$V_L \frac{dC_L}{dt} = K_L a (C_L^* - C_L) V_L$$

The equilibrium oxygen concentration C_L^* is given by the combination of Henry's law and the Ideal Gas Law equation where

$$C_L^* = \frac{RT}{H} C_G$$

and C_L^* is the oxygen concentration in equilibrium with the gas concentration C_G. The above equations can be solved in this form as in simulation example KLAFIT. It is also useful to solve the equations in dimensionless form.

Oxygen Electrode Dynamic Model

The response of the usual membrane-covered electrodes can be described by an empirical second-order lag equation. This consists of two first-order lag equations to represent the diffusion of oxygen through the liquid film on the surface of the electrode membrane and secondly the response of the membrane and electrolyte:

$$\frac{dC_F}{dt} = \frac{C_L - C_F}{\tau_F}$$

and

$$\frac{dC_E}{dt} = \frac{C_F - C_E}{\tau_E}$$

τ_F and τ_G are the time constants for the film and electrode lags, respectively. In non-viscous water phases τ_F can be expected to be very small and the first lag equation can, in fact, be ignored.

Dimensionless Model Equations

Defining dimensionless variables as

$$C'_G = \frac{C_G}{D_{G0}} \qquad C'_L = \frac{C_L}{C_{G0}(RT/H)} \qquad t' = \frac{t}{\tau_G}$$

the component balance equations then become

$$\frac{dC'_G}{dt'} = (1 - C'_G) - K_L a \tau_G \frac{V_L}{V_G} \frac{RT}{H}(C'_G - C'_L)$$

and

$$\frac{dC'_L}{dt'} = K_L a \tau_G (C'_G - C'_L)$$

Initial conditions corresponding to the experimental method are

$$t' = 0; \quad C'_L = C'_G = 0$$

In dimensionless form the electrode dynamic equations are

$$\frac{dC'_F}{dt'} = \frac{C'_L - C'_F}{\tau_F / \tau_G}$$

and

$$\frac{dC'_E}{dt'} = \frac{C'_F - C'_E}{\tau_E / \tau_G}$$

where C'_F is the dimensionless diffusion film concentration.

$$\frac{C_F}{C_{G0}(RT/H)}$$

and C'_E is the dimensionless electrode output

$$\frac{C_E}{C_{G0}(RT/H)}$$

As shown by Dang et al. (1977), solving the model equations by Laplace transformation gives

$$a = \frac{1}{K_L a} + \left(\frac{RT}{H}\frac{V_L}{V_G} + 1\right)\tau_G + \tau_E + \tau_F$$

where a is the area above the C'_E versus t response curve, as shown in Fig. 3.

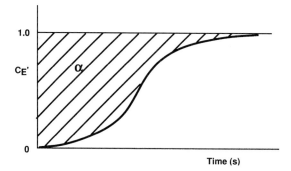

Fig. 3 Determination of the area a above the C'_E versus time response curve.

Program

The program KLADYN can be used to investigate the influence of the various experimental parameters on the method, and is formulated in dimensionless form. The same model, but with dimensions, is used in program KLAFIT and is particularly useful for determining $K_L a$ in fitting experimental data of C_E versus time. A set of experimental data in the text file KLADATA can be used to experiment with the data fitting features of MADONNA.

The program ELECTFIT is used to determine the electrode time constant in the first-order lag model. The experiment involves bringing CE to zero by first purging oxygen from the water with nitrogen and then subjecting the electrode to a step change by plunging it into fully aerated water. The value of the electrode time constant, T_E can be obtained by fitting the model to the set of experimental data in the file ELECTDATA. The value found in this experiment can then be used as a constant in KLAFIT.

Nomenclature

Symbols

C	Oxygen concentration	g/m^3
G	Gas flow rate	m^3/s
H	Henry coefficient	Pa m^3/mol
$K_L a$	Oxygen transfer coefficient	1/s
RT/H	(Gas constant)(Abs. temp.)/Henry coeff.	–
t	Time	s
V	Reactor volume	m^3
a	Area above C'_E-time (s) curve	s
τ	Time constant	s

Indices

E	Refers to electrode
F	Refers to film
G	Refers to gas phase
L	Refers to liquid
′	Prime denotes dimensionless variables

Exercises

4. Compare the values of the time constants (τ_G, $1/K_L a$, τ_F and τ_E) and the dimensionless groups. For the parameter values used, can you estimate whether all processes will be influential? Check your reasoning by changing the parameters in the simulations and observing the plots of C_G, C_L, C_F and C_E versus time.
5. Verify by simulation that $K_L a$ can be estimated from the value of a, as determined by the area under the curve of Fig. 3.
6. The textbook method for calculating $K_L a$ involves plotting $\ln(1/(1-C'_E))$ versus t to obtain $K_L a$ from the slope. What is the reasoning behind this? For what parameter values would it yield accurate $K_L a$ values? Alter the program to study the characteristics of this plot, investigating its validity for a range of parameters.
7. Vary the time constants for transfer ($1/K_L a$) and for measurement ($\tau_E + \tau_F$) to demonstrate when the measurement response is important.
8. Using the simplification of Exercise 5, estimate the time constants and show how their sum can be estimated from the $C'_G = C'_L = C'_E = 0.63$ position on the curves?
9. Determine T_E using the datafile ELECTDATA together with the program ELECTFIT. Use this value in KLAFIT with the datafile KLADATA to determine KLA by fitting the model parameter to the data.
10. Instead of using the previously found value of T_E, determine both T_E and $K_L A$ by fitting to the KLADATA file. Compare the fit of the curve and the values of T_E and $K_L A$ found in Exercise 6.

Results

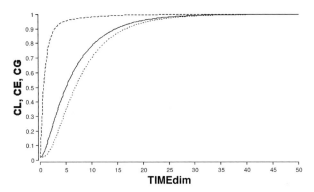

Fig. 4 Response of CG, CL and CE versus TIMEdim from KLADYN.

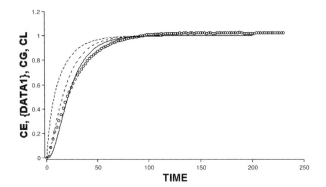

Fig. 5 A fit of experimental data (open circles) as dimensionless CE versus time (s) to determine KLA using KLAFIT, which gives 0.136 1/s. Shown also are CG and CL.

References

Dang, N. D. P., Karrer, D. A. and Dunn, I. J. (1977) Oxygen Transfer Coefficients by Dynamic Model Moment Analysis, Biotechnol. Bioeng. 19, 853.

Ruchti, G., Dunn, I. J. and Bourne J. R. (1981). Comparison of Dynamic Oxygen Electrode Methods for the Measurement of $K_L a$, Biotechnol. Bioeng. 13, 277.

5.9.9
AXDISP – Differential Extraction Column with Axial Dispersion

System

A consideration of axial dispersion is essential in any realistic description of extraction column behaviour. Here a dynamic method of solution is demonstrated, based on a finite differencing of the column height coordinate. Figure 1 below shows the extraction column approximated by N finite-difference elements.

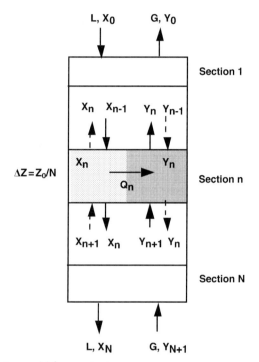

Fig. 1 Axial dispersion model for an extractor.

Model

The detailed derivation of the model equations is described in Sections 4.4.2 and 4.4.3, but are reformulated here in simple concentration terms.

The component balance equations are then:

for the L feed phase

$$\frac{dX_n}{dt} = -E_{1X}(X_{n+1} - X_{n1}) + E_{2X}(X_{n1} - 2X_n + X_{n+1}) - \frac{Q_n}{H_L}$$

5.9 Mass Transfer Process Examples

and for the G phase

$$\frac{dY_n}{dt} = E_{1Y}(Y_{n+1} - Y_{n-1}) + E_{2Y}(Y_{n-1} - 2Y_n + Y_{n+1}) + \frac{Q_n}{H_G}$$

where the factors E_{1X} and E_{2Y} represent the inverse residence times of the phases, and E_{2X} and E_{2Y} represent the inverse diffusion times of the phases, in segment n.

For the L phase

$$E_{1X} = \frac{L}{A_L \Delta Z} = \frac{v_L}{\Delta Z H_L}$$

$$E_{2X} = \frac{D_L A}{\Delta Z^2 A_L} = \frac{D_L}{\Delta Z^2 H_L}$$

where H_L is the fractional holdup of the L phase. Similar equations apply to the G phase.

The column end sections require special treatment since diffusive flux is obviously absent exterior to the column. Thus, the column outlet concentrations X_{N+1} and Y_0 are based on an extrapolation of the concentration profiles over the additional half outlet section of the column, as again described in Sections 4.4.2 and 4.4.3.

Thus for the L and G phase flows leaving end section 8

$$\frac{dX_N}{dt} = E_{1X}(X_{N-1} + X_N - 2X_{N+1}) + E_{2X}(X_{N-1} - X_N) - \frac{Q_N}{H_L}$$

$$\frac{dY_N}{dt} = E_{1Y}(2Y_{N+1} - (Y_{N-1} + Y_N)) + E_{2Y}(Y_{N-1} - Y_N) + \frac{Q_N}{H_G}$$

The balances for the other end section, n=1, take a similar form.

Program

The simulation starts with the initial axial concentration profiles set to zero. Using arrays, the number of sections is set by the parameter N_{slabs}, allowing the axial concentration profile for each phase along the contactor to be demonstrated.

Nomenclature

Symbols

D	Eddy dispersion coefficient	cm²/s
E_1	Inverse residence time	1/s
E_2	Inverse dispersion time	1/s

H	Fractional phase holdup	–
L and G	Superficial phase velocity	cm/s
Q	Mass transfer rate	g/cm^3 s
X and Y	Solute concentration	g/cm^3
Z	Column height	cm

Indices

*	Equilibrium value
n	Stage n
L and G	Liquid phase

Exercises

1. The steady-state outlet values should be checked for the overall balance.
2. The extent of extraction can be investigated as a function of solvent flow rate.
3. The mass transfer coefficient can be increased to obtain an approach to equilibrium conditions.
4. The dispersion coefficients can be increased to observe an approach to ideal mixing. Setting the coefficients to zero should give the best performance.
5. The dynamic responses for changes in inlet conditions can be investigated.

Results

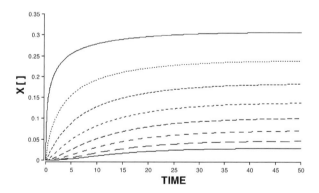

Fig. 2 Time-dependent concentration variations for the L-phase [$D_L = D_G = 100$].

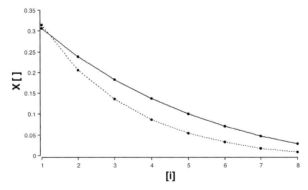

Fig. 3 Steady state concentration profiles for the L-phase [$D_L = D_G = 100$] and [$D_L = D_G = 20$].

5.9.10
AMMONAB – Steady-State Design of a Gas Absorption Column with Heat Effects

System

Ammonia is recovered from an air–ammonia gas mixture by absorption into water, using a countercurrent packed column. The absorption is accompanied by the evolution of heat which causes a rise in the water temperature and hence a change in the equilibrium. The problem and data values are taken from Backhurst and Harker (1990).

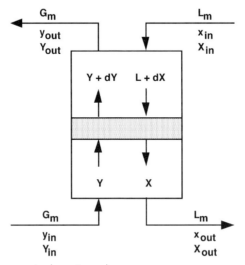

Fig. 1 Steady-state ammonia absorption column.

Model

The gas is at high concentration and therefore the column component balance equations are based on mole ratio concentration units.

The balance equations for countercurrent flow follow Section 4.4.1.

For the gas
$$\frac{dY}{dz} = \frac{K_G a (y - y_e)}{L_m}$$

For the liquid
$$\frac{dX}{dz} = \frac{K_G a (y - y_e)}{G_m}$$

The liquid phase energy balance is given by
$$\frac{dT}{dz} = \frac{K_G a (y - y_e) \Delta H}{L c_p}$$

The temperature in the gas phase is assumed constant with no heat losses and zero gas-liquid heat interchange.

The equilibrium relationship is represented by
$$y_e = \frac{Kx}{P_T}$$

where the equilibrium constant, K, is a function of temperature.

The inlet gas flow is 1.25 m³/m² s measured at NTP containing 5% by volume NH_3. Hence
$$G_m = (1.25)(0.95)/22.4 = 0.053 \, \text{k mol/m}^2 \, \text{s}$$

For Y_{in} equal 0.05
$$Y_{in} = 0.05/0.95 = 0.0526 \, \text{kmol} \, NH_3/\text{kmol air}$$

For 95% ammonia recovery
$$Y_{out} = \frac{0.0526 * 0.05}{1 + 0.0526 * 0.05} = 0.00263 \, \text{kmol} \, NH_3/\text{kmol air}$$

The inlet water flow is 1.95 kg/m² s, hence
$$L_m = 195/18 = 0.108 \, \text{kmol/m}^2 \, \text{s}$$

An overall component balance for the column gives

$$G_m(Y_{in} - Y_{out}) = L_m(X_{out} - X_{in})$$

Calculating the concentration in the outlet liquid

$$X_{out} = \frac{0.053 * (0.0526 - 0.00263)}{0.108} = 0.0245 \text{ kmol NH}_3/\text{kmol water}$$

Backhurst and Harker present the equilibrium data as straight line relationships for the temperatures 293 K, 298 K and 303 K. This data was curve fitted to the form

$$y_e = \frac{Kx}{P_T}$$

where $K = a + bT + cT^2$, and the constants are given by $a = 0.0059$, $b = 0.1227$ and $c = 3.01$.

Program

The program starts from the known concentration and temperature conditions at the top of the column and integrates down the column until the condition Y_{in} (calculated) is equal to the known inlet value Y_{in}. A variable Y_{height} is defined in the program to locate the Y_{IN} position. Note the use of HEIGHT, instead of TIME, as the independent variable in this steady state example.

Nomenclature

Symbols

C_p	Specific heat	kJ/kmol °C
G_m	Molar flow of inert air	kmol/m² s
K	Equilibrium constant	kN/m²
K_Ga	Overall mass transfer capacity coefficient based on the gas phase Δy	kmol/m³ s
L_m	Molar flow of solute-free water	kmol/m² s
P_T	Pressure	kN/m²
ρ	Density	kg/m³
T	Liquid temperature	°C
X	Mole ratio concentration in the liquid phase	–
x	Mole fraction concentration in the liquid phase	–
Y	Mole ratio concentration in the gas phase	–
y	Mole fraction concentration in the gas phase	–
z	Height of packing	m
ΔH	Exothermic heat of adsorption	kJ/kmol

Indices

in, out   Refer to inlet and outlet conditions
e   Refers to equilibrium

Exercises

1. Study the effects of varying liquid rate, inlet water temperature, inlet gas concentration and pressure on the absorption.
3. Modify the program so that you are able to graph the driving force for mass transfer $(y - y_e)$ and the resulting "pseudo" equilibrium curve, i.e., (y_e versus x), allowing for the changing temperature. Note however that the equilibrium relationship is only valid for the range 293 to 303 K.
3. Develop a dynamic model for the column and compare its flexibility with this steady state model.

Results

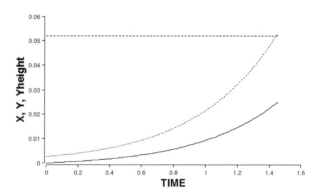

Fig. 2 Gas and liquid concentrations where the required column height is determined as 1.45 m.

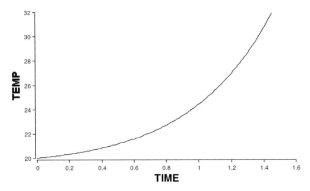

Fig. 3 The corresponding variation in the liquid phase temperature for the column of column height of 1.45 m, as determined in Fig. 2.

Reference

Backhurst, J.R. and Harker, J.H. (1990) Chemical Engineering, Vol. 5, Pergamon.

5.9.11
MEMSEP – Gas Separation by Membrane Permeation

System

An internally staged gas-permeation module is used for the oxygen enrichment of air, using the flow arrangement shown in Fig. 1. Enrichment depends on differing membrane permeabilities for the oxygen and nitrogen to be separated. The permeation rates are proportional to the differences in component partial pressures.

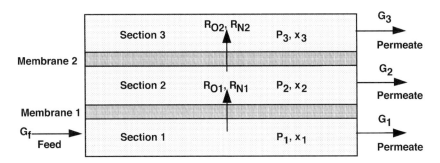

Fig. 1 Configuration of the gas-permeation module.

Model

The module system consists of three well mixed sections, separated by two membranes. Overall steady-state balances can be used to check the computer solution and are given by

Total mass:
$$N_f = N_1 + N_2 + N_3$$

Oxygen:
$$N_f x_{OF} = N_1 x_{O1} + N_2 x_{O2} + N_3 x_{O3}$$

Nitrogen:
$$N_f(1 - x_{OF}) = N_1(1 - x_{O1}) + N_2(1 - x_{O2}) + N_3(1 - x_{O3})$$

Steady-state total mass balances can also be written for Sections 1, 2 and 3, based on the flow rates in each section and the permeation rates between sections. Converting from mol/s to m³/s using the Ideal Gas Law gives

Section 1
$$G_f = G_1 + (R_{O1} + R_{N1})\frac{RT}{P_1}$$

Section 2
$$G_2 = (R_{O2} + R_{N2} - R_{O1} - R_{N1})\frac{RT}{P_2}$$

Section 3
$$G_3 = (R_{O2} + R_{N2})\frac{RT}{P_2}$$

The dynamic component mass balances in terms of concentrations are as follows:

Section 1

Oxygen
$$\frac{dC_{O1}}{dt} = \frac{G_f C_{Of} - G_1 C_{O1} - R_{O1}}{V_1}$$

Nitrogen
$$\frac{dC_{N1}}{dt} = \frac{G_f C_{Nf} - G_1 C_{N1} - R_{N1}}{V_1}$$

Section 2

Oxygen
$$\frac{dC_{O2}}{dt} = \frac{R_{O1} - G_2 C_{O2} - R_{O2}}{V_2}$$

Nitrogen
$$\frac{dC_{N2}}{dt} = \frac{R_{N1} - G_2 C_{N2} - R_{N2}}{V_2}$$

Section 3

Oxygen
$$\frac{dC_{O3}}{dt} = \frac{R_{O2} - G_3 C_{O3}}{V_3}$$

Nitrogen
$$\frac{dC_{N3}}{dt} = \frac{R_{N2} - G_3 C_{N3}}{V_3}$$

The permeation rate is proportional to the partial pressure difference, the membrane permeability, its area and thickness:

Section 1 to Section 2:

For O_2
$$R_{O1} = \frac{k_{pO}}{d}(P_1 x_{O1} - P_2 x_{O2}) A_1$$

For N_2
$$R_{N1} = \frac{k_{pN}}{d}(P_1(1 - x_{O1}) - P_2(1 - x_{O2})) A_1$$

Section 2 to Section 3:

For O_2
$$R_{O2} = \frac{k_{pO}}{d}(P_2 x_{O2} - P_3 x_{O3}) A_2$$

For N_2
$$R_{N2} = \frac{k_{pN}}{d}(P_2(1 - x_{O2}) - P_3(1 - x_{O3})) A_2$$

Program

The program employs an IF THEN ELSE statement to switch the feed rates. An alternative with the Windows version of MADONNA would be to use the CONTINUE feature. For this, the values of N_f would be changed after the STOPTIME was reached and the run continued with a new value of Nf and STOPTIME.

Nomenclature

Symbols

A	Membrane area	m^2
C	Concentration	mol/m^3
d	Membrane thickness	m
G	Flow rate	m^3/s
K_C	Cost constant	$1/Pa^2$
k_p	Permeability coefficient	mol/m s Pa
N	Molar flow rate	mol/s
P	Pressure	Pa
R	Gas constant	m^3 Pa/mol K
R_{in}	Permeation rate	mol/s
V	Section volume	m^3
x	Mole fraction of oxygen	–

Indices

1, 2, 3	Sections 1, 2, 3 and membrane 1, 2
O	Oxygen
N	Nitrogen
f	Feed

Exercises

1. Study the influence of pressures P_1 and P_3 on the separation performance.
2. Introduce new membranes with improved permeabilities to give better separation and test them by simulation.
3. Define a cost function assuming that the cost increases with P_1^2 and G_f and that the enrichment of stream 3 gives profits according to

$$COST = P_1^2 G_f - K_C x_{O3} G_3$$

Use a reasonable value for K_C and then try to find the optimum by varying the pressure in the three stages.

Results

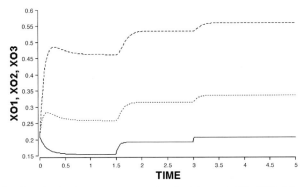

Fig. 2 Mole fractions in the three stages at varying feed flow rate (G_f=0.05, 0.2 and 2 m³/s).

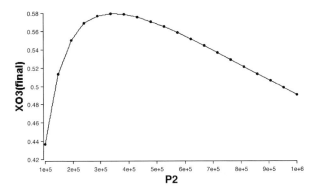

Fig. 3 Influence of P_2 on oxygen mole fractions in the third stage, N_f=50mol/s (range of P_2=0.1 to 1.0 10^6 Pa).

Reference

Li, K., Acharya, R. and Hughes, R. (1991) Simulation of Gas Separation in an Internally Staged Permeator, Trans. IChemE. 69, Part A, 35–42.

5.9.12
FILTWASH – Filter Washing

System

The washing of filter cake is carried out to remove liquid impurities from the valuable solid filter cake or to increase the recovery of valuable filtrate. Wakeman (1990) has shown that the axial dispersion flow model, as developed in

Section 4.3.6, provides a fundamental description of cake washing and can take into account the influence of many phenomena. These include non-uniformities in the liquid flow pattern, non-uniform porosity distributions, the initial spread of washing liquid onto the topmost surface of the filter cake and the desorption of solute from the solid surfaces.

Model

As shown by Wakeman, the solute material balance for the flowing liquid phase, allowing for axial dispersion and desorption of solute is given by the following defining partial differential equation

$$\frac{\partial c}{\partial t} + \frac{(1-\varepsilon)}{\varepsilon}\frac{\partial \eta}{\partial t} = D_L \frac{\partial^2 c}{\partial x^2} - u \frac{\partial c}{\partial x}$$

The concentration of the solute adsorbed on to the particles is related to the concentration of the solute in the liquid by an equilibrium relationship of the form

$$\eta = K_0 + Kc^{1/N}$$

Expressing the model equations in dimensionless form gives

$$\frac{\partial c'}{\partial W}\left[1 + \frac{Kc'^{(1-N)/N}}{N}\left(\frac{1-\varepsilon}{\varepsilon}\right)c'^{(1-N)/N}\right] = P\frac{\partial^2 c'}{\partial x'^2} - \frac{\partial c'}{\partial x'}$$

where

$$c' = \frac{c}{c_0} \quad \text{and} \quad x' = \frac{x}{L}$$

W is the wash ratio given by

$$W = \frac{u_{x=0}t}{L}$$

and P is the Peclet number

$$P = \frac{D_L}{u_{x=0}L}$$

For solution by digital simulation the depth of the filter cake is finite differenced, each element having a dimensionless thickness Δx. For any element n the resulting difference differential equation is given by

$$\frac{dc_n}{dW} = \frac{1}{Z}\left[P\frac{c_{n-1} - 2c_n + c_{n+1}}{\Delta x^2} - \frac{c_{n+1} - c_{n-1}}{2\Delta x}\right]$$

5.9 Mass Transfer Process Examples | 481

where the term Z is given by

$$Z = 1 + \frac{Kc_0^{(1-N)/N}}{N} \frac{1-\varepsilon}{\varepsilon} c_i^{(1-N)/N}$$

In these equations the prime designation for dimensionless concentration, c', has been dropped.

Program

Program FILTWASH models the dimensionless filtration wash curves for the above case of a filter cake with constant porosity, axial dispersion in the liquid flow and desorption of solute from the solid particles of the filter bed.

Limit values on concentration are very important in the program, in order to avoid possible numerical overflow problems, caused by zero concentration values in the denominator of the model equations.

Special relationships apply to the end segments, owing to the absence of axial dispersion exterior to the cake.

Nomenclature

Symbols

c	Solute concentration in liquid	kg/m³
c'	Dimensionless solute concentration	–
c_0	Solute conc. in cake before washing	kg/m³
D_L	Axial dispersion coefficient	m²/s
K_0	Constant in equilibrium isotherm	kg/m³
K	Adsorption equilibrium constant	$\left(\dfrac{\text{kg m}^3}{\text{m}^3 \text{ kg}}\right)^{-1/N}$
L	Cake thickness	m
N	Reciprocal order of the adsorption	–
P	Peclet number per unit volume of voids	–
S	Fractional saturation, volume of liquid	–
t	Time	s
u	Pore velocity	m/s
W	Wash ratio	–
x	Distance measured from filter cloth	m
x*	Dimensionless distance	–
ε	Cake porosity	–
η	Solute concentration on solid	kg/m³

Exercises

1. Use the program to assess the effects of differing degrees of axial dispersion, for values of Peclet number ranging from 0.005 to 0.25. Modify the program to account for zero axial dispersion.
2. Wakeman found that the effect of increasing the equilibrium constant in the description isotherm equation is to increase the wash-ratio required. Is this confirmed by simulation, and what is the explanation of this effect?
3. Increasing the value of N (reciprocal power term) reduces the effect of the desorbed solute and agrees ultimately with the case of zero solute desorption. Try this in a parametric run.
4. In compressible filter cakes, the porosity varies from a minimum next to the filter septum to a maximum at the cake surface. How could you include such a variation within the context of a simulation program? Remember that as the porosity changes, the local fluid velocity will also change as a result of

$$(u\varepsilon)_{x'} = (u_{x'=0})(\varepsilon_{x'=0})$$

A necessary assumption would also be that the initial porosity distribution remains constant and independent of flow rate.
5. The presence of a liquid layer on the surface of the filter cake will cause solute to diffuse from the top layer of cake into the liquid. Also if disturbed, the layer of liquid will mix with the surface layer of filter cake. This effect can be incorporated into a simulation by assuming a given initial depth of liquid as an additional segment of the bed which mixes at time $t=0$ with the top cake segment. The initial concentrations in the liquid layer and in the top cake segment are then found by an initial mass balance. Try altering the model to account for this.

Cases 4 and 5 are solved also by Wakeman (1990).

Results

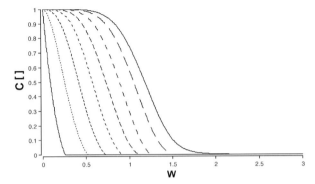

Fig. 1 Solute concentration variations as a function of wash ratio.

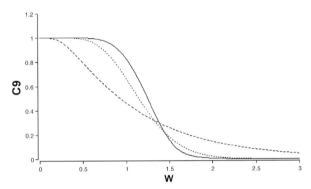

Fig. 2 Variations in outlet solute concentrations [$D_L=8.5E-8$], [$D_L=8.5E-7$] and [$D_L=8.5E-6$].

References

Wakeman, R.J. (1990) Chem. Eng. Res. Des. 68, 161–171.
Wakeman, R.J. (1986) Chem. Eng. Res. Des. 64, 308.
Wakeman, R.J. (1981) in "Solid–Liquid Separation", ed. L. Svarovsky, Butterworths, London.

5.9.13
CHROMDIFF – Dispersion Rate Model for Chromatography Columns

System

As explained in Section 4.4.4, the movement of components through a chromatography column can be modelled by a two-phase rate model, which is able to handle multicomponents with nonlinear equilibria. In Fig. 1 the column with segment n is shown and in Fig. 2 the structure of the model is depicted. This involves the writing of separate liquid and solid phase component balance equations, for each segment n of the column. The movement of the solute components through the column occurs by both convective flow and axial dispersion within the liquid phase and by solute mass transfer from the liquid phase to the solid.

Model

The equations have previously been derived in Section 4.4.4 in a form suitable for programming with MADONNA. Correlations for the column Peclet number are taken from the literature and used to calculate a suitable value for the dispersion coefficient for use in the model.

Fig. 1 Chromatographic column showing finite differencing into column segments.

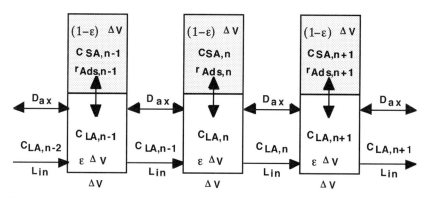

Fig. 2 Structure of the two-phase rate model in finite-difference form.

Program

The model is written in vector form with the last elements taken separately to satisfy the closed end boundary conditions.

Nomenclature

Refer to the nomenclature in Section 4.4.4 and in the program.

Exercises

1. Study the influence of the equilibrium constant ratio K_B/K_A, for the two components A and B, on the resolution of the peaks, as given by the time difference between the peaks.
2. The liquid flow rate directly influences the dispersion coefficient. Investigate its influence on peak width.
3. Correlate the degree of separation obtained as a function of column Peclet number and compare this to the number of theoretical plates required in simulation example CHROMPLATE for a similar degree of separation obtained.
4. Investigate the influence of increasing column length on the degree of separation obtained, by varying the increment number NN, but with constant ΔZ. Keeping NN constant, now vary the length of column by increasing the length of each increment.
5. Amend the program to allow for interacting non-linear Langmuir equilibria for the two components, as given below, using the following numerical equilibrium values $K_A = 3$, $K_B = 1.8$, $b_A = 2$ and $b_B = 4$.

$$C^*_{SA} = \frac{K_A C_{LA}}{1 + b_A C_{LA} + b_B C_{LB}} \quad C^*_{SB} = \frac{K_B C_{LB}}{1 + b_A C_{LA} + b_B C_{LB}}$$

Results

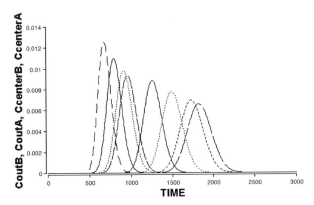

Fig. 3 Decrease of resolution caused by increasing K_B (1.8, 2.3 and 2.8) in 3 runs.

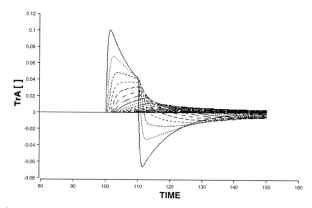

Fig. 4 Transfer rate at the time of the pulse injection, showing direction reversal.

Reference

Strube, J. (1999) Plant Engineering and Design Course Notes, University of Dortmund.

5.9.14
CHROMPLATE – Stagewise Linear Model for Chromatography Columns

System

As explained in Section 4.4.4, there exists an equivalency between tubular dispersion models and stagewise or tank in series models. The stagewise model, used in CHROMPLATE considers the chromatographic column to consist of a large number of well-mixed stirred tanks, arranged in series and thus represents an alternative modelling approach to that of the tubular dispersion model CHROMDIFF. The same two-component separation process is modelled and simulated in both cases.

Each tank contains a known quantity of both solid adsorbant and liquid phase. As depicted in Figs. 1 and 2, only the liquid phase is mobile and the solid phase remains static. Both phases contain solute components A and B, which move along the column, from stage to stage, only by convective flow.

Fig. 1 Chromatographic column showing a mass balance region.

Taking a balance for component A around both phases in stage n gives

$$\begin{pmatrix} \text{Accumulation} \\ \text{in the} \\ \text{liquid phase} \end{pmatrix} + \begin{pmatrix} \text{Accumulation} \\ \text{in the} \\ \text{solid phase} \end{pmatrix} = \begin{pmatrix} \text{Flow} \\ \text{in} \end{pmatrix} - \begin{pmatrix} \text{Flow} \\ \text{out} \end{pmatrix}$$

$$\varepsilon V_{stage} \frac{d(C_{LA,n})}{dt} + (1-\varepsilon)V_{stage} \frac{d(C_{SA,n})}{dt} = L_{in}(C_{LA,n-1} - C_{LA,n})$$

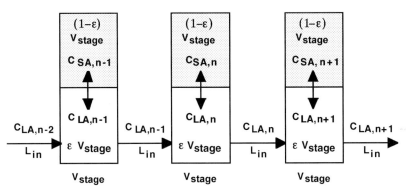

Fig. 2 Plate model with equilibrium stages.

Taking the balance, around both phases, effectively disregards the rate of solute transfer from liquid to solid and, instead, the assumption of a perfect equilibrium stage is employed to provide a relation between the resulting liquid and solid phase concentrations. For the special case of a linear equilibrium

$$C_{LA} = K_A C_{SA}$$

Practical adsorption isotherms are usually more complex, especially at high concentrations. But for the linear equilibrium employed here, the analysis becomes very simple. Combining the equilibrium relationship with the balance equation yields

$$\frac{d(C_{LA,n})}{dt} = \frac{L_{in}(C_{LA,n-1} - C_{LA,n})}{\varepsilon V_{stage} + (1-\varepsilon)V_{stage}K_A}$$

A similar balance equation can be written for the second component B. Thus the resulting model consists of 2*N equations, where N is the number of stages in the cascade. These are then easily solved by simulation.

It should be noted that taking the balance region over both the combined phases, as done in this example, is useful only for linear equilibrium. In general the best method is to use a transfer rate model, as described in Section 4.4.4, and to make balances for each component in each phase separately.

Nomenclature

Symbols

V_{stage}	Volume of a single plate section	cm^3
C_L	Liquid phase concentration	g/cm^3
C_S	Solid phase concentration	g/cm^3
d	Tube diameter	cm
D_p	Diameter of packing particle	cm
ε	Fraction voids	–
k and k_{eff}	Mass transfer rate coefficient	1/s
K	Equilibrium constant	–
L_{in}	Volumetric flow rate	cm^3/s
L_{bed}	Length of column	cm

Subscripts

A	Refers to component
B	Refers to component
in	Refers to inlet
L	Refers to liquid phase
n	Refers to stage number
S	Refers to solid phase adsorbent

Exercises

1. Study the effect of differing number of stages.
2. Investigate the influence of flow rate on the separation.
3. Vary the equilibrium constants to quantify their influence.
4. Choose values of the equilibrium constant so that the separation is difficult. Run simulations with increased column length or number of plates to improve the separation.
5. Write a program for this stagewise model using transfer rates as in CHROM-DIFF.

Results

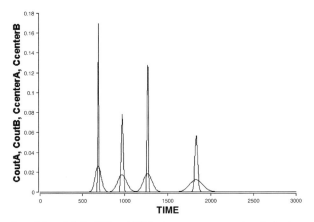

Fig. 3 Separation of A and B for 500 and 5000 plates at center and end of column.

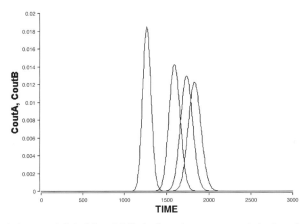

Fig. 4 As K_B is increased (1.8, 2.5 and 2.8) the B peak moves toward the A peak.

Reference

Strube, J. (1999) Plant Engineering and Design Course Notes, University of Dortmund.

5.10
Distillation Process Examples

5.10.1
BSTILL – Binary Batch Distillation Column

The process is as described in Section 3.3.3.2 and consists of a distillation column containing seven theoretical plates, reboiler and reflux drum. Distillation is carried out initially at total reflux in order to first establish the column concentration profile. Distillate removal then commences at the required distillate composition under proportional control of reflux ratio. This model is based on that of Luyben (1973, 1990).

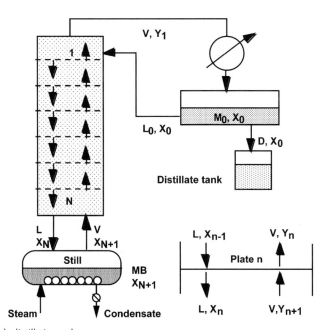

Fig. 1 Batch distillation column.

Model

The component balance for plate n, based on the more volatile component, is

$$M \frac{dX_n}{dt} = L X_{n-1} - L X_n + V (Y_{n+1} - Y_n)$$

For the still

$$\frac{dM_B}{dt} = L - V$$

and
$$\frac{d(M_B X_{N+1})}{dt} = L X_N - V Y_{N+1}$$

where the suffix N refers to the bottom plate and suffix N+1 relates to the still and column base, as represented in Fig. 1.

For the reflux drum and condenser, assuming constant holdup

$$M_0 \frac{dX_0}{dt} = V Y_1 - (L + D) X_0$$

The reflux ratio, R, is given by

$$R = L_0/D = L/D$$

for constant molar overflow, where

$$D = \frac{V}{R+1}$$

and

$$L = V - D$$

Assuming theoretical plate behaviour

$$y_n = \frac{a X_n}{1 + (a-1) X_n}$$

where a is the relative volatility.

The controller equations are given by:

for initial startup
$$R = R_{initial}$$

and for distillate withdrawal
$$R = K_C (X_D - X_{Dset})$$

Program

Start-up is at initially total reflux, with the reflux value set at an arbitrarily high value, $R = R_{init}$. Once the distillate composition, X_0, exceeds the set point composition, X_{0set}, control of reflux ratio is effected by varying the reflux ratio according to the proportional control relationship $R_C = K_C^*(X_{0set} - X_0)$. The change from initial start-up to controlled withdrawal of distillate is effected by means of the logical control variable FLAG.

Nomenclature

Symbols

a	Relative volatility	–
K_C	Controller gain	–
M	Molar holdup	kmol
L	Liquid flow rate	kmol/h
V	Vapour flow rate	kmol/h
D	Distillate rate	kmol/h
X	Liquid phase mole fraction	–
Y	Vapour phase mole fraction	–
R	Reflux ratio	–

Indices

1 to N	Plate number
B	Reboiler
0	Reflux drum
init	initial conditions
n	n-th plate
set	controller set point

Exercises

1. With only seven plates in the column, it is very difficult to maintain the required top product quality for very long. Study the effect of K_C on (i) the time for which required distillate composition can be maintained and (ii) the quantity of the total distillate of required composition obtained.
2. Also, as the problem is set up, further distillation should perhaps be discontinued, once the still composition has reached very low concentration levels. Modify the program to include this effect into the simulation.
3. Increase the number of plates in the column and study this effect on the separation obtained.

4. Modify the program to allow for continuous feeding of fresh feed into the still during the distillation. The feed rate should be adjusted to equal the rate of withdrawal of distillate. What is the effect of feeding on the distillation process?

Results

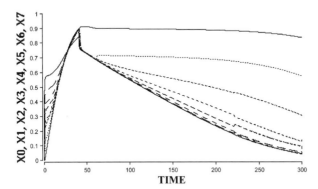

Fig. 2 Composition profiles for distillate withdrawal at controlled reflux ratio, following the initial startup at total reflux. Number of plates = 7.

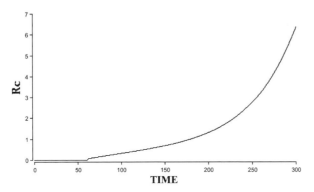

Fig. 3 The corresponding variation in the controlled value of reflux ratio.

Reference

Luyben, W. L. (1973, 1990) Process Modelling, Simulation and Control for Chemical Engineers, McGraw-Hill.

5.10.2
DIFDIST – Multicomponent Differential Distillation

System

A mixture of hydrocarbons A, B, C and D is charged to a batch still and is distilled over without reflux.

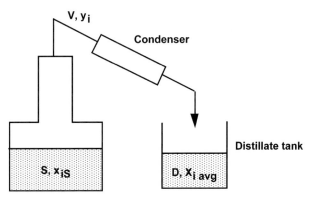

Fig. 1 Single multicomponent batch distillation.

Model

Total mole balance for the still

$$\frac{dS}{dt} = -V$$

Component mole balance

$$\frac{d(Sx_i)}{dt} = -Vy_i$$

The equilibrium is represented by

$$y_i = \frac{a_i x_i}{\Sigma a_i x_i}$$

For the receiver tank

$$\frac{dD}{dt} = V$$

$$\frac{dD_i}{dt} = Vy_i$$

where D is the total distillate in tank and D_i the mole quantity of component i.

Nomenclature

Symbols

S	Still liquid quantity	kmol
V	Distillate removal rate	kmol/h
x	Mole fraction composition in the liquid	–
y	Mole fraction composition in the vapour	–
a	Relative volatility	–

Indices

i	Component i
0	Initial condition
avg	Refers to average values in distillate tank

Exercises

1. Vary the relative volatilities of the mixture and study the effect on the distillation.
2. Vary the initial proportions of the four components.
3. Compare the computer predictions with the Rayleigh equation prediction, where

$$\ln \frac{S}{S_0} = \int_{x_i}^{x_{i0}} \frac{dy}{y^* - x}$$

4. Check the overall material balance in all the above exercises.

Results

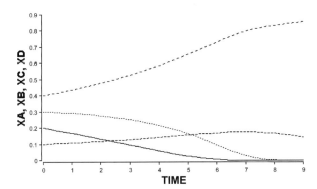

Fig. 2 Liquid compositions as a function of time during the course of the distillation.

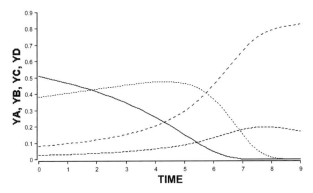

Fig. 3 The corresponding vapour phase compositions.

5.10.3
CONSTILL – Continuous Binary Distillation Column

System

A continuous binary distillation column is represented below. The column is shown in Fig. 1 as consisting of eight theoretical plates, but the number of plates in the column may be changed, since the MADONNA program is written in array form.

5.10 Distillation Process Examples

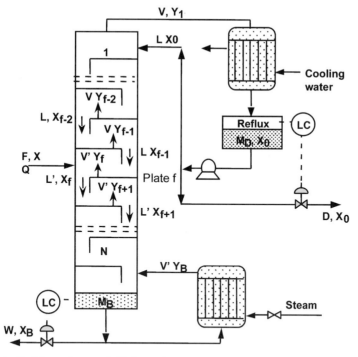

Fig. 1 An N-plate continuous distillation column.

Model

The model equations from Section 3.3.3.3 are

Reflux drum:
$$M_0 \frac{dX_0}{dt} = VY_1 - (L+D)X_0$$

Column enriching section:
$$M \frac{dX_n}{dt} = L(X_{n-1} - X_n) + V(Y_{n+1} - Y_n)$$

Column stripping section:
$$M \frac{dX_n}{dt} = L'(X_{m-1} - X_m) + V'(Y_{m+1} - Y_m)$$

Column feedplate:
$$M \frac{d(X_f)}{dt} = LX_{f-1} - L'X_f + V'Y_{f+1} - VY_f + FX_f$$

Reboiler:

$$M_B \frac{dX_B}{dt} = L'X_N - WX_W - V'Y_B$$

In the enriching section:

$$L = RD$$

and

$$V = (R+1)D$$

In the enriching section:

$$L' = L + qF$$

and

$$V = V' - (1-q)F$$

where for a saturated liquid feed with $q=1$, $L'=L+F$ and $V=V'$.

Assuming theoretical plate behaviour in the column, the corresponding equilibrium vapour compositions are calculated by

$$Y_n = \frac{aX_n}{1 + (a-1)X_n}$$

where a is the relative volatility.

Program

Note that the total number of plates, the feed plate location and reflux ratio can all be varied during simulation. The array form of the program also allows graphing the axial steady state tray-by-tray concentration profile. Part of the program is shown below.

{Example CONSTILL}

{CONDENSER AND REFLUX DRUM - index 1}
d/dt(X[1]) = (V*Y[i+1] - (L+D)*X[i])/MD

{LIQUID PHASE COMPONENT BALANCES FOR PLATES 2 TO Nplates-1 }
d/dt(X[2..Fplate-1]) = (L*(X[i-1]-X[i])+V*(Y[i+1]-Y[i]))/M
{Plates above the feed plate}

d/dt(X[Fplate]) = (F*Xfeed+L*X[i-1]-L1*X[i]+V*(Y[i+1]-Y[i]))/M {FEED PLATE}

```
d/dt(X[Fplate+1..Nplates1]) = (L1*(X(i-1)-X(i))+V*(Y(i-1) -
Y[i]))/M  {Plates below the feed plate}

{REBOILER AND COLUMN BASE - index Nplates}
d/dt(X[Nplates]) = (L1*X[i-1]-V1*Y[i]-W*X[i])/MR

{VAPOUR PHASE EQUILIBRIA}
Y[1..Nplates] = A*X[i]/(1+(A-1)*X[i])

{For plotting purposes}
XFplate=X[Fplate]
Xbottoms=X[Nplates]
Xdist=X[1]
```

Nomenclature

Symbols

D	Distillate rate	kmol/h
F	Feed rate	kmol/h
M	Liquid holdup	kmol
L	Liquid flow rate	kmol/h
q	Feed quality	–
R	Reflux ratio	–
V	Vapour flow rate	kmol/h
x	Liquid phase mole fraction	–
y	Vapour phase mole fraction	–
α	Relative volatility	–

Indices

0	Refers to reflux drum
B	Refers to reboiler
f	Refers to feed plate
n	Refers to plate number
'	Refers to flows below the feed plate

Exercises

1. Study the response of the column to changes in the operating variables feed rate, feed composition and reflux ratio.
2. Verify that the overall steady-state balance is attained.
3. How long does it take for the steady state to be reached?
4. What influence does the reflux ratio have on the separation?
5. What influence does the term q have on the separation?

6. Use sliders for N_{plate}, F_{plate} and R with steady state array plots (choosing [i] as the X axis) to optimise the design of the column in terms of column separation efficiency. Compare your results with those obtained by more conventional calculation methods.

Results

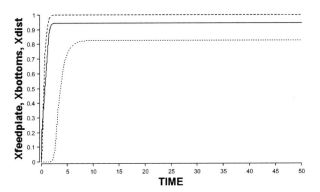

Fig. 2 Top product, feed plate and bottoms product composition variations as a function of time.

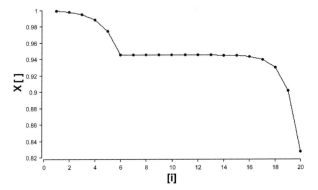

Fig. 3 Column steady state composition profiles. Note that N_{plate}, F_{plate} and R still need further optimisation to ensure good quality product at both ends of the column. Note the pinch region in the vicinity of the feed plate is very apparent.

5.10.4
MCSTILL – Continuous Multicomponent Distillation Column

System

This problem is similar to CONSTILL except that three components benzene, xylene and toluene are considered.

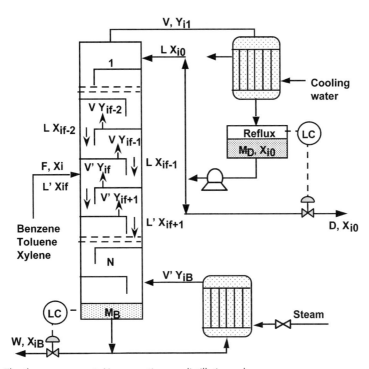

Fig. 1 The three-component, N_{plate}, continuous distillation column.

Model

The component balance equations for reflux drum, column enriching section, feed plate, column stripping section and the column base and reboiler are very similar to those derived previously for the binary distillation example CONSTILL, but are now expressed in multicomponent terms as described in Section 3.3.3.4.

Thus for any component i of the mixture in the enriching section of the column

$$\frac{M_n d(X_{in})}{dt} = L(X_{in-1} - X_{in}) + V(Y_{in+1} - Y_{in})$$

where i=1 to j and where the corresponding vapour equilibrium composition is now represented by

$$Y_{in} = \frac{a_{in} X_{in}}{\sum_{1}^{j} a_{in} X_{in}}$$

Program

The simulation is based on a fixed vapour boilup rate. Component balance equations are represented for "benzene" and "toluene". The "xylene" concentrations are determined by difference, based on the sum of the mole fractions being equal to one.

Note that the plate numbering in the program is slightly different to that shown in Fig. 1, owing to the use of the vector notation to include both the reflux drum and the column base. In the program, index number 1 is used to denote the reflux drum and product distillate, and index $N_{plate}+1$ is used to denote the reboiler and bottoms product. This is convenient in the subsequent plotting of the steady state composition profiles in the column. Both N_{plate} and the feed plate location F_{plate} are important parameters in the simulation of the resulting steady state concentration profiles and the resultant column optimisation.

The column start up is from an initial arbitrarily chosen composition profile. The steady state composition profile obtained from the first run can then be used as the starting profile for subsequent runs. With the MADONNA version for Windows, changes in operating parameters, such as feed rate, can be made with the CONTINUE feature. This can be programmed easily with IF-THEN-ELSE statements. In this way realistic dynamics can be obtained for the column conditions moving from one steady state to another.

Nomenclature

Symbols

D	Distillate rate	kmol/h
F	Feed rate	kmol/h
M	Liquid holdup	kmol
L	Liquid flow rate	kmol/h
q	Feed quality	–
R	Reflux ratio	–
V	Vapour flow rate	kmol/h
x	Liquid phase mole fraction	–
y	Vapour phase mole fraction	–
a	Relative volatility	–

Indices

′	Refers to flows below the feed plate
0	Refers to reflux drum
B	Refers to reboiler
f	Refers to feed plate
i	Refers to the component i
j	Refers to the number of components
n	Refers to plate number

Exercises

1. Study the response of the column to changes in the operating variables feed rate, feed composition and reflux ratio, while keeping the number of plates N_{plate} and feed plate location F_{plate} constant.
2. Verify that the overall steady-state balance is attained.
3. How long does it take for the steady state to be reached?
4. What influence does the reflux ratio have on the separation?
5. Repeat Exercises 1 to 4, but now varying the number of column plates and the feed plate.
6. Use sliders for N_{plate}, F_{plate} and R to determine the optimum design conditions for the column. For this the array index [i] can be chosen for the X axis to obtain steady-state axial profiles.

Results

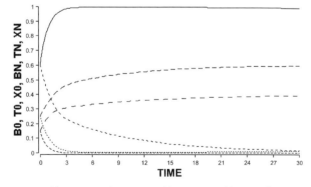

Fig. 2 Column top and bottom product compositions approaching steady state from the rather arbitrary feed composition starting point.

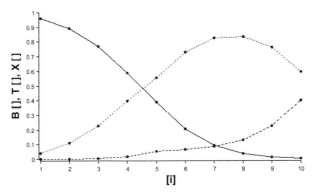

Fig. 3 Steady state column composition profiles for "benzene", "toluene" and "xylene".

5.10.5
BUBBLE – Bubble Point Calculation for a Batch Distillation Column

System

As in example BSTILL, a column containing four theoretical plates and reboiler is assumed, together with constant volume conditions in the reflux drum. The liquid behaviour is, however, non-ideal for this water–methanol system. The objective of this example is to illustrate the need for iterative calculations required for bubble point calculations in non-ideal distillation systems and to show how this can be achieved via the use of simulation languages.

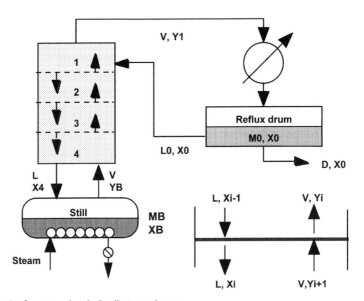

Fig. 1 The four-stage batch distillation column.

Model

Component balances for the more volatile component on any plate n give

$$M \frac{dx_n}{dt} = L(x_{i-1} - x_n) + V(y_{i+1} - y_n)$$

for the still

$$\frac{dM_B}{dt} = L - V$$

$$\frac{d(M_B x_B)}{dt} = Lx_4 - Vy_B$$

where 4 refers to the bottom plate and B to the still and column base.

For the reflux drum and condenser

$$M_0 \frac{d(x_0)}{dt} = Vy_1 - (L + D)x_0$$

where the reflux ratio, R, is given by L_0/D and

$$D = \frac{V}{(R+1)}$$

Throughout the column the total mass balance gives

$$L = V - D$$

Theoretical plate behaviour in the column is assumed, where the liquid–vapour equilibrium is described by

$$p_i = \gamma_i x_i p_{pi}$$

Here p_{pi} is the vapour pressure of the pure fluid i, x_i is its mole fraction in the liquid phase, γ_i is its activity coefficient and p_i is the actual vapour partial pressure.

Vapour pressures p_{pi} are given by

$$\ln p_{pi} = c_i + \frac{d_i}{T_i}$$

For the estimation of activity coefficients the Van Laar equation is used.

$$\gamma_M = \exp\left[\frac{A x_W^2}{\left(x_W + \dfrac{A}{B} x_M\right)^2}\right]$$

$$\gamma_W = \exp\left[\frac{B x_M^2}{\left(x_M + \dfrac{B}{A} x_W\right)^2}\right]$$

These equations represent an implicit loop to satisfy the condition

$$P = p_W + p_M = \gamma_W x_W p_{pW} + \gamma_M x_M p_{pM}$$

Since P is constant, the temperature will adjust at each plate such that the above equation is satisfied.

Program

For convenience, only four stages were used in this model, since MADONNA does not have a ROOTS finder function in an array form. For this reason each stage needs to be programmed separately.

Nomenclature

Symbols

A, B	Coefficient in Van Laar equation	–
C	Constant in the vapour pressure relation	–
D	Distillate rate	kmol/h
d	Constants in the vapour pressure relation	–
L	Liquid flow rate	kmol/h
M	Molar holdup	kmol
R	Reflux ratio	–
p_p	Pure component vapour pressure	bar
P	Total pressure	bar
V	Vapour flow rate	kmol/h
x	Liquid phase mole fraction	–
y	Vapour phase mole fraction	–
α	Relative volatility	–
γ	Activity coefficient	–

Indices

0	Refers to reflux drum
1 to 4 and i	Refers to plate number
B	Refers to reboiler
M	Refers to methanol
W	Refers to water

Exercises

1. Investigate the response of the column to changes in the boilup rate V.
2. The program starts up the column at total reflux (R very high). After steady state is reached on all plates, vary the reflux ratio interactively and attempt to carry out the distillation in minimum time, while attempting to maintain a distillate composition, so far as is possible, that $x_0 > 0.9$ and note the response of the distillate composition x_0.
3. Recharge the column by changing x_B and M_B and note the transients on each plate.
4. Devise a control scheme, such that the column maintains a distillate composition at $x_0 > 0.9$.

Results

Outputs are shown below.

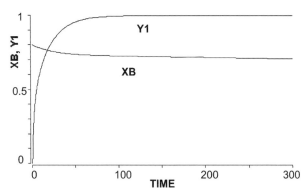

Fig. 2 The column was started with total reflux; the reflux was then reduced to 10 at T=20.

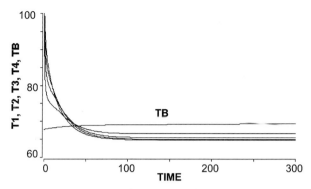

Fig. 3 The temperature profiles here are from the run of Fig. 2. The highest temperature is in the still (T_B) and the lowest is on the top plate (T_1).

Reference

Luyben, W. L. and Wenzel, L. A. (1988) Chemical Process Analysis, Prentice-Hall.

5.10.6
STEAM – Multicomponent, Semi-Batch Steam Distillation

System

This example is based on the theory and model description in Section 3.3.4 and involves a multicomponent, semi-batch system, with both heating and boiling periods. The compositions and boiling point temperatures will change with time. The water phase will accumulate in the boiler. The system simulated is based on a mixture of n-octane and n-decane, which for simplicity is assumed to be ideal but which has been simulated using detailed activity coefficient relations by Prenosil (1976).

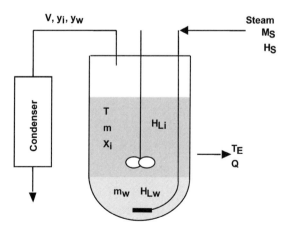

Fig. 1 Schematic of the semi-batch steam distillation.

Model

The model involves first a heating period and, when the boiling temperature is reached, then a distillation period.

All the steam is assumed to condense during the heating period, therefore the steam efficiency, E, equals 1.0. The standard state for enthalpy calculations is considered to be that of the pure liquid components at 0 °C and 1 bar.

Program

The heating period begins with FLAG set initially to zero. When $\Sigma y > 1$ then FLAG becomes 1, and the distillation period begins. At each time interval the ROOT FINDER is used to make the iterative bubble point calculation. The component mass balance determines the removal of volatiles in the vapour, where the total molar flow rate, V, is determined from the energy balance.

Nomenclature

Symbols

c_p	Molar heat capacity	kJ/kmol K
H	Enthalpy	kJ/kmol
M_s	Flow rate of steam	kmol/min
m or M	Mass of organic phase in the still	kmol
m_w or M_w	Mass of water in the still	kmol
P	Total pressure	bar
P^*	Partial pressure	bar
p^0	Vapour pressure	bar
Q	Heat transfer rate	kJ/min
T	Absolute temperature	K
T_E	Absolute ambient temperature	K
U	Overall heat transfer coefficient	kJ/K min
V	Vapour flow rate	kmol/min
x	Mole fraction in the liquid	–
y	Mole fraction in the vapour	–

Indices

0	Refers to initial
1, 2, i	Refers to organic components
Dist	Refers to distillate
L	Refers to liquid
S	Refers to steam
V	Refers to vapour
W	Refers to water
L	Refers to liquid

Exercises

1. Investigate the response of the system to changes in steam rate M_S.
2. Change the initial composition by varying the mole fractions X_{01} and X_{02}.

Results

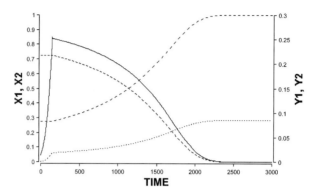

Fig. 2 The dynamic changes of the organic mole fractions in the gas and liquid phases are seen here.

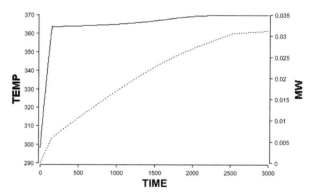

Fig. 3 Temperature and water mass variations in the still during the heating and distillation periods.

Reference

Prenosil, J. E. (1976) Multicomponent Steam Distillation: A Comparison between Digital Simulation and Experiment, Chem. Eng. J. 12, 59–68, Elsevier.

5.11
Heat Transfer Examples

5.11.1
HEATEX – Dynamics of a Shell-and-Tube Heat Exchanger

System

A shell-and-tube heat exchanger is to be investigated for both its dynamic and steady-state behaviour.

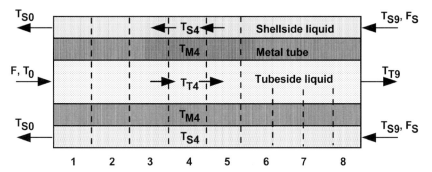

Fig. 1 Finite differencing the heat exchanger for dynamic modelling.

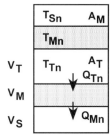

Fig. 2 Detail of section n showing heat flows, volumes and transfer areas.

Model

The model equations follow from Section 4.5.2.2. For section n, the rate of heat transfer from the tube contents to the metal wall is given by

$$Q_{Tn} = U_{TM} A_T (T_{Tn} - T_{Mn})$$

The rate of heat transfer from metal wall to shell-side content is

$$Q_{Mn} = U_{MS} A_M (T_{Mn} - T_{Sn})$$

For the tube-side fluid

$$c_{pT}\rho_T V_T \frac{dT_{Tn}}{dt} = F_T c_{pT}(T_{Tn-1} - T_{Tn}) - Q_{Tn}$$

For the metal wall

$$c_{pM}\rho_M V_M \frac{dT_{Mn}}{dt} = Q_{Tn} - Q_{Mn}$$

For the shell-side fluid

$$c_{pS}\rho_S V_S \frac{dT_{Sn}}{dt} = F_S c_{pT}(T_{Sn+1} - T_{Sn}) + Q_{Mn}$$

The respective volumes and area are calculated by

$$V_T = \Delta L \pi D_T^2 \; ; \; A_T = \Delta L \pi D_T$$

$$V_M = \Delta L \pi (D_M^2 - D_T^2), \; A_M = \Delta L \pi D_M$$

$$V_S = \Delta L \pi (D_S^2 - D_M^2), \; A_S = \Delta L \pi D_S$$

where D_T, D_M, D_S are the respective diameters and A_T, A_M, A_S are the areas.

The tube inlet and outlet end sections are modelled by averaging the temperatures entering the section.

For example, for the shell-side section 1:

Entering shell-side temperature (averaged) = $\dfrac{(T_{S2} + T_{S1})}{2}$

Leaving shell-side temperature = T_{S0}

Thus

$$c_{pS}\rho_S V_S \frac{dT_{S1}}{dt} = F_S c_{pS}\left(\frac{(T_{S2} + T_{S1})}{2} - T_{S0}\right) + Q_{M1}$$

Program

The model is programmed using arrays. This permits varying the number of sections. As usual with such finite-differenced models, the end sections are treated separately. Part of the program is shown below.

```
{Example HEATEX}
{SHELL AND TUBE HEAT EXCHANGER}

{ELEMENT 1}
QT[1]=UTM*(TT[1]-TM[1])
```

```
QM[1]=UMS*(TM[1]-TS[1])
TS0=TS[1]+0.5*(TS[1]-TS[2])

d/dt(TT[1])=ZT*(TT0-0.5*(TT[1]+TT[2]))-QT[1]/YT
d/dt(TM[1])=(QT[1]-QM[1])/YM
d/dt(TS[1])=ZS*(0.5*(TS[1]+TS[2])-TS0)+QM[1]/YS

{ELEMENTS 2 TO ELEMENT Nelem-1}
{TUBE HEAT TRANSFER RATES, kJ/s}
QT[2..Nelem-1]=UTM*(TT[i]-TM[i])

{SHELL SIDE HEAT TRANSFER RATES, kJ/s}
QM[2..Nelem-1]=UMS*(TM[i]-TS[i])

{TUBE SIDE HEAT BALANCES}
d/dt(TT[2..Nelem-1])=ZT*(TT[i-1]-TT[i])-QT[i]/YT

{METAL WALL HEAT BALANCES}
d/dt(TM[2..Nelem-1])=(QT[i]-QM[i])/YM

{SHELL SIDE HEAT BALANCES}
d/dt(TS[2..Nelem-1])=ZS*(TS[i+1]-TS[i])+QM[i]/YS

{ELEMENT Nelem}
QT[Nelem]=UTM*(TT[i]-TM[i])
QM[Nelem]=UMS*(TM[i]-TS[i])
TTout=TT[Nelem]-0.5*(TT[Nelem-1]-TT[Nelem])
d/dt(TT[Nelem])=ZT*(0.5*(TT[i-1]+TT[i])-TTout)-QT[i]/YT
d/dt(TS[Nelem])=ZS*(TSIN-0.5*(TS[i-1]+TS[i]))+QM[i]/YS
d/dt(TM[Nelem])=(QT[i]-QM[i])/YM
```

Nomenclature

Symbols

A	Transfer area ($\Delta A = \pi D \Delta L$)	m²
C	Specific heat	kJ/kg K
D_M	Outside diameter of metal wall	m
D_T	Internal diameter of tube wall	m
F	Mass flow velocity	m³/s
H_{MS}	Film heat transfer coefficient from the wall to the shell	kJ/m² K s
H_T	Film heat transfer coefficient, tube-side	kJ/m² K s
Q	Heat transfer rate	kJ/s
T_{Mn}	Metal wall temperature in element n	K
T_{Sn}	Shell-side temperature in element n	K
T_{Tn}	Temperature in element n, tube-side	K

| V | Volumes of shell-side fluid, tube-side fluid and metal wall in length ΔL | m^3 |
| ρ | Density of the shell-side fluid, tube-side fluid and metal wall | kg/m^3 |

Indices

S	Refers to shell
M	Refers to metal
T	Refers to tube
1, 2, ..., n	Refers to section

Exercises

1. Establish that the overall steady-state balance is satisfied.
2. Investigate the response of the system to a change in entering temperatures.
3. Increase the heat transfer coefficients and observe the resulting trend in the temperatures T_T, T_M, and T_S.
4. Increase the flow of shell-side fluid and observe the effect on the inlet shell-side and outlet tube-side temperatures.
5. Increase the mass of the wall and observe the influence on the dynamics.
6. Derive a dimensionless form of the equations and thus obtain the important dimensionless groups governing the dynamic behaviour of the heat exchanger.

Results

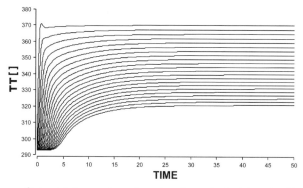

Fig. 3 Response of tube-side temperatures for a step change in the inlet tube-side temperature.

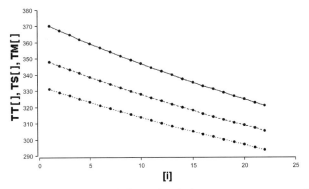

Fig. 4 Steady-state axial profile for the shell, metal and tube temperature corresponding to Fig. 3.

5.11.2
SSHEATEX – Steady-State, Two-Pass Heat Exchanger

System

A purified water stream for a food processing plant is raised to required temperature by heat exchange with warm untreated process water. A conventional two-pass heat exchanger is to be used. This problem is based on the formulation of Walas (1991) and data taken from Backhurst, Harker and Porter (1974).

In the first pass, both the hot and cold fluids flow in cocurrent flow through the heat exchanger, whereas in the second pass the cold fluid now flows countercurrent to the hot shell-side fluid. Half the heat exchange area is therefore in cocurrent flow and half in countercurrent flow.

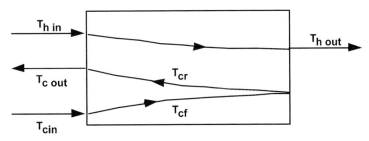

Fig. 1 Two-pass heat exchanger schematic, showing two passes for the cold fluid.

Model

The derivation of the model equations follows that of Section 4.5.2.1.

For the hot shell-side fluid

$$(Wc_p)_h \frac{dT_h}{dA} = -\frac{U(T_h - T_{cf})}{2} - \frac{U(T_h - T_{cr})}{2}$$

For the cold-side fluid in forward cocurrent flow

$$(Wc_p)_c \frac{dT_{cf}}{dA} = \frac{U(T_h - T_{cf})}{2}$$

and for the cold-side fluid in reverse countercurrent flow

$$(Wc_p)_c \frac{dT_{cr}}{dA} = \frac{U(T_h - T_{cr})}{2}$$

For a given inlet and outlet cold water temperature, the outlet hot water temperature is determined by an overall heat balance, where

$$(W)_h(T_{h\,in} - T_{h\,out}) = (Wc_p)_c(T_{c\,out} - T_{c\,in})$$

Program

Starting at the given temperature conditions $T_{c\,in}$, $T_{c\,out}$ and $T_{h\,in}$, the model equations are integrated to the condition that $T_{cf} = T_{cr}$, which must be true when the fluid reverses directions. It should be noted that a solution exists only for certain values of initial conditions and flow rates.

Nomenclature

Symbols

A	Area	m^2
c_p	Specific heat capacity	kJ/kg s
L	Length	m
T	Temperature	K
t	Time	s
U	Overall heat transfer coefficient	kJ/m^2 s K
W	Mass flow rate	kg/s

Indices

h, c	Refer to hot and cold fluids
f, r	Refer to the cold fluid in forward (cocurrent) and reverse (countercurrent) flow

Exercises

1. Study the effects of varying flow rates of both hot and cold fluids on the heat transfer area required.
2. Vary the hot fluid temperature in the range 383 to 353 K and study the effect on the area.
3. The model assumes a constant specific heat for water. If this varies with temperature, according to $c_p = a + bT$, how could this effect be implemented in the program, and what additional difficulties does it cause?
4. Is any advantage to be gained by changing the hot fluid to the two-pass arrangement?

Results

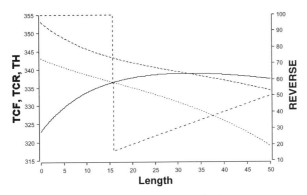

Fig. 2 The steady state axial profiles for the cold and hot fluids. The variable REVERSE marks the length of exchanger needed, as shown by the vertical line.

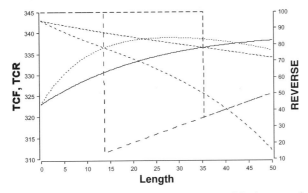

Fig. 3 Using an overlay plot gives the solutions for two values of the heat transfer coefficient.

References

Walas, S. M. (1991) "Modeling with Differential Equations in Chemical Engineering", Butterworth.
Backhurst, J. R., Harker, J. H. and Porter, J. E. (1974) "Problems in Heat and Mass Transfer", Edward Arnold.

5.11.3
ROD – Radiation from Metal Rod

System

A metal rod of radius r is in contact with a constant temperature source at each end. At steady state the heat conducted towards the center is balanced by the heat loss by radiation. This leads to a symmetrical temperature profile in the rod, as shown.

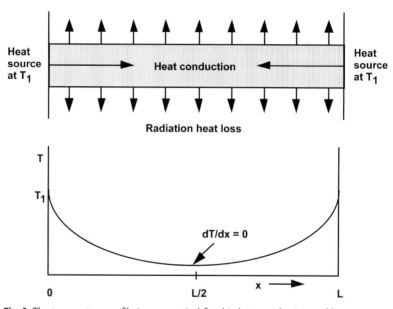

Fig. 1 The temperature profile is symmetrical for this heat conduction problem.

Model

The model is based on the steady-state energy balance combined with Fourier's law which gives

$$\frac{d^2T}{dx^2} = \frac{2\sigma}{rk} \cdot T^4$$

or

$$\frac{d^2T}{dx^2} = aT^4$$

where $a = 1.25 \times 10^{-8}$ (1/m² K³).

Boundary conditions are

$$T_1 = 2000 \text{ K} \quad \text{at} \quad x = 0$$

At the midpoint of the rod

$$\frac{dT}{dx} = 0 \quad \text{at} \quad x = \frac{L}{2}$$

This boundary condition is deduced from the symmetrical nature of the rod profile L=0.5 m.

Program

Knowledge of the derivative at the centre of the rod requires a solution involving repeated estimates of the temperature gradient at the rod end. The integration proceeds from the end to the rod centre. The required trial and error iteration can be easily done with a slider to adjust the parameter SLOPE.

Nomenclature

Symbols

T	Temperature	K
σ	Emissivity	kJ/s m² K⁴
k	Conductivity	kJ/s m K
x	Distance along rod	m
a	Constant	1/K³ m²
r	radius of rod	m

Exercises

1. Experiment with the various parameters of the model to see their influence.
2. Alter the program so that the conductivity and radius can be varied. Using various metals show how the value of k influence the solution.
3. Reformulate the problem in a dynamic form to be solved with respect to time and distance changes along the rod.

Results

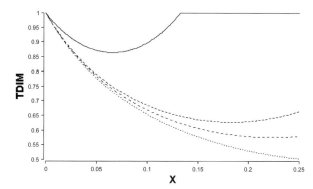

Fig. 2 Using a slider for SLOPE the correct value can be found to achieve a zero slope at the center of the rod at X=0.25.

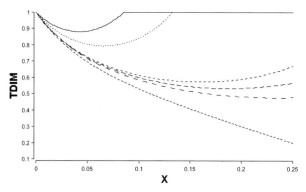

Fig. 3 Doubling Alpha compared with the run in Fig. 2 requires an increase in SLOPE.

5.12
Diffusion Process Examples

5.12.1
DRY – Drying of a Solid

System

A thin slab of solid material resting on a surface dries first by evaporation from the top surface and then by diffusion from the interior of the solid. The water movement towards the top surface is approximated by the diffusion equation

$$\frac{\partial C}{\partial t} = D\frac{\partial^2 C}{\partial X^2}$$

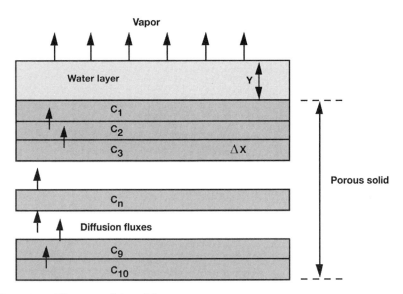

Fig. 1 Finite-differencing of the solid.

Model

Surface Drying

Water is removed from the top surface where $X = 0$.

Drying rate

$$M_1 = UA\frac{T_g - T_w}{\Delta H}$$

Water mass evaporated

$$W_1 = \int_0^t M_1 \, dt$$

Depth of water at time t

$$Y = \frac{W - W_1}{\rho A}$$

where the initial water mass in the layer is

$$W = V_0 \rho$$

Solid Drying

At the instant that the depth of the water layer is reduced to zero, the drying of the solid begins. As shown in Fig. 1, the solid is considered as 10 finite difference elements, each of length ΔX.

For the water balance of element n and applying Fick's law

$$V_n \frac{dC_n}{dt} = D\left(\frac{C_{n+1} - C_n}{\Delta X}\right) A - D\left(\frac{C_n - C_{n-1}}{\Delta X}\right) A$$

with

$$V_n = A \Delta X$$

giving

$$\frac{dC_n}{dt} = D\left(\frac{C_{n+1} - 2C_n + C_{n-1}}{\Delta X^2}\right)$$

C_n represents the concentration at the midpoint of element n, and the gradient dC/dX is approximated by $\Delta C/\Delta X$.

The boundary conditions are as follows:

$$X = 0, \quad C_0 = \text{constant}$$

Corresponding to zero flux at the lower surface

$$X = L, \quad \frac{dC}{dX} = 0$$

This can be approximated by setting $C_{11} = C_{10}$.

The drying rate is obtained by approximating the concentration gradient at the surface

$$M_2 = DA \frac{C_1 - C_0}{\Delta X}$$

where C_0 is the water concentration of the outer solid in equilibrium with the air. The total water removed from the solid is

$$W_2 = \int_0^t M_2 dt$$

Total water removed in both stages of drying

$$W_3 = W_1 + W_2$$

The water remaining is

$$W_4 = W + (ALC) - W_3$$

The percent dryness is

$$P = \frac{(W_3 \cdot 100)}{(W + A \cdot L \cdot C)}$$

Program

The program is written with two logical control variable flags, one takes the value of 1.0 to control the surface evaporation and the other a value of 1.0 to start the diffusional drying process. Array notation for the number of solid segments is used to allow plotting the concentration versus distance at any desired time.

Nomenclature

Symbols

A	Area	m^2
A	Surface area	m^2
C	Water concentration initially	$kmol/cm^3$
C_0	Surface equilibrium concentration	$kmol/cm^3$
C_n	Water concentration of solid	$kmol/cm^3$
D	Diffusion coefficient	m^2/s
ΔH	Latent heat of evaporation	$kJ/kmol$
ΔX	Depth of finite element	m

L	Depth of solid	m
L	Thickness	m
M_1	Drying rate of water layer	kmol/s
M_2	Drying rate of solid	kmol/s
P	Percent dryness	–
ρ	Water density	kmol/m^3
T_g	Air temperature	K
T_w	Water temperature (wet bulb)	K
U	Heat transfer coefficient	kJ/m^2 K s
V	Volume of water layer	m^3
W	Initial mass of water layer	kmol
W_1	Water mass evaporated	kmol
W_2	Water removed from solid	kmol
W_3	Total water removed	kmol
W_4	Remaining water	kmol
X	Length variable	m
Y	Water layer depth	m

Exercise

1. Run the program and graph the water concentrations in the solid versus time. Graph also the rate R1 versus percent dryness P.
2. The drying time for the solid can be characterised by L^2/D, a diffusion time. Verify this by simulation by varying D and L separately.
3. Improve the model by assuming a mass transfer rate at the solid surface, such that $M_2 = K_T A C_1$. Assume a suitable value for K_T.

Results

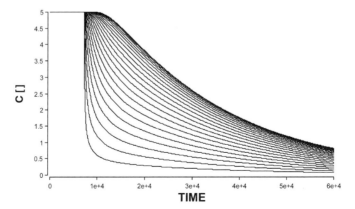

Fig. 2 Concentration-time profiles in the solid.

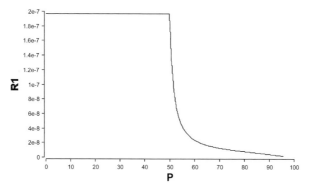

Fig. 3 Drying curve plotted as drying rate versus percent water remaining.

5.12.2
ENZSPLIT – Diffusion and Reaction: Split Boundary Solution

System

A rectangular slab of porous solid supports an enzyme. For reaction, reactant S must diffuse through the porous lattice to the reaction site, and, as shown in Fig. 1, this occurs simultaneously from both sides of the slab. Owing to the decreasing concentration gradient within the solid, the overall rate of reaction will generally be lower than that at the exterior surface. The magnitude of the concentration gradient determines the effectiveness of the catalyst.

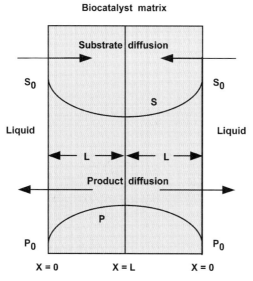

Fig. 1 Symmetrical concentration gradients for substrate and product.

Model

Under steady-state conditions

$$\begin{pmatrix} \text{Rate of diffusion of} \\ \text{reactant into the slab} \end{pmatrix} = \begin{pmatrix} \text{Rate at which reactant is} \\ \text{consumed by reaction} \end{pmatrix}$$

$$D_S \frac{d^2 S}{dX^2} = R$$

A quasi-homogeneous form for the reaction term is assumed.

The boundary conditions are given by

At $X = L$

$$S = S_0, \quad P = P_0$$

At $X = 0$

$$\frac{dS}{dX} = \frac{dP}{dX} = 0$$

The external concentration is known, and the concentration profile throughout the slab is symmetrical.

The reaction rate for this enzyme kinetics example is expressed by the Michaelis–Menten equation and with product inhibition.

$$R = \frac{kES}{K_M(1 + P/K_I + S)}$$

where k, K_M and K_I are kinetic constants and E and P are the enzyme and product concentrations. At steady state, the rate of diffusion of substrate into the slab is balanced by the rate of diffusion of product out of the slab.

Assuming the simple stoichiometry $S \rightarrow P$,

$$D_S \frac{dS}{dX} = -D_P \frac{dP}{dX}$$

which on integration gives

$$P = (S_0 - S) \frac{D_S}{D_P}$$

where P is assumed zero at the exterior surface.

Defining dimensionless variables

$$S' = \frac{S}{S_0}, \quad P' = \frac{P}{S_0} \quad \text{and} \quad X' = \frac{X}{L}$$

Substituting gives

$$\frac{d^2 S'}{dX'^2} - \frac{L^2 R'}{D_S S_0} = 0$$

where

$$R' = \frac{kES'}{(K_M/S_0)(1 + (S_0 P'/K_I)) + S'}$$

and

$$P' = (1 - S')\frac{D_P}{D_S}$$

The boundary conditions are then given by

$$X' = 1, \quad S' = 1 \quad \text{and} \quad X' = 0, \quad \frac{dS'}{dX'} = 0$$

The catalyst effectiveness may be determined from

$$\eta = \frac{D_S S_0 (dS'/dX')_{X=1}}{L^2 R_0}$$

where R_0 is the reaction rate determined at surface conditions

$$R_0 = \frac{kES_0}{K_M(1 + P_0/K_I + XS_0)}$$

Program

The dimensionless model equations are used in the program. Since only two boundary conditions are known, i.e., S at X′=1 and dS′/dX′ at X′=0, the problem is of a split-boundary type and therefore requires a trial and error method of solution. Since the gradients are symmetrical, as shown in Fig. 1, only one-half of the slab must be considered. Integration begins at the center, where X′=0 and dS′/dX′=0, and proceeds to the outside, where X′=1 and S′=1. This value should be reached at the end of the integration by adjusting the value of S_{guess} at X=0 with a slider.

The MADONNA software allows an automatic, iterative solution of boundary value problems. Selecting **Model/Modules/Boundary Value ODE** prompts for the boundary condition input: Set S=1 at X=1 with unknowns S_{guess}. Allowing

S_{guess} to vary between 0.5 and 0.6 gives a solution that approaches $S_{guess} = 0.562$. After an additional RUN command the solution is shown. It is also instructive to use the manual slider method, as explained above.

The Thiele modulus is also calculated and explained in the program.

Nomenclature

Symbols

D	Diffusion coefficient	m²/h
E	Enzyme concentration	mol/m³
K	Kinetic constant	kmol/m³
k	Reaction rate constant	1/h
L	Distance from slab centre to surface	m
P	Product concentration	kmol/m³
R	Reaction rate	kmol/m³ h
S	Substrate concentration	kmol/m³
X	Length variable	m
η	Effectiveness factor	–

Indices

I	Refers to inhibition
M	Refers to Michaelis–Menten
P	Refers to product
S	Refers to substrate
′	Refers to dimensionless variables
0	Refers to bulk concentration
guess	Refers to assumed value

Exercises

1. Vary the parameters L, D_S, S_0, k and E, but maintaining the Thiele parameter [L^2 k E/D_S S_0] at a constant value less than 0.1.
2. Vary the above parameters separately, noting the value of the Thiele parameter (L^2 k E/D_S S_0) and its influence.
3. Vary K_I and K_M and note their influence on the gradients and on η.
4. Alter the program for no reaction by setting E=0 and note the results. Run with various bulk concentrations.
5. Alter the program for nth-order kinetics.

Results

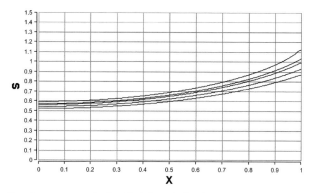

Fig. 2 A few iterations were required to find the solution by varying S_{guess} with a slider to reach $S=1$ at $X=1$.

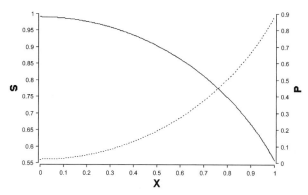

Fig. 3 The P and S profiles for the solution found in Fig. 2.

5.12.3
ENZDYN – Dynamic Diffusion with Enzymatic Reaction

System

This example involves the same diffusion-reaction problem as in the example, ENZSPLIT, except that here a dynamic solution is obtained, using the method of finite differencing. The substrate concentration profile in the porous biocatalyst is shown in Fig. 1.

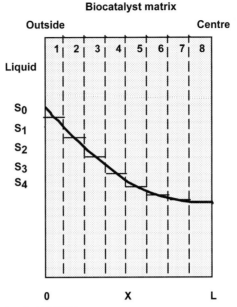

Fig. 1 Finite differencing for ENZDYN.

Model

With complex kinetics, a steady-state split boundary problem as solved in the example ENZSPLIT may not converge satisfactorily. To overcome this, the problem may be reformulated in the more natural dynamic form. Expressed in dynamic terms, the model relations become

$$\frac{\partial S}{\partial t} = D_s \frac{\partial^2 S}{\partial X^2} - R$$

$$\frac{\partial P}{\partial t} = D_p \frac{\partial^2 P}{\partial X^2} + R$$

where at the centre of the slab

$$\frac{dS}{dX} = \frac{dP}{dX} = 0$$

The kinetic equation used here is an enzymatic Michaelis–Menten form with product inhibition

$$R = \frac{kES}{S + K_M \frac{1+P}{K_I}}$$

This behaves as a zero-order reaction if $S \gg K_M$ and $P \ll K_I/K_M$.

Using finite-differencing techniques for any given element n, these relations may be expressed in semi-dimensionless form using the definitions

$$S'_n = S_n/S_0, \quad P'_n = P_n/S_0 \quad \text{and} \quad \Delta X' = X/L$$

Here S_0 is the external substrate concentration and $\Delta X'$ is the length of the finite difference element.

This gives,

$$\frac{dS'_n}{dt} = \frac{D_S}{L^2}\left(\frac{S'_{n-1} - 2S'_n + S'_{n+1}}{\Delta X'^2}\right) - \frac{R'_n}{S_0}$$

$$\frac{dP'_n}{dt} = \frac{D_P}{L^2}\left(\frac{P'_{n-1} - 2P'_n + P'_{n+1}}{\Delta X'^2}\right) + \frac{R'_n}{S_0}$$

and

$$R'_n = \frac{kES'_n}{(K_M/S_0)(1 + S_0 P'_n/K_I) + S'_n}$$

Boundary conditions are given by the external concentrations S_0 and P_0 and by setting $S_{N+1} = S_N$ and $P_{N+1} = P_N$, at the slab centre.

Catalyst effectiveness may be determined from the slope of the reactant concentration, S, at the solid surface

$$\eta_1 = \frac{D_s S_0}{L^2 R_0}\left(\frac{1 - S'_1}{\Delta X'}\right)$$

where

$$R_0 = \frac{kES_0}{K_M(1 + P_0/K_I + S_0)}$$

Program

The number of increments can be changed using the parameter ARRAY. The boundary conditions dictate the formulation of the first and last segments as follows: Center of slab (segment ARRAY), dS/dX=0 and dP/dX=0; surface conditions (segment 0), P_0 and S_0. With the same parameter values, the results of example ENZSPLIT should be exactly the same as the steady-state of ENZDYN.

Nomenclature

The nomenclature is the same as ENZSPLIT except as follows:

Symbols

ΔX   Increment length   m

Indices

n   Refers to segment n

Exercises

1. Vary K_I and K_M and note the inhibition effects. It might be useful to plot the rates of reaction to see when inhibition becomes important.
2. Vary E and L and note the effect on the gradients and η.
3. Experiment with parametric changes, while keeping $(D_S S_0/L^2 \, k \, E)$ constant. How is η influenced?
4. Vary L^2/D_S, and note that this influences the time to reach steady state.
5. Compare the steady-state results with those of ENZSPLIT.
6. Verify the form of the dimensionless equations. Rewrite the program in dimensional form.
7. Revise the program for a simple nth-order reaction.

Results

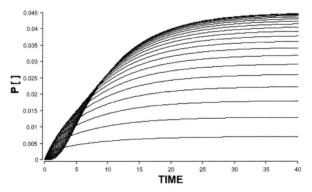

Fig. 2 A plot of P versus time for each segment.

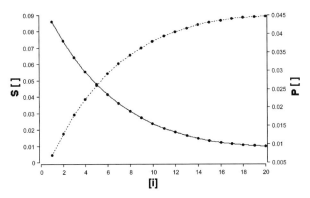

Fig. 3 The substrate and product profiles plotted versus segment position for the steady state.

References

Blanch, H. W. and Dunn, I. J. (1973) Modelling and Simulation in Biochemical Engineering. In: Advances in Biochemical Engineering, Vol. 3, Eds. T. K. Ghose, A. Fiechter, N. Blakebrough, Vol. 3, Springer.

Goldman, R., Goldstein, L. and Katchalski, C. I. (1971) in Biochemical Aspects of Reactions on Solid Supports, Ed. G. P. Stark, Academic Press.

5.12.4
BEAD – Diffusion and Reaction in a Spherical Catalyst Bead

System

This example treats a diffusion-reaction process in a spherical biocatalyst bead. The original problem stems from a model of oxygen diffusion and reaction in clumps of animal cells by Keller (1991), but the modelling method also applies to bioflocs and biofilms, which are subject to potential oxygen limitation. Of course, the modelling procedure can also be applied generally to problems in heterogeneous catalysis.

Diffusion and reaction takes place within a spherical bead of volume $=\frac{4}{3}\pi R_P^3$ and area $=4\pi R_P^2$. It is of interest to find the penetration distance of oxygen for given specific activities and bead diameters. As shown, the system is modelled by dividing the bead into shell-like segments of equal thickness. The problem is equivalent to dividing a rectangular solid into segments, except that here the volumes and areas are a function of the radial position. Thus each shell has a volume of $\frac{4}{3}\pi(r_n^3 - r_{n-1}^3)$. The outside area of the nth shell segment is $4\pi\, r_n^2$ and its inside area is $4\pi\, r_{n-1}^2$.

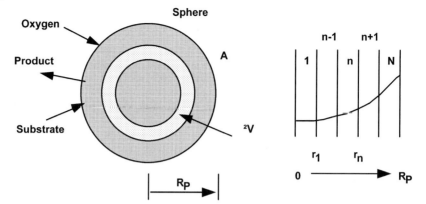

Fig. 1 The finite differencing of the spherical bead geometry.

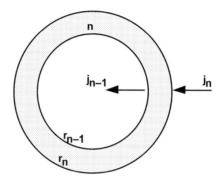

Fig. 2 The diffusion fluxes entering and leaving the spherical shell with outside radius r_n and inside radius r_{n-1}.

Model

Here the single limiting substrate S is taken to be oxygen.

The oxygen balance for any element of volume ΔV is given by

$$\frac{4}{3}\pi(r_n^3 - r_{n-1}^3)\frac{dS_n}{dt} = j_n 4\pi r_n^2 - j_{n-1} 4\pi r_{n-1}^2 + R_{sn}\frac{4}{3}\pi(r_n^3 - r_{n-1}^3)$$

The diffusion fluxes are

$$j_n = D_S \frac{S_{n+1} - S_n}{\Delta r}$$

$$j_{n-1} = D_S \frac{S_n - S_{n-1}}{\Delta r}$$

Substitution gives

$$\frac{dS_n}{dt} = \frac{3 \cdot D_S}{\Delta r(r_n^3 - r_{n-1}^3)}[r_n^2(S_{n+1} - S_n) - r_{n-1}^2(S_n - S_{n-1})] + R_{sn}$$

The balance for the central increment 1 (solid sphere not a shell) is

$$\frac{4}{3}\pi(r_1)^3 \frac{dS_1}{dt} = j_1 4\pi r_1^2 + R_{sn}\frac{4}{3}\pi(r_1)^3$$

Since $r_1 = \Delta r$, this becomes

$$\frac{dS_1}{dt} = \frac{3 D_S}{\Delta r^2}(S_2 - S_1) + R_{s1}$$

The reaction rate is expressed by a Monod-type equation

$$R_{Sn} = -OUR_{max} X \frac{S_n}{K_S + S_n}$$

where X is the biomass concentration (cell number/m^3) in the bead, OUR is the specific oxygen uptake rate (mol/cell s) and S_n is the oxygen concentration (mol/m^3) in shell n.

Program

The segments are programmed using the array facility of MADONNA, numbered from the outside to the centre. The effectiveness factor, expressing the ratio of the reaction rate to its maximum, is calculated in the program, part of which is shown below.

```
{Example BEAD}

INIT S[1..Array]=0
deltar=Radius/Array
r[1..Array]=((Array+1)-i)*deltar
RS[1..Array]=0

{SUBSTRATE OXYGEN BALANCES FOR EACH DELTA R ELEMENT}
{Outer shell}
d/dt(S[1])=3*D* ( ((r[1]**2) * (S0-S[1])) - ((r[2]**2) *
(S[1]-S[2])))/(deltar* ((r[1]**3) - (r[2]**3)) )+RS[1]

{Shells 2 to Array-1}
d/dt(S[2..(Array-1)])=3*D* ( ((r[i]**2) * (S[i-1]-S[i])) -
((r[i+1]**2) *(S[i]-S[i+1])) )/(deltar* ((r[i]**3) -
(r[i+1]**3)) )+RS[i]
```

```
{Inner spherical section}
d/dt(S[Array])=3*D*(S[(Array-1)]-
-S[Array])/(deltar**2)+RS[Array]
Sinner=S[array] {Defined for graphing}
```

Nomenclature

Symbols

D	Diffusion coefficient	m^2/s
K_S	Saturation constant in Monod equation	mol/m^3
OUR (max)	Maximum specific oxygen uptake rate	mol/cell s
r	Radius at any position	m
R_P	Outside radius of bead	m
R_S	Reaction rate in the Monod equation	$mol/s\ m^3$
S	Oxygen substrate concentration	mol/m^3
X	Biomass concentration	$cells/m^3$
Δr (Deltar)	Increment length, r/Array	m

Indices

1	Refers to segment 1
2	Refers to segment 2
n	Refers to segment n
P	Refers to particle
S	Refers to substrate

Exercises

1. Investigate the response of the bead to changes in the outside to the bulk concentrations. Can you relate the time constant to D/r^2?
2. Note the position of zero oxygen when changing D_{O2} and R_P.
3. Change the program to calculate the effectiveness factor.
4. Write the model in dimensionless form. What are the governing parameters for first and zero-order kinetics? Verify by simulation.
5. Develop a model and program for a rectangular slab geometry with diffusion from both sides and taking the thickness as $V_{bead}/A_{bead}=R_P$. Compare the results for the slab with those for the bead.

Results

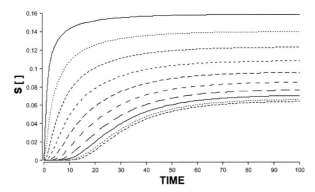

Fig. 2 Profiles of oxygen concentrations versus time for each shell.

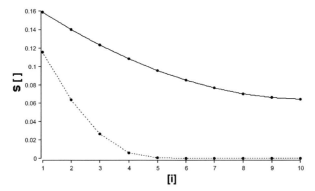

Fig. 3 Doubling the bead radius causes oxygen deficiency inside the bead (lower curve) as these radial profiles show.

Reference

Keller, J. (1991) PhD Dissertation No. 9373, ETH Zurich.

5.13
Biological Reaction Examples

5.13.1
BIOREACT – Process Modes for a Bioreactor

System

The general balance equations for stirred tank bioreactor systems can be derived by referring to Section 3.2. Here the three principle modes of operation are considered, i.e., batch, fed batch and chemostat operation, as represented in Fig. 1. The important variables are biological dry mass or cell concentration, X, substrate concentration, S, and product concentration, P. The reactor volume V is assumed to be well mixed, and cell growth is assumed to follow Monod kinetics, based on one limiting substrate, as described in Section 1.3.6. Substrate consumption is related to cell growth by means of a constant yield factor $Y_{X/S}$. Product formation is the result of both growth and non-growth associated rates of production. The lag and decline phases of cell growth are not included in the model.

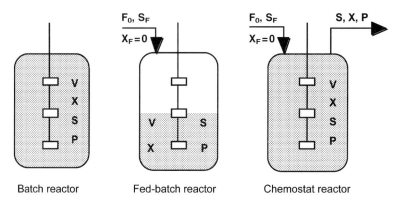

Fig. 1 Three modes of operation for a biological reactor.

Simple batch fermentation system is simulated within the program by setting the values of the input and outlet flow rates to zero.

For fed-batch fermentation, the model equations need to include the continuous feeding of sterile substrate to the fermenter, but zero outflow. The increase in volume (total accumulation of mass) that occurs in the fermenter due to the feeding is represented by a total mass balance relationship.

A continuous fermenter with sterile feed is referred to as a chemostat. For constant volume operation, the inlet volumetric flow rate is equal to that at the output. With this model chemostat start-up, resultant steady state behaviour and cell washout phenomena are easily investigated by simulation.

Model

The generalised material balances for a well-stirred tank fermentation can be represented as:

(Accumulation rate) = (In by flow) − (Out by flow) + (Production rate)

Total balance

$$\frac{dV}{dt} = F_0 - F_1$$

Cell balance

$$\frac{d(VX)}{dt} = -F_1 X + r_X V$$

Substrate balance

$$\frac{d(VS)}{dt} = F_0 S_F - F_1 S + r_S V$$

Product balance

$$\frac{d(VP)}{dt} = -F_1 P + r_P V$$

Knowing the mass quantities, VX, VS, and VP, the corresponding concentration terms are determined as

$$X = \frac{VX}{V}, \quad S = \frac{VS}{V}, \quad P = \frac{VP}{V}$$

The kinetic equations for the model are given by

$$r_X = \mu X$$

with μ being given by the Monod equation:

$$\mu = \frac{\mu_m S}{(K_S + S)}$$

and the rate of consumption of the limiting substrate

$$r_S = -\frac{r_X}{Y_{X/S}}$$

The rate of product formation is assumed to follow both a growth related and non-growth related kinetic form

$$r_P = (k_1 + k_2\mu)X$$

where k_1 is the non-growth associated coefficient and k_2 is the coefficient associated with growth.

Program

The MADONNA program, is designed to simulate the three reactor modes of operation by enabling the setting of appropriate values of the input and output flow rates.

Nomenclature

Symbols

D	Dilution rate ($=F_0/V$)	1/h
F_0	Volumetric feed rate	m^3/h
F_1	Volumetric outlet rate	m^3/h
k_1 and k_2	Product formation constants	1/h and kg/kg
K_S	Saturation constant	kg/m^3
M	Maintenance coefficient	kg/kg h
P	Product concentration	kg/m^3
r	Reaction rate	kg/m^{3}h
S	Substrate concentration	kg/m^3
V	Reactor volume	m^3
X	Biomass concentration	kg/m^3
Y	Yield coefficient	kg/kg
μ and U	Specific growth rate	1/h
τ	Time lag constant	h

Indices

1	Refers to non-growth association rate
2	Refers to growth-association rate
F	Refers to feed
m	Refers to maximum
P	Refers to product
S	Refers to substrate
X	Refers to biomass

Exercises for Batch Operation

1. Vary the kinetic parameters K_S and μ_m and observe the effects for the case of a batch reactor.
2. Vary the product kinetics constants (k_1 and k_2) and observe the effects. Note the shape of the P versus time curve when S reaches zero.
3. Vary the initial amount of biomass and note its influence on the batch performance.

Exercises for Chemostat Operation

1. Increase the volumetric flow rate successively until washout is obtained.
2. Change the feed concentration of substrate. Does this alter the steady state value of S in the reactor outlet? Why?
3. Under steady state conditions, the specific growth rate becomes equal to the dilution rate, $\mu = D$. Using the Monod equation, calculate a set of steady state S versus D values. Verify this result by simulation.
4. What is the effect of operating the chemostat at a dilution rate of $D > \mu_m$?
5. Operate the chemostat initially as a batch reactor with $D=0$, and then switch to chemostat operation with $D < \mu_m$. Note the effect on the startup time.

Exercises for Fed-batch Operation

1. Start as a batch reactor and then switch to fed batch operation. Explain why this procedure is preferable to starting initially as a fed batch?
2. Operate the fed batch with high values of SF and low values of F. Note that it is possible to obtain a quasi steady state condition for which μ becomes equal to dilution rate $D = F/V$. It is seen from the simulations that this results in constant values of X but low and slowly decreasing values of S.
3. Operate at low SF and note the slope of the X versus time curve, which will eventually become linear, and note also that the slope is directly related to the feed rate. Why should this be so?

Results

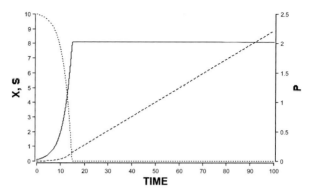

Fig. 2 Concentration profiles for a batch reaction.

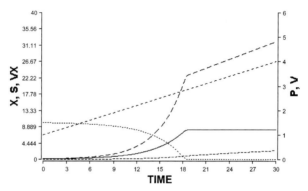

Fig. 3 Results from a fed batch operation. Here both the biomass and the biomass concentration are plotted.

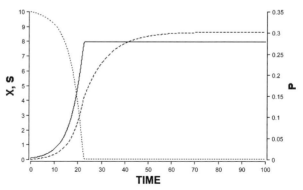

Fig. 4 Results from a chemostat run, where the steady state values of the concentrations can be observed.

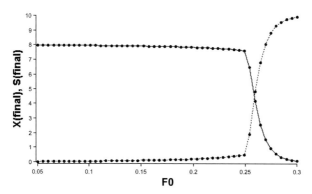

Fig. 5 A parametric run for a chemostat varying F_0 versus final steady state values for X and S. Washout is shown in the region of UM. Here $V=1$ and therefore $D=F_0$.

5.13.2
INHIBCONT – Continuous Bioreactor with Inhibitory Substrate

System

Inhibitory substrates at high concentrations reduce the specific growth rate below that predicted by the simple Monod equation. The inhibition function may often be expressed empirically as

$$\mu = \frac{\mu_m S}{K_S + S + S^2/K_I}$$

where K_I is the inhibition constant (kg m^{-3}).

In batch cultures the term S^2/K_I may be significant during the early stages of growth, even for higher values of K_I, owing to the initially high values of S. If substrate concentrations are low, the inhibition function reduces to the Monod equation. It can be shown that the value of μ passes through a maximum at $S_{max} = (K_S K_I)^{0.5}$.

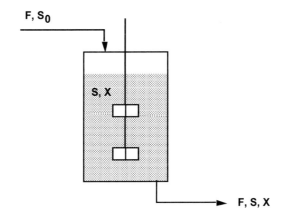

Fig. 1 Model variables for a continuous bioreactor.

Model

A continuous bioreactor with its variables is represented in Fig. 1.

Biomass balance

$$V\frac{dX}{dt} = \mu VX - FX$$

or

$$\frac{dX}{dt} = (\mu - D)X$$

where D is the dilution rate (= F/V). Thus steady-state behaviour, where $dX/dt = 0$, is represented by the conditions that $\mu = D$.

Substrate mass balance

$$V\frac{dS}{dt} = F(S_0 - S) - \frac{\mu XV}{Y}$$

or

$$\frac{dS}{dt} = D(S_0 - S) - \frac{\mu X}{Y}$$

where Y is the yield factor for biomass from substrate. Also from this equation at steady state, since $\mu = D$ and $dS/dt = 0$, the steady-state cell concentration is given by

$$X = Y(S_0 - S)$$

A continuous inhibition culture will often lead to two possible steady states, as defined by the steady-state condition $\mu = D$, as shown in Fig. 2.

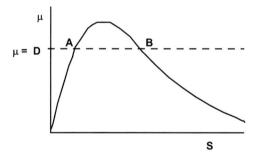

Fig. 2 Steady states A (stable) and B (unstable) for a chemostat with inhibition kinetics.

One of these steady states (A) can be shown to be stable and the other (B) to be unstable by reasoning somewhat similar to that described in Section 3.2.7. Thus only state A and the washout state ($S=S_0$) are possible. Using simulation techniques, the behaviour of the dynamic approach to these states can be investigated.

Program

When the system equations are solved dynamically, one of two distinct steady-state solutions is obtained, i.e., the reactor passes through an initial transient but then ends up under steady-state conditions either at the stable operating condition, represented by A, or the washout condition, for which $X=0$. The initial concentrations for the reactor will influence the final steady state obtained. A PI controller has been added to the program, and it can be used to control a substrate setpoint below S_{max}. The controller can be turned on setting by $K_P > 0$. The control constants K_P, τ_I and the time delay τ_F can be adjusted by the use of sliders to obtain the best results. Appropriate values of control constants might be found in the range 0.1 to 10 for K_P and 0.1 to 10 for τ_I. Note that the control does not pass S_{max} even though the setpoint may be above S_{max}. Another feature of the controller is a time delay function to remove chatter. The program comments should be consulted for full details.

Nomenclature

Symbols

D	Dilution rate	1/h
F	Flow rate	m³/h
K_I	Inhibition constant	kg/m³
K_P	Controller constant	kg/m³
K_S	Saturation constant	m⁶/kg h
S	Substrate concentration	kg/m³
V	Volume	m³
X	Biomass concentration	kg/m³

Y	Yield coefficient	kg/kg
μ (U)	Specific growth rate coefficient	1/h
τ_I	Controller time constant	h
τ_F	Time constant controller delay	h

Indices

0	Refers to inlet
I	Refers to initial value
m, max	Refer to maximum

Exercises

1. Vary K_S and K_I to observe the changes in the inhibition curve by plotting versus S. Note especially the value of S at the maximum in μ.
2. Start the reactor with the initial concentration conditions of S=0, X=2 and D<μ_m, and observe the approach to steady state. Did washout occur?
3. Start the reactor with S>S_{max} and X=0.1. What is the steady-state result?
4. Run a simulation under the conditions of Exercise 3 and program to stop the feed (D=0) just before washout. Allow S to return to near zero. Plot μ versus S.
5. Rerun Exercise 2 with S_0=20. Stop the flow when S>S_{max}, and allow the system to return to S<S_{max} before starting again.
6. Operate the model to include feedback control of the feed flow rate and experiment with various values of K_P, τ_I and τ_F.
7. Make suitable changes in the initial conditions of X and S, and plot the phase plane diagram X versus S. By making many runs at a range of initial conditions, the washout region can be identified.

Results

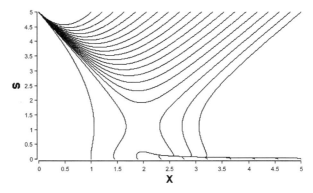

Fig. 3 A phase plane made by varying the initial conditions for S and X, with two Multiple Runs on an overlay plot.

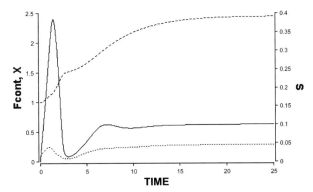

Fig. 4 A control simulation of the process with the setpoint below S_{max}.

Reference

Edwards, V. H., Ko, R. C. and Balogh, S. A. (1972) Dynamics and Control of Continuous Microbial Propagators Subject to Substrate Inhibition, Biotechnol. Bioeng. 14, 939–974.

5.13.3
NITBED – Nitrification in a Fluidised Bed Reactor

System

Nitrification is the sequential oxidation of NH_4^+ to NO_2^- and NO_3^- which proceeds according to the following reaction sequence:

$$NH_4^+ + \frac{3}{2}O_2 \rightarrow NO_2^- + H_2O + 2H^+$$

$$NO_2^- + \frac{1}{2}O_2 \rightarrow NO_3^-$$

The overall reaction is thus

$$NH_4^+ + 2O_2 \rightarrow NO_3^- + H_2O + 2H^+$$

Both steps are influenced by dissolved oxygen and the corresponding substrate concentration. The nitrification as a wastewater treatment process benefits greatly from biomass retention, owing to the relatively slow growth rates of the nitrifiers.

In this example, a fluidised biofilm sand bed reactor for nitrification, as investigated by Tanaka et al. (1981), is modelled as three tanks-in-series with a recycle loop (Fig. 1). With continuous operation, ammonium ion is fed to the reactor, and the products nitrite and nitrate exit in the effluent. The bed expands in volume because of the constant circulation flow of liquid upwards through the bed. Oxygen is supplied external to the bed in a well-mixed gas-liquid absorber.

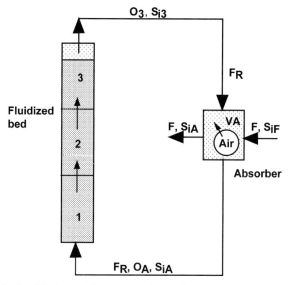

Fig. 1 Biofilm fluidised-bed recycle loop reactor for nitrification.

Model

The model balance equations are developed by considering both the individual tank stages and the absorber. Component balances are required for all components in each section of the reactor column and in the absorber, where the feed and effluent streams are located. Although the reaction actually proceeds in the biofilm phase, a homogeneous model apparent kinetics model is employed, which is justified by its simplicity.

In the absorber, oxygen is transferred from the air to the liquid phase. The nitrogen compounds are referred to as S_1, S_2, and S_3, respectively. Dissolved oxygen is referred to as O. Additional subscripts, as seen in Fig. 1, identify the feed (F), recycle (R) and the flows to and from the tanks 1, 2 and 3, each with volume V, and the absorption tank with volume V_A.

The fluidised bed reactor is modelled by considering the component balances for the three nitrogen components (i) and also for dissolved oxygen. For each stage n, the component balance equations have the form

$$\frac{dS_{in}}{dt} = \frac{F_R}{V}(S_{i,n-1} - S_{i,n}) - r_{Sin}$$

$$\frac{dO_n}{dt} = \frac{F_R}{V}(O_{i,n-1} - O_{i,n}) - r_{on}$$

Similarly for the absorption tank, the balance for the nitrogen containing components include the input and output of the additional feed and effluent streams, giving

$$\frac{dS_{iA}}{dt} = \frac{F_R}{V_A}(S_{i,3} - S_{iA}) + \frac{F}{V_A}(S_{iF} - S_{iA})$$

The oxygen balance in the absorption tank must account for mass transfer from the air, but neglects the low rates of oxygen supply and removal of the feed and effluent streams. This gives

$$\frac{dO_A}{dt} = \frac{F_R}{V_A}(O_3 - O_A) + K_L a(O_A^* - O_A)$$

For the first and second biological nitrification rate steps, the reaction kinetics for any stage n are given by

$$r_{1n} = \frac{v_{m1} S_{1n}}{K_1 + S_{1n}} \frac{O_n}{K_{O1} + O_n}$$

$$r_{2n} = \frac{v_{m2} S_{2n}}{K_2 + S_{2n}} \frac{O_n}{K_{O2} + O_n}$$

The oxygen uptake rate is related to the above reaction rates by means of the constant yield coefficients, Y_1 and Y_2, according to

$$r_{On} = -r_{1n} Y_1 - r_{2n} Y_2$$

The reaction stoichiometry provides the yield coefficient for the first step

$$Y_1 = 3.5 \text{ mg } O_2/(\text{mg } N_{NH_4})$$

and for the second step

$$Y_2 = 1.1 \text{ mg } O_2/(\text{mg } N_{NO_2})$$

Nomenclature

Symbols

F	Feed and effluent flow rate	L/h
F_R	Recycle flow rate	L/h
$K_L a$	Transfer coefficient	h
K	Saturation constants	mg/L
K_1	Saturation constant for ammonia	mg/L
K_2	Saturation constant for ammonia	mg/L
O	Dissolved oxygen concentration	mg/L
O_S and O^*	Oxygen solubility, saturation conc.	mg/L
Our	Oxygen uptake rate	mg/L
r	Reaction rate	mg/L h
S	Substrate concentration	mg/L
V	Volume of one reactor stage	L
V_A	Volume of absorber tank	L

v_m	Maximum velocity	mg/L h
Y	Yield coefficient	mg/mg

Indices

1, 2	Refer to oxidation rates of ammonia and nitrite, resp.
1, 2, 3	Refer to ammonia, nitrite and nitrate, resp.
1, 2, 3	Refer to stage numbers
A	Refers to absorption tank
F	Refers to feed
ij	Refers to substrate i in stage j
m	Refers to maximum
O1 and O2	Refer to oxygen in first and second reactions
S1, S2	Refer to substrates ammonia and nitrite
S and *	Refer to saturation value for oxygen

Exercises

1. Assuming maximum reaction rates, choose a value of F and S_{1F}, and estimate the outlet substrate concentration at steady state. What oxygen supply rate would be required? Verify by simulation.
2. Vary $K_L a$ to give a low sufficient oxygen transfer capacity and note that low dissolved oxygen concentrations result in high nitrite concentrations. Investigate the reasons for this.
3. Vary the operating parameters and interpret the results in terms of the axial oxygen gradients in the reactor column.
4. Vary the rates of substrate and oxygen supply and note the interaction of both these factors on the outlet values of S_{13} and O_3.
5. Operate the reactor continuously with a $K_L a$ value of 40 1/h. With an ammonia feed concentration of 100 mg/L find the feed flow rate which will give 99% conversion to nitrate.

Results

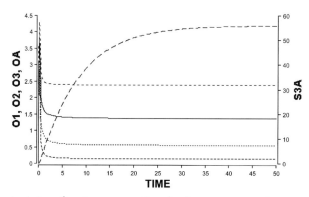

Fig. 2 Dynamic startup of continuous operation showing oxygen concentrations and nitrate in the effluent.

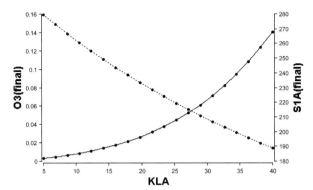

Fig. 3 Parametric run of continuous operation showing oxygen at the column top and ammonia in the effluent versus $K_L a$.

5.13.4
BIOFILM – Biofilm Tank Reactor

System

Active biomass can be retained effectively within continuous reactors by providing a carrier surface upon which a biofilm can develop. Examples are trickling filters (see BIOFILT) and fluidised bed reactors (see NITBED). In this example, biomass is retained within an aerated continuous tank reactor as a biofilm (Fig. 1). Wastewater entering the reactor is mixed throughout the tank and the pollutant S diffuses into the biofilm, along with oxygen, where it is degraded. The overall reaction rate in such a case will depend on the concentration gradients within the film. A diffusion model can be used to describe the biofilm diffusion-reaction process. As shown in Fig. 2, the biofilm is divided into segments for simulation purposes. Mass balances are written in terms of diffusion fluxes for each segment, and the diffusion fluxes at each position are expressed in terms of the corresponding concentration driving forces. An appropriate kinetic model is used for the local reaction rate, which depends on the local concentration within the segment. The bulk phase is coupled to the biofilm by means of an additional relationship representing the diffusion flux at the liquid-biofilm interface.

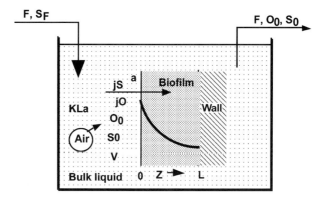

Fig. 1 Continuous tank reactor with biofilm.

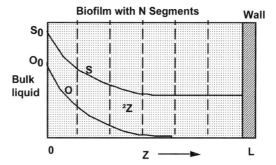

Fig. 2 Finite differencing of the concentration profiles within the biofilm into segments 1 to N.

Model

Multicomponent reaction within a biofilm can be described by diffusion-reaction equations. A component mass balance is written for each segment and for each component, respectively, where

$$\begin{pmatrix} \text{Accumulation} \\ \text{rate} \end{pmatrix} = \begin{pmatrix} \text{Diffusion} \\ \text{rate in} \end{pmatrix} - \begin{pmatrix} \text{Diffusion} \\ \text{rate out} \end{pmatrix} + \begin{pmatrix} \text{Production rate} \\ \text{by reaction} \end{pmatrix}$$

$$A \Delta Z \frac{dS_n}{dt} = j_{n-1} A - j_n A + r_{Sn} A \Delta Z$$

Using Fick's law, the diffusional flux can be described by

$$j_{n-1} = D_S \frac{S_{n-1} - S_n}{\Delta Z}$$

giving

$$\frac{dS_n}{dt} = D_S \frac{(S_{n-1} - 2S_n + S_{n+1})}{\Delta Z^2} + r_S$$

Thus N dynamic equations are obtained for each component at each position, within each segment. The equations for the first and last segment must be written according to the boundary conditions. The boundary conditions for this case correspond to the following: the bulk tank concentration is S_0 at the external surface of the biofilm where $Z=0$; a zero flux at the biofilm on the wall means that $dS/dZ=0$ at $Z=L$.

The kinetics used here consider carbon-substrate inhibition and oxygen limitation. Thus,

$$r_S = -v_m \frac{S}{K_S + S + (S^2/K_I)} \frac{O}{K_O + O}$$

A constant yield coefficient Y_{OS} describes the oxygen uptake rate

$$r_O = Y_{OS} r_S$$

For the well-mixed continuous-flow liquid phase shown in Fig. 1, the balance equations for oxygen and substrate must account for the supply of each component both by convective flow and by gas-liquid transfer, as well as by the diffusion rate into the biofilm.

For substrate with continuous inflow and outflow from the reactor:

$$\frac{dS_0}{dt} = \frac{F}{V}(S_F - S_0) - aD_S \frac{S_0 - S_1}{\Delta Z}$$

For oxygen transferred from the gas phase

$$\frac{dO_0}{dt} = K_L a(O_S - O_0) - aD_O \frac{O_0 - O_1}{\Delta Z}$$

In the above equations, symbol "a" represents the area of biofilm per unit volume of bulk liquid. The diffusion rates in and out of the biofilm are driven by concentration differences between the bulk liquid (0) and the outer biofilm segment (1).

Nomenclature

Symbols

a	Specific area perpendicular to the flux	1/m
A	Area perpendicular to diffusion flux	m^2
D	Diffusion coefficient	m^2/h

F	Volumetric flow rate	m³/h
j	Diffusion flux	g/m² h
K_S	Saturation constant	g/m³
K_I	Inhibition constant	g/m³
$K_L a$	Oxygen transfer coefficient	1/h
L	Biofilm thickness	m
O	Dissolved oxygen concentration	g/m³
O_S	Saturation concentration for oxygen	g/m³
R, r	Reaction rate	g/m³h
S	Substrate conc. of carbon source	g/m³
V	Volume of tank	m³
v_m	Maximum reaction rate	g/m³h
Y_{OS}	Yield for oxygen uptake	–
Z	Length of element	m

Indices

0	Refers to bulk liquid
1–N	Refer to sections 1–N
I	Refers to inhibition
O	Refers to oxygen
S	Refers to carbon source
n	Refers to section n
F	Refers to feed

Exercises

1. Observe the response time of the reactor for changes in the operating parameters S_F, F and $K_L a$. Is the biofilm always the slowest to respond? Relate the response to the diffusion time constant L^2/D.
2. Experiment by varying the magnitude of the parameter K_I.
3. Vary the ratio of substrate and oxygen in the bulk phase, O_0/S_0, above and below the value of Y_{OS} for a series of runs. Note that the steady-state penetration limitation of one or the other substrates depends on whether the bulk ratio is greater or equal to the stoichiometric requirement, Y_{OS}.
4. Vary the amount of biomass by changing the numerical value of v_m and note the influence on the reaction rate.
5. Use the model to perform steady-state experiments to determine the apparent kinetics. Calculate the rates by sampling the substrate in the feed and effluent. Avoid the inhibition effect by setting K_I very high, while keeping S_0 high enough to maintain nearly zero order throughout the biofilm. By varying $K_L a$, run the reactor at various bulk dissolved oxygen concentrations. Plot the results as substrate uptake rate versus O_0 to determine the apparent K_S value for oxygen.
6. Repeat Exercise 5 for substrate kinetics, while operating at constant O_0.

Results

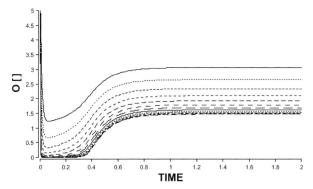

Fig. 3 Response of the oxygen in the 12 segments of the biofilm.

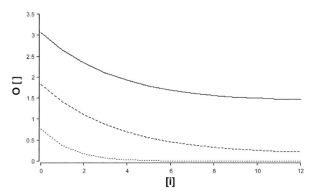

Fig. 4 Oxygen concentrations with distance from the outside (left side) to the inside of the film for three values of K_La.

5.13.5
BIOFILT – Biofiltration Column for Removing Ketone from Air

System

Biofiltration is an environmental process for treating contaminated air streams. Moist air is passed through a packed column, in which the pollutants in the contaminated air are adsorbed onto the moist packing. There the resident population of organisms oxidizes the pollutants biologically. Such columns can be run with a liquid phase flow (bio-trickling filter) or only with moist packing (biofilter). The work of Deshusses et al. (1994, 1995) investigated the removal of two ketones, methyl isobutyl ketone (MIBK) and methyl ethyl ketone (MEK) in such a biofilter The kinetics of this multi-substrate system is especially interesting since both substances exhibit mutual inhibitory effects on their rates of degradation. Here the original dynamic model is greatly simplified in this example to calculate the steady-state concentration profiles in the column for the removal of only one component, MEK. The modelling concept is similar to the example of countercurrent gas-liquid absorption (Section 5.9.10, AMMONAB).

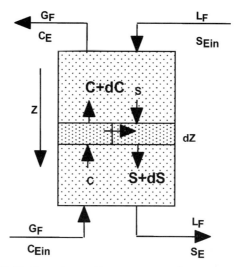

Fig. 1 Steady-state biofiltration column with countercurrent flow.

Model

The kinetics can be described approximately according to the Monod-type kinetics.

For MEK

$$r_{SE} = -\frac{v_{MSE}}{S_E + K_{SE}}$$

The steady-state mass balances for countercurrent flow can be formulated by considering changes in the component mass rates over a differential length of column. In this case, the length is measured from the top of the column ($Z=0$) to the bottom, as shown in Fig. 1. The column is assumed to consist of a moist solid or liquid phase with volume V_S, and a gas phase of volume V_G. The actual support volume is included in V_S.

Thus for MEK in the gas phase

$$\frac{dC_E}{dZ} = \frac{1}{v_{ZG}} K_L a (S_{ES} - S_E)$$

5.13 Biological Reaction Examples

Thus in these equations for the steady-state, the convective terms are balanced by the transfer terms. Here the K_La value is based on the total volume V_T and is assumed to have the same value for each component. The linear velocities, expressed in terms of the volumetric flow rates and the empty tube cross-section, are for the gas and liquid phases

$$v_{ZG} = \frac{G_F}{A_T}$$

$$v_{ZL} = \frac{L_F}{A_T}$$

Assuming the reaction to occur within the wet packing volume V_S, the liquid phase balance for MEK is

$$\frac{dS_E}{dZ} = \frac{1}{v_{ZL}} K_La(S_{ES} - S_E) + r_{SE}\frac{1}{v_{ZL}}(1-\varepsilon)$$

The gas-liquid equilibrium relationship for the MEK is represented by

$$S_{ES} = M_E C_E$$

Very low liquid flow rates correspond to the case of biofilter operation and higher rates to the case of a trickling filter. The formulation of this problem for the two ketones would best be done with a tanks in series dynamic model. An alternative iterative steady state solution is given in Snape et al. (1995) for the two-component case with recycle.

Program

The program BIOFILT solves the steady-state countercurrent biofiltration operation for MEK removal. The integration is started at the top of the column by assuming a desired concentration in the outlet air. The required column length is determined when the inlet gas concentration exceeds the actual inlet value.

Nomenclature

Symbols

A_T	Cross-sectional area of empty tube	m^2
C	Concentration in gas phase	kg/m^3
D	Diameter	m
G_F	Gas flow rate	m^3/s
K_La	Mass transfer coefficient	1/s
K_S	Monod coefficient	kg/m^3

L_F	Liquid flow rate	m³/s
M	Partition coefficient	–
r	Reaction rate	kg/m³s
S	Concentration in moist or liquid phase	kg/m³
v_M	Maximum reaction velocity	kg/m³s
v_{ZG}	Velocity of the gas phase	m/s
v_{ZL}	Velocity of the liquid phase	m/s
V_G	Volume of liquid phase	m³
V_S	Volume of solid or liquid phase	m³
Z	Length or height	m
ε	Bed porosity	–

Indices

E	Refers to MEK
G	Refers to gas
L	Refers to liquid
M	Refers to maximum
S	Refers to saturation
Z	Refers to axial direction

Exercises

1. Investigate the influence of the entering gas flow rate and entering concentration on the column length required to obtain 90% removal of MEK.
2. Alter the program to describe a cocurrent column. Compare the results with the countercurrent case for various extremes of liquid flow rate. Using the length found in Exercise 1, operate the column cocurrently to see whether the same degree of removal can be obtained. Show that the comparison depends on whether the column performance is reaction rate or transfer rate controlled.
3. The activity of the organisms v_M and the value of the mass transfer coefficient are important to the process. Vary these to determine their influence on the column length required.
4. Operate BIOFILT as a trickling filter by setting reasonably high values of liquid flow and investigate the influence of liquid flow rate.
5. Develop a model and program this for the case of a countercurrent column removing two components. Use a dynamic model to avoid the difficulty of making iterations for steady state.

Results

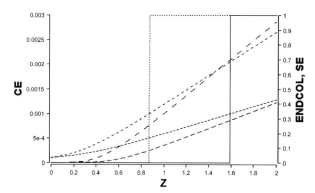

Fig. 2 Using two gas flow rates gave the respective column lengths required.

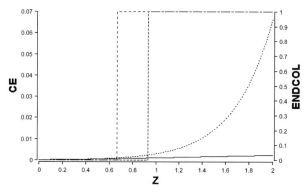

Fig. 3 Using very low and very high liquid flow rates gave these column lengths for the desired elimination.

References

Deshusses, M.A. (1994) ETH-Zurich, PhD Thesis, No. 10633.

Deshusses, M.A., Hamer, G. and Dunn, I.J. (1995) Part I, Behavior of Biofilters for Waste Air Biotreatment: Part I, Dynamic Model Development and Part II, Experimental Evaluation of a Dynamic Model, Environ. Sci. Technol. 29, 1048–1068.

5.14
Environmental Examples

5.14.1
BASIN – Dynamics of an Equalisation Basin

System

The flow rate and concentration of wastewater do not remain constant but vary during the course of the day and are also dependent on the time of year. If the flow rate is too high, loss of micro-organisms by washout may occur in secondary treatment processes. If the flow rate is too low, then the lack of nutrients will lead to a reduction of the micro-organism population. Wastewaters entering a treatment plant usually flow first into an equalisation basin, so that the flow rate out of the basin is maintained constant, or between prescribed limits, to protect the subsequent processes. The equalisation tank also reduces the effect of toxic shocks on the biological processes within the main treatment plant.

Model

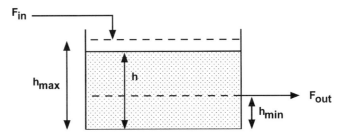

Fig. 1 Basin for process flow equalisation.

The effluent flow rate is assumed to show a diurnal variation with maximum values occurring in the morning and early evening. This variation can be approximated by a sine wave function:

$$F_{in} = F_{av} + F_{amp} \sin\left(\frac{2\pi t}{12} - \pi\right)$$

where t is the time in hours, F_{av} the average flow rate and F_{amp} the amplitude of the variation in flow rate. This function gives maximum flow rates at 9 am and 9 pm, assuming that t=0 at midnight and accounts for the average time taken for water to travel from household to the treatment plant. A random component may be added to the flow rate to give more realistic variations in flow rate. Variations in the concentration of organics (S), both in the wastewater and in the tank, are not considered in this simplified example.

5.14 Environmental Examples

The purpose of the tank (Fig. 1) is to ensure that the flow rate out of the tank, F_{out}, is kept constant at a value equal to F_{con} for as long as the depth of the liquid in the basin does not exceed some maximum value or fall below a minimum value, i.e.,

$$F_{out} = F_{con} \quad \text{for} \quad h_{min} < h < h_{max}$$

If the liquid depth exceeds the maximum depth of the tank, then the outlet flow F_{out} will equal the inlet flow F_{in}, so long as F_{in} exceeds the required flow rate to the plant F_{con}. If F_{in} falls below the value of F_{con}, the outlet flow F_{out} will be maintained equal to F_{con} for as long as possible. This prevents the tank from overflowing and allows the liquid level to fall when the inlet flow is reduced.

$$F_{out} = F_{in} \quad \text{for} \quad h > h_{max} \quad \text{and} \quad F_{in} > F_{con}$$

$$F_{out} = F_{con} \quad \text{for} \quad h > h_{max} \quad \text{and} \quad F_{in} < F_{con}$$

The magnitude of the inlet flow similarly regulates the outlet flow rate whenever the water level drops below the minimum depth. This prevents the tank from running dry and allows the tank to fill up again when the inlet flow is increased. In this case, the regulation of outlet flow is expressed by

$$F_{out} = F_{in} \quad \text{for} \quad h < h_{min} \quad \text{and} \quad F_{in} < F_{con}$$

$$F_{out} = F_{con} \quad \text{for} \quad h < h_{min} \quad \text{and} \quad F_{in} > F_{con}$$

The tank is assumed to have vertical sides so that the depth of liquid can be related to the volume (V) and tank cross-sectional area (A) by

$$V = Ah$$

A total mass balance for the tank gives the relationship

$$\frac{dV}{dt} = F_{in} - F_{out}$$

i.e., the rate of change in the volume of the tank contents with respect to time is equal to the difference in the volumetric flow rate to and from the tank.

Substitution gives

$$\frac{d(Ah)}{dt} = F_{in} - F_{out}$$

which for constant tank cross-sectional area this becomes

$$\frac{dh}{dt} = \frac{F_{in} - F_{out}}{A}$$

Nomenclature

A	Cross-sectional area	m²
F_{in}	Volumetric flow rate into basin	m³ h⁻¹
F_{av}	Daily average volumetric flow rate	m³ h⁻¹
F_{amp}	Daily variation in flow rate	m³ h⁻¹
F_{con}	Constant flow rate	m³ h⁻¹
F_{out}	Flow rate out of basin	m³ h⁻¹
F_{storm}	Storm flow rate	m³ h⁻¹
L	Length of basin	m
h_{max}	Maximum depth of water in basin	m
h_{min}	Minimum depth of water in basin	m
h	Depth of water at time t	m
Q_{av}	Annual average volumetric flow rate	m³ h⁻¹
Q_{amp}	Annual variation in flow rate	m³ h⁻¹
S_{in}	Concentration of organics in inflow	g m⁻³
S	Concentration of organics in tank	g m⁻³
t	Time	h
V	Volume of basin	m³
W	Width of basin	m

Exercises

1. During a storm the flow rate of wastewater will increase significantly. This effect can be simulated by programming conditional (IF.THEN.ELSE) changes in F_{STORM} at certain times. Simulate storms of varying duration and severity and study how the basin can cope with these varying demands while maintaining a constant outlet flow rate.

2. One problem with the above control strategy is that when the tank fails there is a rapid change in the flow rate entering the main treatment plant. Modify the program so that the flow rate leaving the tank is controlled such that the flow is proportional to the depth of liquid in the tank. The result of this should be that although oscillations in the flow rate are damped and not eliminated entirely, the risk of complete and sudden failure is reduced.

3. Modify the program to allow for inlet variation in the concentration of organics. This can be simulated by inclusion of the component balance equation

$$\frac{d(VS)}{dt} = F_{in} S_{in} - F_{out} S$$

where S = VS/V. The magnitude of the inlet organic concentration S_{in} will need to be defined.

Is constant flow rate the best strategy to avoid variations to subsequent processes?

4. As well as daily variations in flow rate and concentration there are also annual variations. These can be simulated by assigning a variable function to the value of F_{av}. For instance, the following sine function could be used:

$$F_{av} = Q_{av} + Q_{amp} \sin\left(\frac{2\pi t}{365 \times 24}\right)$$

where Q_{av} and Q_{amp} are the annual average flow rate and amplitude of the annual variations in flow rate, respectively. Include this in the model and see how this additional factor affects the operation of the basin.

5. In practice not all basins are rectangular in shape and the cross-sectional area of the basin can vary with height. Include this effect in the model assuming a basin with a trapezoidal cross-sectional area, where the height and volume of liquid are related by the following formula:

$$V = h(L + 2h)(W + 2h)$$

where L and W are the length and width of the base of the basin, respectively (i.e., a basin for which the walls are at 45° to the vertical).

6. Use a conditional statement to simulate the effect of a shock load on the system: i.e., change the value of S_{in} for a short time and follow the resulting variations in the value of S.

7. Modify the program to operate at constant organic concentration rather than constant flow rate and repeat the Exercises 1 to 6. Which strategy gives the minimum variation in load to subsequent processes? Devise a strategy that takes into account both changes in flow rate and concentration.

8. Modify the program so that the organic concentration is reduced in the basin by the biological action of bacteria in the wastewater. Assume that the basin remains well mixed so that the concentration in the outflow is the same as that in the basin.

Results

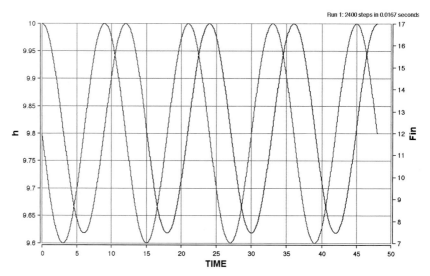

Fig. 2 Variation in height with inlet flow, maintaining outlet flow constant.

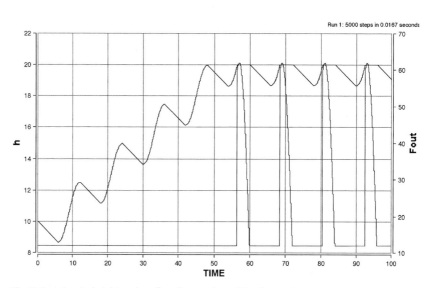

Fig. 3 Variation in height and outflow for very large inlet fluctuations.

5.14.2
METAL – Transport of Heavy Metals in Water Column and Sediments

System

Heavy metals find their way into the water cycle via the natural processes of erosion, weathering and volcanic activity. In addition the human activities of mining, smelting and burning fossil fuels contribute significantly to the metals found in natural waters. These metals can be accumulated by some micro-organisms and thus enter into the food chain, so presenting a possible toxic risk to wildlife and humans. Rivers passing through industrial and mining areas transport metals, partly as metal in solution and partly as metal adsorbed to suspended material. This suspended material sediments out in lakes and estuaries and there accumulates in the sediment. The model presented here, based on van de Vrie (1987), aims at predicting the concentration of zinc, cadmium and lead in the sediment and water of a lake in the southwest of the Netherlands.

Model

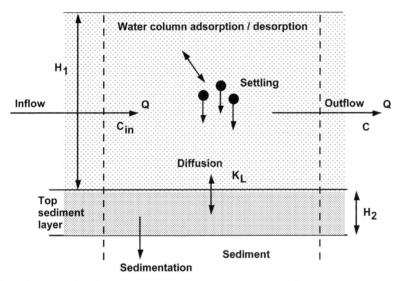

Fig. 1 Balance regions for heavy metal transport for a lake in the Netherlands.

The whole lake is divided into a number of vertical elements, each comprising a column of water and the top layer of sediment. In this simulation only one such element is considered in order to improve the speed of execution. The general principle however can be extended to many elements.

It is assumed that the metals are distributed homogeneously in the water column as well as in the upper sediment layer. In both compartments, the metals can exist in either a dissolved form or as adsorbed to solid particles.

$$C_T = C_d + C_a$$

$$C_d = f_d C_T$$

The dissolved fraction, f_d, is given by the distribution coefficient, K_D:

$$f_d = \frac{1}{1 + mK_D}$$

where m is the total concentration of the suspended particles.

The suspended particles settle with a velocity v_s, and enter the top sediment layer, at a rate given by

$$J_s = v_s A C_a = v_s A (1 - f_d) C_T$$

The particles in the top layer of sediment also settle until they reach a certain depth at which they are no longer influenced by the water column.

There is a diffusion exchange between the two compartments owing to differences in the dissolved metal concentration in the water column and in the top layer of the sediment. The rate of this exchange is proportional to the concentration difference and the area of the sediment-water interface and is defined by proportionality constant K_L.

$$J_d = K_L A (C_{d1} - C_{d2})$$

Water flows into the water column at a volumetric flow rate Q containing a total metal concentration C_{in}. It is assumed that water flows out from the column at the same rate and at the same concentration as in the water column. Mass balances on the water column and top sediment layer give

$$V_1 \frac{dC_{T1}}{dt} = Q(C_{in} - C_{T1}) - J_{s1} - J_d$$

$$V_2 \frac{dC_{T2}}{dt} = J_{s1} + J_d - J_{s2}$$

where V_1 and V_2 are the volumes of the water column and top layer of sediment, respectively, and H_1 and H_2 are the corresponding heights.

Substituting and rearrangement gives

$$\frac{dC_{T1}}{dt} = Q(C_{in} - C_{T1}) - \frac{v_s}{H_1}(1 - f_d)C_T - \frac{K_L}{H_1}(C_{d1} - C_{d2})$$

$$\frac{dC_{T2}}{dt} = \frac{V_{s1}}{H_1}(1 - f_{d1})C_{T1} + \frac{K_L}{H_2}(C_{d1} - C_{d2}) - \frac{V_{s2}}{H_2}(1 - f_{d2})C_{T2}$$

Nomenclature

Symbols

A	Area of water-sediment interface in each element	m^2
C	Concentration of metal	$g\,m^{-3}$
f_d	Fraction of metal that is dissolved	–
H_1	Height of water column	m
H_2	Depth of top sediment layer	m
J	Fluxes of metals	$g\,m^{-2}\,day^{-1}$
K_D	Distribution coefficient	g^{-1}
K_L	Diffusion exchange coefficient	$m\,day^{-1}$
m	Concentration of suspended particles	$g\,m^{-3}$
Q	Volumetric flow rate of water into water column	$m^3\,day^{-1}$
t	Time	days
V_1	Volume of water column	m^3
V_2	Volume of top sediment layer	m^3
v_s	Sedimentation velocity	$m\,day^{-1}$

Indices

1	Water column
2	Top sediment layer
a	Adsorbed
d	Dissolved
s	Settling
T	Total

Exercises

1. Run the model with the following parameters and compare the relative rates of accumulation for the different metals.

Parameters for METAL:

Parameter	Zinc	Cadmium	Lead
Thickness of top sediment layer (m)	0.39	0.37	0.26
Distribution coefficient water column (g^{-1})	114	81	309
Distribution coefficient sediment (g^{-1})	18.8	22.0	38.0
Diffusion exchange coefficient ($m\,day^{-1}$)	1.0	8.1	0.88

Results

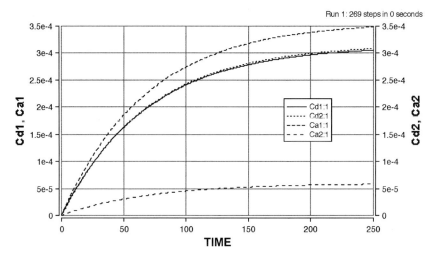

Fig. 2 Variation in adsorbed (A) and dissolved (D) metal both in the water column (1) and in the sediment (2).

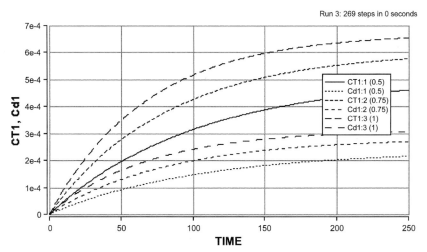

Fig. 3 Dissolved and total metals concentration at three flow rates (1.0, 0.75 and 0.5 m^3 d^{-1}).

References

van de Vrie, E. M. (1987) Modelling and Estimating Transport and Fate of Heavy Metals in Water Column and Sediment Layer in Some Enclosed Branches of the Sea in the SW Netherlands, In: Dynamical Systems and Environmental Models (eds. Bothe, H. G., Ebeling, W., Kurzhanski, A. B. and Peschel, M.), Akademie, Berlin.

Salomons, W. and Förstner, U. (1984) Metals in the Hydrocycle, Springer Verlag, Berlin.

5.14.3
OXSAG – Classic Streeter-Phelps Oxygen Sag Curves

System

When wastewaters are discharged into rivers the degradable organic compounds are utilised by micro-organisms, causing the dissolved oxygen concentration to be reduced. Simultaneously, oxygen is transferred from the air and at a certain distance downstream the transfer rate will equal the oxygen uptake rate; the dissolved oxygen level will be at a minimum. When the organics are used up, further downstream, the dissolved oxygen will start to increase to its saturation value. Streeter and Phelps in 1925 developed a simple model for this system. The plot of dissolved oxygen concentration versus the distance along the river is known as the sag curve, which is shown in below.

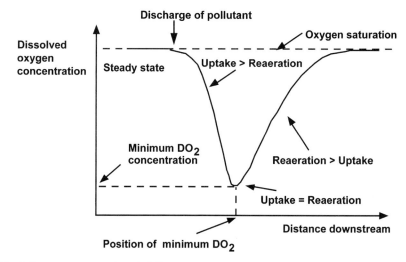

Fig. 1 Oxygen sag curve characteristics.

Model

The Streeter Phelps equation relates the dissolved oxygen concentration (DO) at any point in the river to the rate of aeration (R_{air}) and the rate of utilisation of oxygen by the river micro-organisms (R_{bio}):

$$\frac{dDO}{dt} = R_{air} + R_{bio}$$

The rate of reaeration is assumed to be a first order process, which is proportional to the oxygen deficit (i.e., the saturation dissolved oxygen concentration minus the actual dissolved oxygen concentration.):

$$R_{air} = K_{air}(DO_{sat} - DO)$$

The rate of substrate utilisation is assumed to follow Monod kinetics as

$$\frac{dS}{dt} = R_S$$

where

$$R_S = -R_{max} \frac{S}{K_S + S}$$

and

$$R_{bio} = \frac{R_S}{Y_{SO}}$$

Here Y_{SO} is a yield coefficient relating the rate of substrate utilisation to the rate of oxygen uptake.

Assuming plug flow, the distance downstream from the point of discharge, Z, is related to the average river velocity, v, and the time of travel, t, by

$$Z = vt$$

Nomenclature

DO	Dissolved oxygen concentration	g m^{-3}
DO$_{sat}$	Saturated dissolved oxygen conc.	g m^{-3}
K$_{air}$	Reaeration coefficient	h^{-1}
K$_S$	Monod constant	g m^{-3}
R$_{air}$	Rate of reaeration	g m^{-3} h^{-1}
R$_{bio}$	Rate of biological oxidation	g m^{-3} h^{-1}
R$_{max}$	Maximum reaction rate	h^{-1}

S	Substrate concentration	g m^{-3}
t	Time	h
v	Average river velocity	m h^{-1}
Y_{SO}	Yield coefficient, substrate/oxygen	–
z	Distance along river	m

Exercises

1. Vary the concentration of the discharge between 10 and 100 g m^{-3} and see how the distance downstream and the value of the minimum dissolved oxygen concentration vary with the concentration of the discharge.
2. Vary the velocity of the river between 200 and 1000 m h^{1} and see how this influences the distance downstream and the value of the minimum dissolved oxygen concentration.
3. Vary the reaeration coefficient K_{air} and see how this influences the dissolved oxygen profiles.

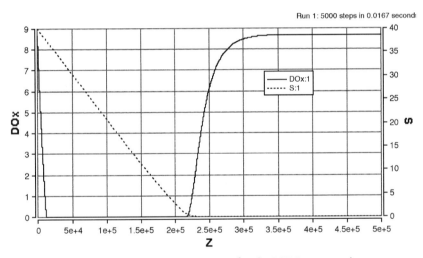

Fig. 2 Typical oxygen sag curve. Reaeration occurs after the BOD is consumed.

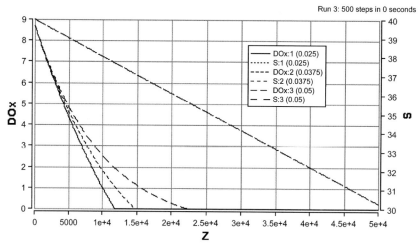

Fig. 3 Oxygen profiles for 3 values of K_{air}.

Reference

Streeter, H. W. and Phelps, E. B. (1925) A Study of the Pollution and Natural Purification of the Ohio River, Public Health Bulletin 146, US Public Health Service, Washington D. C.

5.14.4
DISCHARGE – Dissolved Oxygen and BOD Steady-State Profiles Along a River

System

Wastewater treatment plants and most factories discharge water into a river system. The wastewater will differ from the river water in temperature, BOD and dissolved oxygen concentration and nitrogen content. If the wastewater is discharged at an approximately constant rate then it may be assumed that steady state will be attained at some point along the length of the river with respect to temperature, BOD and dissolved oxygen. The minimum dissolved oxygen concentration likely to be encountered in the river and its position downstream from the point of the discharge can therefore be predicted.

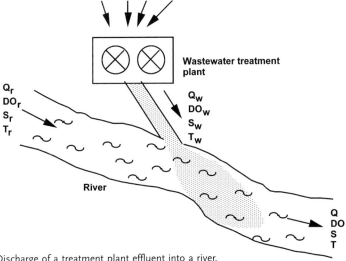

Fig. 1 Discharge of a treatment plant effluent into a river.

Model

Wastewater flows into a river and is immediately mixed with the river water. The flow rate of the river is increased as a result, where

$$Q = Q_r + Q_w$$

Similarly, the dissolved oxygen concentration (DO), BOD (S), nitrates (N) and temperature of the river will also be effected according to

$$DO = \frac{DO_r Q_r + DO_w Q_w}{Q}$$

$$S = \frac{S_r Q_r + S_w Q_w}{Q}$$

$$Temp = \frac{Temp_r Q_r + Temp_w Q_w}{Q}$$

$$N = \frac{N_r Q_r + N_w Q_w}{Q}$$

The dissolved oxygen deficit can be defined as the difference between the actual dissolved oxygen concentration and the saturation concentration for oxygen in the river at the same conditions.

$$\varepsilon = DO_{sat} - DO$$

The saturated dissolved oxygen concentration can be related to the temperature:

$$DO_{sat} = 14.652 - 0.41\,\text{Temp} + 0.008\,\text{Temp}^2$$

Oxygen is consumed by micro-organisms as they utilise the substrate and also as a result of nitrification, but oxygen is gained by the river by processes of surface aeration, so that

$$\frac{d\varepsilon}{dt} = aK_n N + K_s S - K_a \varepsilon$$

The rate of nitrification is assumed to be first order with respect to nitrate concentration, and the value of the rate constant, K_n, is a function of temperature.

$$K_n = K_{n20}\, 1.07^{(\text{Temp}-20)}$$

$$\frac{dN}{dt} = -K_n N$$

The rate of pollutant degradation is also first order and temperature dependent:

$$K_s = K_{s20}\, 1.05^{\text{Temp}-20}$$

$$\frac{dS}{dt} = -K_s S$$

The rate of reaeration is dependent on the velocity, v, of the river and its depth D, where

$$K_{a20} = \frac{2.26v}{D^{0.667}}$$

The reaeration rate is also temperature dependent:

$$K_a = \frac{K_{a20}}{24} e^{0.024(\text{Temp}-20)}$$

Assuming steady-state conditions, the distance downstream (km) can be related to the time of passage along the river (hours) by

$$z = 3.6vt$$

The velocity of the river is related to the volumetric flow rate and the river depth:

$$v = \frac{Q}{3600\,DW}$$

where this assumes approximately rectangular cross-sectional area of the river, W being the width.

Nomenclature

Symbols

ε	Oxygen deficit	mg L^{-1}
a	Stoichiometric parameter	–
D	Depth of river	m
DO	Dissolved oxygen concentration	mg L^{-1}
DO$_{sat}$	Saturated dissolved oxygen concentration	mg L^{-1}
K$_a$	Reaeration coefficient	h^{-1}
K$_n$	Nitrification rate coefficient	h^{-1}
K$_s$	Biodegradation rate coefficient	h^{-1}
N	Nitrate concentration	mg L^{-1}
Q	Volumetric flow rate	m^3 h^{-1}
S	BOD substrate concentration	mg L^{-1}
t	Time	h
Temp	Temperature	°C
v	Velocity of river	m s^{-1}
W	Width of river	m
z	Distance downstream	km

Indices

r	River
w	Wastewater
20	Reference temperature (°C)

Exercises

1. Compare the results obtained with those for the model OXSAG. What assumptions are appropriate to each particular model, and how valid are these likely to be in reality?
2. Investigate the discharge of a pollutant into rivers of different sizes, flow rates and velocities. What is the best and the worst type of river for the discharge of industrial wastewaters (from the point of view of minimising environmental damage)?
3. Model seasonal variations in temperature, flow rate and river depth. At what time of year is the river most at risk?
4. Is there any difference between the discharge of a small volume of highly concentrated waste and a large volume of more dilute waste if the overall amount of pollutant is the same?
5. Is it more worthwhile to reduce the temperature of the wastewater by 10% or the concentration by 10%? Is this always the case or does it depend on the absolute values?

Results

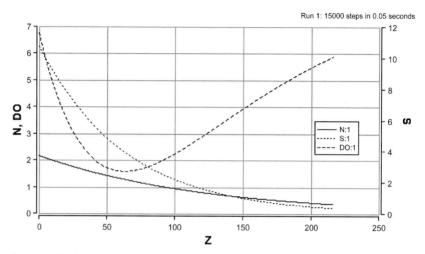

Fig. 2 Dissolved oxygen, BOD and nitrate concentration-time profiles (25 °C wastewater).

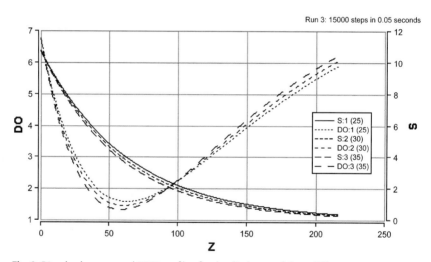

Fig. 3 Dissolved oxygen and BOD profiles for the discharge of three different wastewaters (15, 25 and 35 °C). Note that the higher temperature leads to a lower minimum dissolved oxygen concentration that occurs nearer to the discharge point, but also recovers more quickly.

Reference

Jorgensen, S. E. (1986) Fundamentals of Ecological Modelling, Elsevier.

5.14.5
ASCSTR – Continuous Stirred Tank Reactor Model of Activated Sludge

System

The activated sludge process, depicted in Fig. 1, involves basically the aeration and agitation of an effluent in the presence of a flocculated suspension of micro-organisms which are supported on particulate organic matter. After a predetermined residence time (usually several hours) the effluent is passed to a sedimentation tank where the flocculated solids are separated from the treated liquid. A reduction of BOD from 250–350 mg L^{-1} to a final value of 20 mg L^{-1} is achieved under typical operating conditions. Part of the settled sludge is usually recycled to the aeration tank in order to maintain biological activity.

In this model the aeration tank is modelled as a continuous stirred tank reactor.

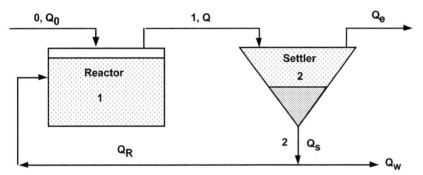

Fig. 1 Continuous stirred tank reactor and settler with sludge recycle.

Model

The total volumetric flow rate entering and leaving the reactor, Q, is the sum of the influent and recycle flow rates:

$$Q = Q_0 + Q_R$$

The recycle ratio is defined as

$$R = \frac{Q_R}{Q_0}$$

Combining the above equations gives

$$Q = (1 + R)Q_0$$

The reactor residence time is defined by the volume of the reactor divided by the volumetric influent flow rate:

$$\theta = \frac{V}{Q_0}$$

The concentration factor β is defined as the ratio of the recycle biomass to effluent biomass concentrations:

$$\beta = \frac{X_R}{X} \quad \beta > 1$$

The growth of biomass in the reactor is assumed to follow Monod kinetics with a first-order death rate. A mass balance on the biomass in the reactor yields the following differential equation (assuming that no biomass enters the reactor in the feed):

$$\frac{dX}{dt} = \frac{X}{\Theta}(\beta R - 1 - R) + \mu_m \frac{XS}{K_s + S} - K_d X$$

Similarly, assuming there is only negligible substrate in the recycle, a mass balance on the substrate (BOD) in the reactor gives

$$\frac{dS}{dt} = \frac{S_0 - S}{\Theta} - \frac{\mu_m}{Y}\frac{XS}{(K_s + S)} - K_d X$$

Various operating parameters can be calculated from the above equations as follows:

(1) Substrate to micro-organism ratio or process load factor:

$$\text{Load} = \frac{\text{Substrate consumed per day}}{\text{Biomass in the reactor}}$$

$$\text{Load} = \mu_m \frac{\frac{XSV}{Y(K_s + S)}}{XV}$$

$$\text{Load} = \mu_m \frac{S}{Y(K_s + S)}$$

(2) Mean solids residence time in the reactor θ_m:

$$\theta_m = \frac{\text{Solids in the reactor}}{(\text{Rate of biomass synthesis}) + (\text{Solids input to reactor})}$$

(3) The sludge age θ_c:

$$\theta_c = \frac{\text{Biomass in the reactor}}{\text{Rate of biomass generation}}$$

$$\theta_c = \frac{XV}{Q_w X_R + Q_e X_e}$$

Nomenclature

Symbols

K_d	Death rate constant	h^{-1}
K_s	Monod constant	$g\ m^{-3}$
Load	Loading factor	h^{-1}
Q	Volumetric flow rate	$m^3\ h^{-1}$
R	Recycle ratio	–
S	Substrate concentration	$g\ m^{-3}$
V	Volume	m^3
X	Biomass concentration	$g\ m^{-3}$
Y	Yield biomass/substrate	–
β	Sludge concentration factor	–
μ_m	Maximum growth rate	h^{-1}
θ	Reactor residence time	h
θ_c	Sludge age	h
θ_m	Mean solids residence time	h

Indices

0	Refers to inlet
e	Refers to effluent
R	Refers to recycle
w	Refers to waste

Exercises

1. Vary the recycle ratio and see how this influences the process performance.
2. Change the model so that the reactor is modelled as two or more tanks in series and compare the results with those obtained for the single tank model.
3. Simulate a shock loading by changing the values of either the flow rate or the effluent concentration interactively.
4. Add a calculation of the operating parameters substrate consumption rate to micro-organisms ratio (Load), solids residence time (θ_m) and sludge age (θ_c) to the program. Experiment with variations in the values of these parameters under various conditions of flow and feed concentration to test their value as process control parameters.

Results

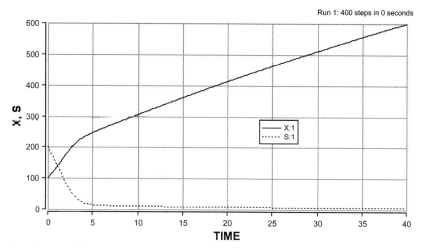

Fig. 2 Startup of the continuous activated sludge reactor with recycle ratio R=0.3.

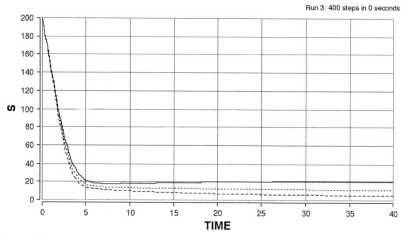

Fig. 3 Substrate concentration profiles for different values of the recycle ratio.

Reference

Grady, C. P. L. and Lim, H. C. (1980) Biological Wastewater Treatment: Theory and Applications. Pollution Engineering and Technology 12, Chapter 16.

5.14.6
DEADFISH – Distribution of an Insecticide in an Aquatic Ecosystem

System

The environmental impact of a new product needs to be assessed before it can be released for general use. Chemicals released into the environment can enter the food chain and be concentrated in plants and animals. Aquatic ecosystems are particularly sensitive, in this respect, since chemicals, when applied to agricultural land, can be transported in the ground water to rivers and then to the lakes, where they can accumulate in fish and plant life. The ecokinetic model presented here is based on a simple compartmental analysis and is based on laboratory ecosystem studies (Blau et al., 1975). The model is useful in simulating the results of events, such as the accidental spillage of an agrochemical into a pond, where it is not ethical to perform actual experimental studies.

The distribution of the insecticide Dursban in a simple aquatic environment consisting of water, fish, soil and plants is simulated, as shown in Fig. 1. The kinetic constants were obtained from radio label experiments in a test aquarium.

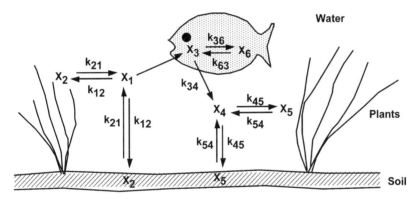

Fig. 1 Kinetic model of the water-fish-plant-soil interactions.

Model

It is assumed that at time zero a defined quantity of Dursban enters a lake and is distributed subsequently between the fish, the soil and the plants. The soil and plants are lumped together as one compartment. The quantities in each compartment are expressed as a percentage of the initial contamination.

Dursban in the pond water can be taken up by the soil and plants and also by the fish in the pond, and it is assumed that there is an equilibrium between the amount of insecticide in the plants and soil and in the pond water. Inside the fish, Dursban is metabolised to pyridinol and then excreted. Some insecticide is stored separately within the fish. The released pyridinol is taken up by the soil and plants but not again by the fish.

In this model only the quantities of the components in the different compartments are considered, and therefore no assumptions have to be made about the sizes of the different compartments.

Writing mass balances for compartments 1 to 6 and referring to Fig. 1 we obtain:

For the Dursban in the water

$$\frac{dX_1}{dt} = -k_{13}X_1 - k_{12}X_1 + k_{21}X_2$$

For the Dursban in the plants and soil

$$\frac{dX_2}{dt} = -k_{21}X_2 + k_{12}X_1$$

For the Dursban in the fish

$$\frac{dX_3}{dt} = k_{13}X_1 + k_{63}X_6 - k_{36}X_3 - k_{34}X_3$$

For the Pyridinol in the water

$$\frac{dX_4}{dt} = k_{34}X_3 - k_{45}X_4 + k_{54}X_5$$

For the Pyridinol in the plants and soil

$$\frac{dX_5}{dt} = k_{45}X_4 - k_{54}X_5$$

For the stored Dursban in the fish

$$\frac{dX_6}{dt} = k_{36}X_3 - k_{63}X_6$$

Nomenclature

Symbols

k_{ij}	Rate constant for reaction from compartment i to compartment j	h^{-1}
t	Time	h
X_i	Percentage insecticide in compartment i	%

Indices

1	Dursban in water
2	Dursban in plants and soil
3	Dursban in fish
4	Pyridinol in water
5	Pyridinol in plants and soil
6	Stored Dursban in fish

Exercises

1. Compare the accumulation of the insecticide in the fish in systems with and without the plant and soil compartments.
2. Modify the model to simulate the slow release of insecticide (for instance via runoff) into the lake.

Results

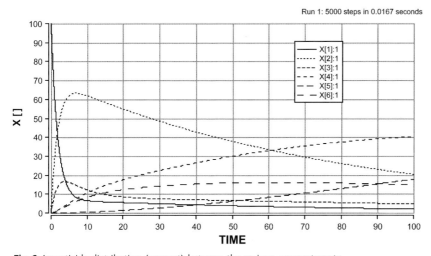

Fig. 2 Insecticide distribution (percent) between the various compartments as a function of time.

584 | 5 Simulation Tools and Examples of Chemical Engineering Processes

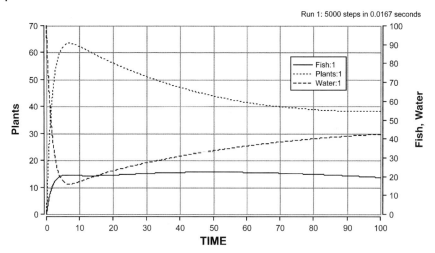

Fig. 3 For the same run, insecticide plus metabolite distribution (percent) between the fish, plants and water.

References

Blau, G. E., Neely, W. B. and Branson, D. R. (1975) Ecokinetics: A Study of the Fate and Distribution of Chemicals in Laboratory Ecosystems, AIChE J. 21, 5, 854–861.

Butte, W. (1991) Mathematical Description of Uptake, Accumulation and Elimination of Xenobiotics in a Fish/Water System, In: Bioaccumulation in Aquatic Systems: Contributions to the Assessment (eds. Nagel, R. and Loskill, R.), VCH, Weinheim.

5.14.7
LEACH – One-Dimensional Transport of Solute Through Soil

System

Groundwater is a major source of drinking, industrial and agricultural water. Its quality is a primary environmental concern, not only for health reasons, but also because of the decrease in crop productivity caused by pollutants. The ability to model the migration of pollutants through the vadose soil layer plays an important part in the combat against the degradation of groundwater.

Model

In this model from Corwin et al. (1991), the transport of a solute such as a pesticide through the soil is simulated from ground level to the water table. As shown in Fig. 1, the soil column is divided into N elements each of thickness z, with element 1 at the surface and element N directly above the water table.

Hence

$$z = \frac{W_{table}}{N}$$

The water fraction for any element n is expressed as the ratio of the volume of water to the total volume of the element:

$$F_n = \frac{W_n}{V_n}$$

The water fraction of each element of soil can not exceed a certain maximum value, known as the field capacity. If the water content is less than the field capacity, then no water is able to leave the element. If the water content is equal to or greater than the field capacity, then water is able to leave the element at the same rate as water enters. The water is also taken up by the roots of plants. Here it is assumed that the uptake rate of water is proportional to the length of root. If the water content falls below a critical value, the plants can no longer take up water, and they wilt and eventually die.

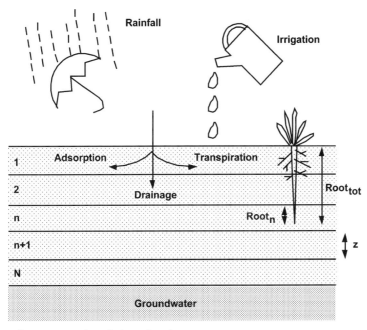

Fig. 1 Solute transport through the vadose layer.

A mass balance based on the volume of water in each element leads to the following equation:

$$\frac{dW_n}{dt} = Q_{n-1} - Q_n - ET_n$$

where Q_{n-1} and Q_n are the volumetric flow rates of water in and out of element n, and ET_n is the rate of assimilation by the plant or evapotranspiration rate.

The limiting behaviour on the rate of evapotranspiration is expressed by the logical condition, that with

$$ET_n = 0 \quad \text{when } F_n < F_{min}$$

$$ET_n = ET\,Root_n \quad \text{when } F_n > F_{min}$$

and F_{min} is the minimum water content for plant growth. The water leaving any element is described by the following logical condition statements:

$$Q_n = 0 \quad \text{when } F_n < FC$$

$$Q_n = Q_{n-1} \quad \text{when } F_n > FC$$

where FC is the field capacity value.

The roots are assumed to grow at a constant rate (GR), so that the length of the root at any time t is given by

$$Root_{tot} = Root_0 + GR\,t$$

where $Root_{tot}$ is the total root length at time t, and $Root_0$ is the root length at time t=0.

The growth stops if the water content is too low, given by the condition

$$GR = 0 \quad \text{if } F_n < F_{min}$$

If the water content is less than the minimum then growth stops. Thus the root length in any element n is given by

$$Root_n = z \quad \text{for } Root_{tot} > nz$$

Thus the root occupies the total length of the element for those elements less than the total root length. When the root penetrates only part way into the soil element, as shown in Fig. 1,

$$Root_n = Root_{tot} - (n-1)z \quad \text{for } (n-1)z < Root_{tot} < nz$$

and

$$Root_n = 0 \quad \text{for } Root_{tot} < (n-1)z$$

The input of water to the top element is by rain or via human irrigation. A water balance for the top element 1 therefore leads to the mass balance

$$\frac{dW_1}{dt} = A(\text{Rain} + \text{Irrig}) - Q_1 - ET_1$$

where Rain and Irrig are specific rates given in cm day^{-1}, and the cross-sectional area, A, of the soil column is related to the height, z, and the volume, V, of each element by

$$A = \frac{V}{z}$$

If a pesticide, or any other solute, is sprayed onto the field it will travel in the water through the soil, but may also be adsorbed onto the soil. It is assumed that the process of adsorption occurs almost instantaneously, as compared to the slower rate of transport through the soil. Only solute dissolved in the mobile water column is transported from one element to the next.

A mass balance on the total solute concentration, C, including both adsorbed and mobile solute for element n yields the following relationship:

$$V_n \frac{dC_n}{dt} = Q_{n-1} Cs_{n-1} - Q_n Cs_n$$

where Cs, the concentration in the soil water, is related to the total and adsorbed solute concentrations by the water fraction, F_n, i.e.,

$$C_n = F_n Cs_n + (1 - F_n) C_{ads}$$

An assumed instantaneous distribution between the adsorbed and free solute is modelled by a Langmuir-type adsorption isotherm, where

$$C_{ads} = C_{max} \frac{Cs}{Cs + K}$$

C_{max} is the maximum concentration of solute that can be adsorbed by the soil, and K is the Langmuir adsorption coefficient. Combining the above two equations, a quadratic equation can be derived in terms of Cs, which on solving for the positive root gives

$$Cs = \frac{-b + (b^2 - 4as)^{0.5}}{2a}$$

where

a = F
b = F K−C + C_{max} (1−F)
s = −C K

Hence the corresponding values for C, Cs and C_{ads} can be determined for each element.

It is assumed that the plants do not take up solute and that the solute is not biodegraded. As water is taken up by the plants at a greater rate than the rate of supply, the value of F will decrease, and the solute will become more concentrated.

Program

Five elements are used to model the soil column.

Nomenclature

a	Constant in quadratic equation	–
A	Cross-sectional area of soil column	cm^2
b	Constant in quadratic equation	–
C	Solute concentration in irrigation water	$g\ cm^{-3}$
C_n	Total solute concentration in element	$g\ cm^{-3}$
C_{ads}	Concentration of adsorbed solute	$g\ cm^{-3}$
Cs	Concentration of solute in liquid phase	$g\ cm^{-3}$
C_{max}	Maximum adsorbed solute conc.	$g\ cm^{-3}$
ET_n	Evapotranspiration rate in element n	$cm^3\ day^{-1}$
F_n	Water fraction in element n	$cm^3\ cm^{-3}$
FC	Field capacity fraction	–
F_{min}	Minimum moisture content for plant growth	–
GR	Plant root growth rate	$cm\ day^{-1}$
Irrig	Specific rate of human irrigation	$cm\ day^{-1}$
K	Langmuir adsorption coefficient	$g\ cm^{-3}$
N	Number of elements in the soil model	–
Q_n	Volumetric flow rate from element n	$cm^3\ day^{-1}$
Rain	Specific rainfall rate	$cm\ day^{-1}$
$Root_n$	Length of root in element n	cm
$Root_{tot}$	Total length of root	cm
s	Constant in quadratic equation	–
t	Time	day
V	Volume of one element	cm^3
W_n	Water content in element n	cm^3
W_{table}	Depth of water table	cm
z	Height of one soil element	cm

Exercises

1. Run the simulation under the following conditions: (a) Dry conditions time 0 to day 2, (b) spraying day 2 to day 5 and (c) rain day 8 to day 13. See how the concentrations of adsorbed and unadsorbed solute vary with time and depth. Vary the time of spraying to before, during and after rainfall.
2. Change the chemical equilibrium parameters (K and C_{max}) to see how they effect the distribution and transport of the solute (see Lyman et al. (1982) for a comprehensive set of data). Experiment by using a different model for the adsorption of the solute. e.g., Freundlich linear adsorption.
3. In practise some water will always be transported through the soil even if the local water content is less than the field capacity owing to preferential flow through cracks and large pores in the soil structure. This can be modelled simply by defining a bypass parameter to account for the fraction of water that can pass through each element. Modify the model to include bypass and see how this influences the solute profiles (see Corwin et al. (1991) for further details).
4. Different soil types have differing field capacities and minimum water contents.

   Typical values are:

Soil type	Field capacity ($cm^3\ cm^{-3}$)	Minimum water content ($cm^3\ cm^{-3}$)
Sandy soil	0.08	0.05
Clay soil	0.35	0.16

   Vary the parameters within the above ranges. How does the soil type affect the transport of solutes? Modify the model so that the soil characteristics vary with depth, e.g., a sandy soil near the surface becoming clay deeper below the surface.
5. The assumption of linear water uptake along the length of the plant roots is not usually true in practise. Often the uptake is divided 40:30:20:10 i.e. 40% of water is taken up by the top quarter of the roots, 30% by the second quarter and so on. Modify the model to take this into account. Try other models of water uptake (e.g., 44:31:19:6).
6. Modify the model for a solute that is capable of being biodegraded or being assimilated by the plants. An appropriate kinetic relationship must be assumed.

Results

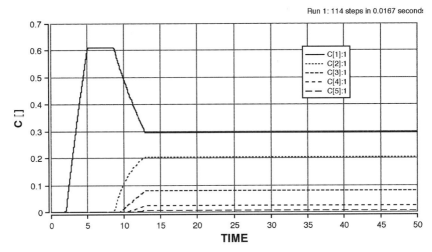

Fig. 2 Time profiles for the overall concentration of solutes at five different depths. Irrigation between days 2 and 5. Rain on days 8 to 13.

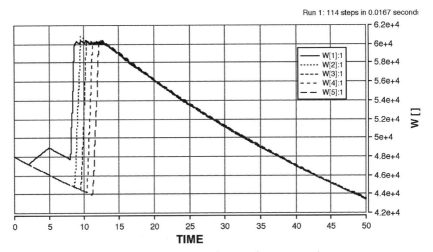

Fig. 3 Water fractions for the same run. The influence of irrigation and rain is clearly seen near the surface at depth 1.

References

Corwin, D. L., Waggoner, B. L. and Rhoades, J. D. (1991) A Functional Model of Solute Transport that Accounts for Bypass, J. Environ. Qual. 20, 647–658.

Lyman, W. J., Reehl, W. F. and Rosenblatt, D. H. (1982) Handbook of Chemical Property Estimation Methods, Chapter 4, Adsorption Coefficients for Soils and Sediments, McGraw-Hill.

5.14.8
SOIL – Bioremediation of Soil Particles

System

Soil that has been contaminated with an organic compound can be cleaned by naturally occurring soil bacteria. These bacteria are present as suspended forms existing within the liquid in the pores of the soil particles as well as existing as microcolonies attached to the solid surface. The micro-organisms utilise the organic compounds as a carbon source but also require oxygen. Oxygen and the contaminant are transported by diffusion within the liquid phase of the soil particles. The time required for the contaminated soil to be cleaned, or the bioremediation time, is an important parameter in influencing the course of action to minimise the environmental impact of the contamination.

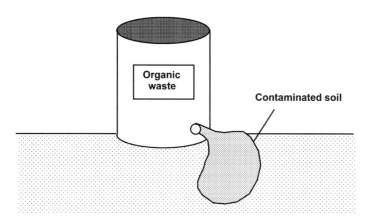

Fig. 1 Leaching of an organic compound into the soil.

Model

The soil aggregates are assumed to be spherical in form and to have constant temperature and to contain initially uniform distributions of substrate (contaminant) and biomass. The external concentrations of biomass and substrate are assumed to be zero and the external oxygen concentration is constant. Substrate is adsorbed onto the solid phase to an extent determined by an equilibrium partition coefficient.

Thus for the substrate

$$S_{sol} = S_{liq} K_{ps}$$

and for the biomass

$$X_{sol} = X_{liq} K_{px}$$

Substrate is present in both the liquid and soil phases of the soil particle. A factor F_s can be defined such that all the substrate can be considered to be found in the total volume V, where

$$S_{liq} F_s V = S_{liq} V E_a + S_{sol} V (1 - E_a)$$

This can be simplified to

$$F_s = E_a + (1 - E_a) \frac{S_{sol}}{S_{liq}}$$

which is

$$F_s = E_a + (1 - E_a) K_{ps}$$

A similar expression can be derived for biomass, which gives

$$F_x = E_a + (1 - E_a) K_{px}$$

where E_a is the volumetric fraction of liquid, and K_{ps} and K_{px} are the partition coefficients. The factors F_s and F_x are introduced to avoid having to use separate differential biomass and substrate balances for both the solid and liquid phases.

Oxygen is assumed to be present only in the liquid phase. The biomass is assumed to follow Monod growth kinetics, depending on both the oxygen and substrate concentrations in the liquid phase and to decline according to a first order decay term, where

$$\frac{R_x}{VF_x} = \mu_{max} X \frac{S}{K_s + S} \frac{O}{K_o + O} - K_d X$$

The rates of uptake of substrate and oxygen are related to the biomass growth rate by appropriate yield constants:

$$\frac{R_s}{VF_x} = -\frac{\mu_{max} X}{Y_s} \frac{S}{K_s + S} \frac{O}{K_o + O}$$

$$\frac{R_o}{VF_x} = -\frac{\mu_{max} X}{Y_o} \frac{S}{K_s + S} \frac{O}{K_o + O}$$

Oxygen, substrate and biomass are all transported by diffusion within the liquid phase contained in the aggregate. The modelling of this process is achieved via the use of a finite differencing technique. In this, the spherical aggregate is divided into a number of shells, as seen in Fig. 1.

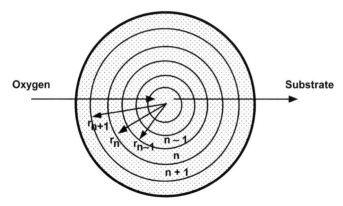

Fig. 2 Finite differencing of a soil particle.

The outer diameter of the nth shell is r_n and its inner diameter is r_{n-1}. Therefore the volume of shell n is given by

$$V_n = \frac{4}{3}\pi(r_n^3 - r_{n-1}^3)$$

and the outside area of the shell by

$$A_n = 4\pi r_n^2$$

Mass balances for element n give

$$F_x V_n \frac{dX}{dt} = J_{x,n} + R_{x,n}$$

$$F_s V_n \frac{dS}{dt} = J_{s,n} + R_{s,n}$$

$$E_a V_n \frac{dO}{dt} = J_{o,n} + R_{o,n}$$

where J_x, J_s and J_o are the respective net rates of diffusion of biomass, substrate and oxygen into the nth element which are approximated by

$$J_{x,n} = D_x^* \frac{A_n(X_{n+1} - X_n) - A_{n-1}(X_n - X_{n-1})}{r_n}$$

$$J_{s,n} = D_s^* \frac{A_n(S_{n+1} - S_n) - A_{n-1}(S_n - S_{n-1})}{r_n}$$

$$J_{o,n} = D_o^* \frac{A_n(O_{n+1} - O_n) - A_{n-1}(O_n - O_{n-1})}{r_n}$$

D_x^*, D_o^*, and D_s^* are the diffusion coefficients for biomass, oxygen and substrate in the liquid phase of the aggregates. It is assumed that the rates of diffusion in the solid phase are negligible compared to those in the liquid phase. The biomass diffusion coefficient also incorporates the effect of cell motility. The rate of diffusion through the aggregate will depend on the volume fraction of liquid in the aggregates and the degree of tortuosity of the pore space. The diffusion constants in the aggregates can be related to the diffusion constants in free liquid by the following relationship

$$D_i^* = D_i \frac{E_a}{\tau}$$

where i is substrate, biomass or oxygen, E_a is the liquid volume fraction in the aggregate and τ is a measure of the degree of tortuosity of the pore space.

Program

The model equations are very stiff and require a suitable integration method.

Nomenclature

Symbols

A	Outside area of shell	cm^2
D	Diffusion coefficient	$cm^2 \, s^{-1}$
DO	Dissolved oxygen concentration	$g \, cm^{-3}$
E_a	Liquid fraction in aggregate	–
F_x	Biomass concentration correction factor	–
F_s	Substrate concentration correction factor	–
J	Diffusion rate	$g \, s^{-1}$
K_d	First order cell death constant	s^{-1}
K_s	Monod constant for substrate	$g \, cm^{-3}$
K_o	Monod constant for oxygen	$g \, cm^{-3}$
K_{ps}	Partition coefficient substrate	–
K_{px}	Partition coefficient biomass	–
r	Radius of aggregate	cm
R	Reaction rates	$g \, s^{-1}$
S	Substrate (contaminant) concentration	$g \, cm^{-3}$
t	Time	s
V	Volume of aggregate	cm^3
X	Biomass concentration	$g \, cm^{-3}$
Y_o	Yield oxygen/biomass	–
Y_s	Yield substrate/biomass	–
μ_{max}	Maximum growth rate	s^{-1}
τ	Tortuosity of the pore space	–

Indices

ex	External
liq	Liquid phase
n	Element number (1 to 6)
o	Oxygen
sol	Solid phase
s	Substrate
x	Biomass
in	Initial

Exercises

1. Vary the radius of the soil aggregates in the range 0.1 to 10 cm and see how this influences the bioremediation time.
2. Vary the substrate partition coefficient in the range 1 to 1500 and see how this also influences the bioremediation time.
3. Also vary the initial substrate concentration, i.e. the degree of contamination in the range 1.0×10^{-6} to 0.1, and see how this influences the bioremediation time.
4. Compare the bioremediation time for a small, highly contaminated particle to that of a larger, less contaminated particle having an equal initial mass of contaminant.

Results

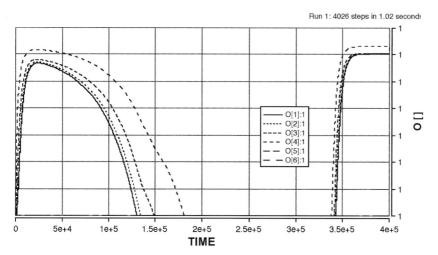

Fig. 3 Variation in oxygen concentrations with time.

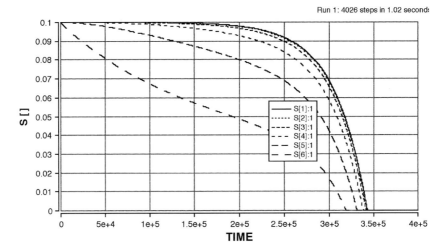

Fig. 4 Variation in substrate concentrations with time for the same run as in Fig. 3.

Reference

Dhawan, S., Fan, L.T., Erickson, E. and Tuitemwong, P. (1991) Modeling, Analysis and Simulation of Bioremediation of Soil Aggregates, Environmental Progress 10, 4, 251–260.

Appendix: Using the MADONNA Language

1 A Short Guide to MADONNA

Computer Requirements

Two unregistered versions of MADONNA are supplied with this book on the CD, one for Windows 95, NT 4.0 and later and one for the Power Macintosh, including OSX. More information with free downloads of this software can be found on the following MADONNA website: http://www.berkeleymadonna.com. The unregistered version does not allow saving new programs but will run any existing program.

Installation

The files are compressed on the CD in the same form as they are available on the Internet. Information on registering MADONNA is contained in the files. Registration is optional since all the examples in the book can be run with the unregistered version. Registration makes available a detailed manual and is necessary for anyone who wants to develop his or her own programs.

Running Programs

To our knowledge, MADONNA is by far the easiest simulation software to use, as can be seen on the Screenshot Guide in this Appendix. Running an example typically involves the following steps:
- Open MADONNA and open a prepared program file.
- Go to Model/Equations on the menu and study the equations and program logic.
- Go to Parameters/Parameter Window on the menu and see how the values are set. They may be different than on the program listing. Those with a* can be reset to the original values. Also, if necessary, here the integration

Chemical Engineering Dynamics: An Introduction to Modelling and Computer Simulation, Third Edition
J. Ingham, I. J. Dunn, E. Heinzle, J. E. Prenosil, J. B. Snape
Copyright © 2007 WILEY-VCH Verlag GmbH & Co. KGaA, Weinheim
ISBN: 978-3-527-31678-6

method and its parameters (D_T, $D_{T_{min}}$ and $D_{T_{max}}$, Stoptime, etc.) values can be changed.
- Decide which plot might be interesting, based on the discussion in the text.
- Go to Graph/New Window and then Graph/Choose Data to select data for each axis. All calculated results on the left side of the equations are available and can be selected.
- Run the program and make a graph by clicking Run on the graph window.
- Adjust the graph by setting the legend with the legend button. Perhaps put one of the variables on the right side of the graph with Graph/Choose Data.
- Possibly select the range of the axes with Graph/Axis Settings. Choose colours or line types with the buttons.
- Decide on further runs. It is most common to want to compare runs for different values of the parameters. This is usually done with Parameter/Batch Runs and also with Model/Define Sliders. If the overlay button is set, then more than one set of runs can be graphed on top of the first run. Sometimes more than one parameter needs to be set; this is best done with changes done in the Parameters/Parameter Window, with an overlay graph if desired.

As seen at the end of the Screenshot Guide, Parametric Plots are very useful to display the steady-state values as a function of the values of one parameter. For this, one needs to be sure that the Stoptime is sufficiently long to reach steady-state for all the runs.

When running a program with arrays, as found in the finite-differenced examples, the X axis can be set with [i] and the Y axes with the variables of interest. The resulting graph is a plot of the variable values at the Stoptime in all of the array sections. For equal-sized segments, this is the equivalent of a plot of the variables versus distance. If the steady-state has been reached then the graph gives the steady-state profile with distance. More on running programs is found in Section 1.1.7 and in Section 2 of the Appendix.

Special Programming Tips

MADONNA, like all programming languages, has certain functions and characteristics that are worth noting and that do not appear elsewhere in this book.

Editing Text

The very convenient built-in editor is usually satisfactory. Also the program can be written with a word processor and saved as a text-only file with the suffix ".mmd". MADONNA can then open it.

Finding Programming Errors. Look at a Table Output of the Variables

Sometimes programs do not run because of errors in the program that cause integration problems. Some hint as to the location of the error can often be found by making an output table of all the calculated variables. This is done by going to Graph/New Window and then Graph/Choose Data to select all the variables. Then the program is run and the tabular data button is chosen. Inspection of all the values during the first one or two time intervals will usually lead to an isolation of the problem for those values that are marked in red with NAN (not a number). Also, values going negative can be found easily here and often indicate an integration error.

The Program Does Not Compile. Is a Bracket Missing?

Sometimes variables are not found when the program is compiling. Often this can be due to one of the comment brackets missing, usually on the right side. MADONNA does not presently test for bracket pairs, and a missing bracket means that the whole program will be read as a comment. The error message is usually "Undetermined comment" or expect "=".

Setting the Axes. Watch the Range of Values

Remember that each Y axis can have only one range of values. This means that you must choose the ranges so that similarly sized values are located on the same axis. Choose the right axes in Graph/Choose Variables.

Are There Bugs or Imperfections in MADONNA?

Yes, there are some that we are aware of. You may find some or you may have some special wishes for improvements. The MADONNA developers would be glad to receive your suggestions. See the homepage for the email address.

Making a Pulse Input to a Process

This can be done in two ways: Either use the pre-programmed PULSE function (see the program CSTRPULSE for an example) or use an IF-THEN-ELSE statement to turn a stream on and off (see example CHROM).

Making a More Complex Conditional Control of a Program

In general the IF-THEN-ELSE conditional statement form is used, combined with the inequality and/or possibilities as found in the HELP. This can involve a switching from one equation to another within this statement. Another way is to use flags or constants that take values of 0 or 1 and are multiplied by terms in the equations to achieve the desired results. The programs RUN and RELUY contain such programming. As shown in RUN, nesting of multiple IF statements is possible:

```
V = IF (Disk < 1 AND P > 1.9)
      THEN 0.85*KV*P/SQRT(TR+273)
           ELSE IF (Disk < 1 AND P < = 1.9 AND P > 1.1)
                THEN KV*P/SQRT(TR+273)*SQRT(1+ (1/P)*(1/P))
```

Parameter Estimation to Fit Parameters to Data

For fitting sets of data to one or more parameters the data can be imported as a text file (Import Data Set) found in the window of Parameters/Curve Fitting. The Preferences/Graph Windows provide the possibility of having the data as open circles. See the programs KLAFIT and ESTERFIT for examples of this. Normally this data would be time in the first column and the measured variable in the second column. If the data is available at equal time increments then two sets of data can be used. See also Help/How Do I.

Optimisation of a Variable

There is a real optimisation routine available in MADONNA under Parameters/Optimize. It is actually a minimization and the function needs to be written in the window. It is also useful to check the results by making runs. If it is something simple with one or two parameters, then sliders can be effectively used. If the value of a maximum is sought as a function of a single parameter value, then the Parametric Plot for maximum value should be used.

Finding the Influence of Two Parameters on the Steady State?

A Parametric Plot choosing the "final" value can be used to find the influence of one variable on the steady state. The second parameter can be changed in the Parameter Window and additional parametric runs made and plotted with an overlay plot. Thus it is possible to obtain a sort of contour plot with a series of curves for values of the second parameter. Unfortunately, no automatic contour plot is yet possible.

Nice Looking Results Are Not Always Correct

A warning! It is possible to obtain results from a program that at first glance seems OK. Always make sure that the same results are obtained when D_T is reduced by a factor of 10 or when a different integration method is used. Plotting all the variables may reveal oscillations that indicate integration errors. These may not be detectable on plots of the other variables.

Setting the Integration Method and Its Parameters?

It is recommended to choose the automatic step-size method AUTO and to set a small value of $D_{T_{min}}$ and a large value for $D_{T_{max}}$. If the results are good, try to improve the speed by increasing both parameters. Sometimes the resulting curves are not smooth if $D_{T_{max}}$ is too high. In most cases, good results are obtained with AUTO and $D_{T_{min}}$ set to about 1/1000 of the smallest time constant.

If no success is found with AUTO, then try Rosenbrock (STIFF) and adjust by the same procedure. Oscillations can sometimes be seen by zooming in on a graph; often these are a sign of integration problems. Sometimes some variables look OK but others oscillate, so look at all of them if problems arise. Unfortunately there is not a perfect recipe, but fortunately MADONNA is very fast so the trial-and-error method usually works out.

Checking Results by Mass Balance

For continuous processes, checking the steady-state results is very useful. Algebraic equations for this can be added to the program, such that both sides became equal at steady state. For batch systems, all the initial mass must equal all the final mass, not always in mols but in kg. Expressed in mols the stoichiometry must be satisfied.

What Is a "Floating Point Exception"?

This error message comes up when something does not calculate correctly, such as dividing by zero. This is a common error that occurs when equations contain a variable in the denominator that is initially zero. Often it is possible to add a very small number to it, so that the denominator is never exactly zero. These cases can usually be located by outputting a table of all the variables.

Plotting Variables with Distance and Time

Stagewise and finite-differenced models involve changes with time and distance. When the model is written in array form the variable can be plotted as a function of the array index. This is done by choosing an index variable for the Y axis and the [ ] symbol for the X axis. The last value calculated is used in the plot, which means that if the steady-state has been reached then it is a steady-state profile with distance. An example is given in the "Screenshot Guide" in Section 2 of the Appendix and in many other simulation examples.

Writing Your Own Plug-in Functions or Integration Methods

Information on using C or C++ for this can be obtained by email from Tim Zahnley, the MADONNA programmer-developer, at madonna@kagi.com.

2 Screenshot Guide to BERKELEY-MADONNA

This guide is intended as a supplementary introduction to MADONNA.

```
donna  File  Edit  Flowchart  Model  Compute  Graph  Parameters  Window  Help
○ ○ ○                    BATSEQ.MMD - Equations
 Run

METHOD AUTO
DT = 0.1
STOPTIME= 500  {Length of simulation, sec.}

K1 = 0.01  {First order rate constant, 1/s}
K2 = 0.01  {First order rate constant, 1/s}
K3 = 0.05  {First order rate constant, 1/s}
K4 = 0.02  {First order rate constant, 1/s}

INIT CA=1  {Initial concentration, kmoles/m3}
INIT CB=0  {Initial concentration, kmoles/m3}
INIT CC=0  {Initial concentration, kmoles/m3}
INIT CD=0  {Initial concentration, kmoles/m3}

{Batch mass balances for components A, B, C and D }
d/dt(CA)= rA
d/dt(CB)= rB
d/dt(CC)= rC
d/dt(CD)= rD

{Kinetic rate terms for components A, B, C and D}
rA=-K1*CA
rB=K1*CA-K2*CB+K3*CC-K4*CB
rC=K2*CB-K3*CC
rD=K4*CB

MOLE = CA+CB+CC+CD  {Sum of moles}
```

Fig. 1 The example BATSEQ equation window has been opened with Model/Equations.

Fig. 2 The example BATSEQ has been run with the Colors, Data Points and Grid selected. From left: Lock, Overlay, Table, Legend, Parameters, Colors, Dashed Lines, Data Points, Grid, Value Output and Zoom.

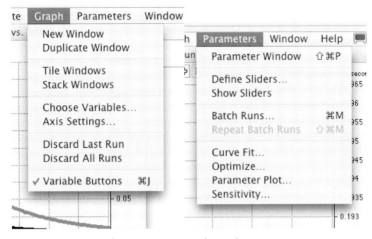

Fig. 3 The menus GRAPH and PARAMETERS are shown above.

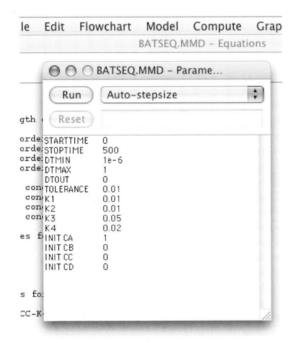

Fig. 4 The MODEL/EQUATIONS was chosen. Seen here is also the PARAMETER WINDOW.

Appendix: Using the MADONNA Language

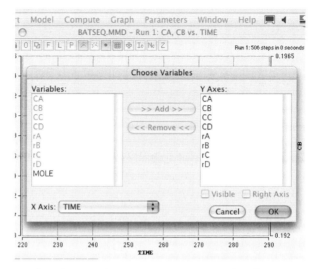

Fig. 5 If a new graph is chosen under GRAPH/NEW WINDOW then the data must be selected under GRAPH/CHOOSE VARIABLES.

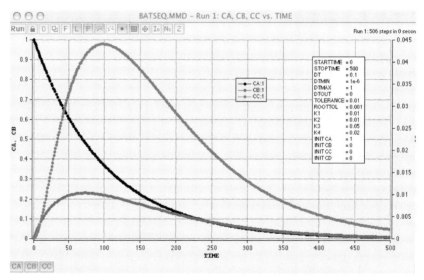

Fig. 6 A graph window with CC on the right-side Y-axis with Legend, Parameter and Data buttons selected.

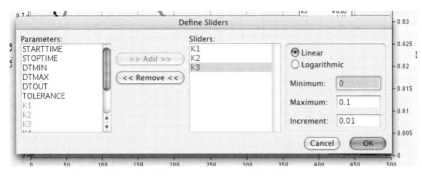

Fig. 7 The window to define the sliders (Parameters/Define Sliders).

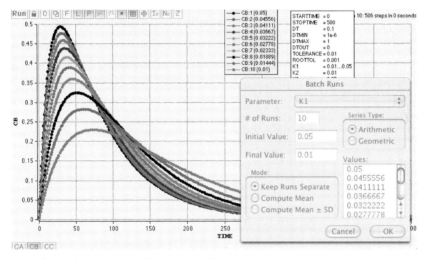

Fig. 8 The Batch Runs window for 10 values of k_1 as shown.

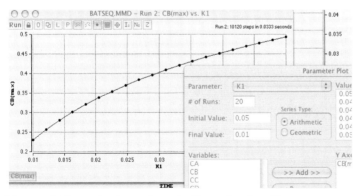

Fig. 9 A Parametric Plot to give the maximum in CB as a function of k_1.

3 List of Simulation Examples

AMMONAB	Steady-State Absorption Column Design 471
ANHYD	Oxidation of O-Xylene to Phthalic Anhydride 324
ASCSTR	Continuous Stirred Tank Reactor Model of Activated 577
AXDISP	Differential Extraction Column with Axial Dispersion 468
BASIN	Dynamics of an Equalisation Basin 560
BATCHD	Dimensionless Kinetics in a Batch Reactor 235
BATCOM	Batch Reactor with Complex Reaction Sequence 240
BATEX	Single Solute Batch Extraction 442
BATSEG, SEMISEG and COMPSEG	Mixing and Segregation 394
BATSEQFL	Example of Flowchart Programming 227
BATSEQ and BATSEQFL	Complex Reaction Sequence 232
BEAD	Diffusion and Reaction in a Spherical Catalyst Bead 533
BENZHYD	Dehydrogenation of Benzene 320
BIOFILM	Biofilm Tank Reactor 551
BIOFILT	Biofiltration Column for Removing Ketone from Air 555
BIOREACT	Process Modes for a Bioreactor 538
BSTILL	Binary Batch Distillation Column 490
BUBBLE	Bubble Point Calculation for a Batch Distillation Column 504
CASCSEQ	Cascade of three Reactors with Sequential Reactions 276
CASTOR	Batch Decomposition of Acetylated Castor Oil 243
CHROMDIFF	Dispersion Model for Chromatography Columns 483
CHROMPLATE	Stagewise Model for Chromatography Columns 486
COMPREAC	Complex Reaction 237
CONFLO1, CONFLO2 and CONFLO3	Continuous Flow Tank 406
CONSTILL	Continuous Binary Distillation Column 496
CONTUN	Controller Tuning Problem 427
COOL	Three-Stage Reactor Cascade with Countercurrent Cooling 287
CSTRCOM	Isothermal Reactor with Complex Reaction 265
CSTRPULSE	Continuous Stirred-Tanks, Tracer Experiment 273
DEACT	Deactivating Catalyst in a CSTR 268
DEADFISH	Distribution of an Insecticide in an Aquatic Ecosystem 581
DIFDIST	Multicomponent Differential Distillation 494
DISCHARGE	Dissolved Oxygen and BOD Profiles in a River 572

3 List of Simulation Examples

DISRE	Isothermal Reactor with Axial Dispersion 335
DISRET	Non-Isothermal Tubular Reactor with Axial Dispersion 340
DRY	Drying of a Solid 521
DSC	Differential Scanning Calorimetry 258
ENZDYN	Dynamic Diffusion with Enzymatic Reaction 529
ENZSPLIT	Diffusion and Reaction: Split Boundary Solution 525
EQBACK	Multistage Extractor with Backmixing 453
EQEX	Simple Equilibrium Stage Extractor 447
EQMULTI	Continuous Equilibrium Multistage Extraction 449
ESTERFIT	Esterification of Acetic Acid with Ethanol, Fitting 261
EXTRACTCON	Extraction Cascade, Backmixing and Control 456
FILTWASH	Filter Washing 479
GASLIQ1/GASLIQ2	Gas–Liquid Mixing and Mass Transfer 385
HEATEX	Dynamics of a Shell-and-Tube Heat Exchanger 511
HMT	Semi-Batch Manufacture of Hexamethylene-triamine 353
HOLDUP	Transient Holdup Profiles in an Agitated Extractor 459
HOMPOLY	Homogeneous Free-Radical Polymerisation 310
HYDROL	Batch Reactor Hydrolysis of Acetic Anhydride 247
INHIBCONT	Continuous Bioreactor with Inhibitory Substrate 543
KLADYN, KLAFIT and ELECTFIT	Dynamic Oxygen Electrode 462
LEACH	One-Dimensional Transport of Solute Through Soil 584
MCSTILL	Continuous Multicomponent Distillation Column 501
MEMSEP	Gas Separation by Membrane Permeation 475
METAL	Transport of Heavy Metals in Water and Sediment 565
MIXFLO1/MIXFLO2	Residence Time Distribution Studies 381
NITBED	Nitrification in a Fluidised Bed Reactor 547
NITRO	Conversion of Nitrobenzene to Aniline 329
NOCSTR	Non-Ideal Stirred-Tank Reactor 374
OSCIL	Oscillating Tank Reactor Behaviour 290
OXIBAT	Oxidation Reaction in an Aerated Tank 250
OXSAG	Classic Streeter-Phelps Oxygen Sag Curves 569
REFRIG1/REFRIG2	Auto-Refrigerated Reactor 295
RELUY	Batch Reactor of Luyben 253
REVREACT	Reversible Reaction with Temperature Effects 305
REVTEMP	Reversible Reaction with Variable Heat Capacities 299
REXT	Reaction with Integrated Extraction of Inhibitory Product 280
ROD	Radiation from Metal Rod 518
RUN	Relief of a Runaway Polymerisation Reaction 355
SELCONT	Optimized Selectivity in a Semi-Continuous Reactor 362

SEMIEX	Temperature Control for Semi-Batch Reactor 430
SEMIPAR	Parallel Reactions in a Semi-Continuous Reactor 347
SEMISEQ	Sequential-Parallel Reactions in a Semi-Continuous Reactor 350
SOIL	Bioremediation of Soil Particles 591
SPBEDRTD	Spouted Bed Reactor Mixing Model 390
SSHEATEX	Steady-State, Two-Pass Heat Exchanger 515
STEAM	Multicomponent, Semi-Batch Steam Distillation 508
SULFONATION	Space-Time-Yield and Safety in a Semi-Continuous Reactor 365
TANK and TANKDIM	Single Tank with nth-Order Reaction 270
TANKBLD	Liquid Stream Blending 409
TANKDIS	Ladle Discharge Problem 412
TANKHYD	Interacting Tank Reservoirs 416
TEMPCONT	Control of Temperature in a Water Heater 420
THERM and THERMPLOT	Thermal Stability of a CSTR 283
THERMFF	Feedforward Control of an Exothermic CSTR 437
TRANSIM	Transfer Function Simulation 435
TUBDYN	Dynamic Tubular Reactor 332
TUBE and TUBEDIM	Tubular Reactor Model for the Steady-State 315
TUBEMIX	Non-Ideal Tube-Tank Mixing Model 378
TUBETANK	Design Comparison for Tubular and Tank Reactors 317
TWOEX	Two-Solute Batch Extraction with Interacting Equilibria 444
TWOTANK	Two Tank Level Control 424
VARMOL	Gas-Phase Reaction with Molar Change 344

Subject Index

a
absorption 471
acceleration of gravity 414, 417
accumulation term energy balance 26
accumulation term material balance 11
ACSL-OPTIMIZE 82
activated sludge continuous reactor 577
activation energy 38
activity coefficient 167, 505 f.
– relations 508
adaptive control 78
adiabatic tank 407
adsorption 587
advanced control strategies 76
aerated tank with oxygen electrode 462
agitated extraction columns 459
ammonium 471, 547
Antoine equation 100, 172
aquatic ecosystem 581
area for heat transfer 100
array facility 535
array form 498
Arrhenius equation 38
Arrhenius relation 104
auto-refrigerated reactor 295
automatic control 68
average holdup time 108, *see also* residence time
axial concentration profiles 469
axial dispersion 468, 480 f.
– coefficient 481
– flow model 479
– model 194, 208, 210, 468
axial mixing liquid–liquid extraction columns 205

b
backmixing 149, 454 f.
backmixing flow rates 137
balance region 6, 10
– for distributed parameter systems 8
– for lumped parameters 8
balances for the end sections 454
balancing procedures 6
batch continuous stirred tanks 93
batch distillation 494
– column 158, 490, 504
batch extraction 130
batch extractor 444
batch fermentation 126
batch fermenter 126
batch liquid–liquid extraction 442
batch operation 94, 541
batch reactor 102, 399
– analogy 189
– examples 232
batch two-phase extraction system 442
benzene 501
binary batch distillation 158
– column 490
bio-trickling filter 555
biofilm 551
– diffusion-reaction process 551
biofilter 555
biofiltration 555
biological flocs and films 533
biological reaction examples 538
biological reactors, tank-type 124 f.
biomass 126
– retention 547
bioremediation 591
– time 591
BOD profile 572
boiling period 508
bottom plate 491
boundary conditions 192, 195, 197, 204, 207, 218, 519, 522, 526–527
boundary value problems 527
bubble point calculations 504

Chemical Engineering Dynamics: An Introduction to Modelling and Computer Simulation, Third Edition
J. Ingham, I. J. Dunn, E. Heinzle, J. E. Prenosil, J. B. Snape
Copyright © 2007 WILEY-VCH Verlag GmbH & Co. KGaA, Weinheim
ISBN: 978-3-527-31678-6

c

cadmium 565
cake washing 480
calorimetry-differential scanning 258
capacity 64
cascade control 76
cascade of equilibrium stages 449
cascade of three reactors 276
cascades with side streams 137
catalyst effectiveness 527, 531
cell concentration 124
chemical engineering modelling 1
chemical equilibrium 39
chemical kinetics 35
chemical reaction time constant 65
chemical reactor safety 177
chemical reactor waste minimisation 120 ff.
chemostats 126
chromatographic processes 207
cocurrent flow 516 f.
Cohen–Coon controller settings controller tuning 75
Cohen–Coon settings 422
coils or jackets 96 f.
column hydrodynamics 152
column plate 158 ff.
common time constants 64
complex batch reaction sequence 232, 355, 538
complex column simulations 167
complex reaction 237
complex sequential-parallel reaction 276
component balance equation 95
component balances 5, 18
concentration driving forces 48, 131
concentration profiles 44
– for countercurrent 45
concentration response 54
condenser 491, 505
conditional control 599
CONFLO, see also continuous flow tank 406
constant molal overflow 160
contaminant removal in soil 591
continuous binary distillation 162
– column 496
continuous column 123
continuous distillation column 163, 497, 501
continuous equilibrium multistage extraction 449

continuous equilibrium stage extraction 133
continuous-flow 123
– equilibrium stage 448
– reactor 96
– stirred tank 52, 420
– tank 406
continuous heating tank 27
continuous multicomponent distillation column 501
continuous operation 94
continuous phase 460
continuous stirred-tank reactor 7, 32, 106, 400
– cascade 93
continuous tank reactor 34, 265
continuous tubular 123
control of extraction cascades 141
control of temperature 99, 420
control of time 76
controller 424
– action 69
– adjustment 434
– equation 117, 428
– tuning 73
– tuning problem 427
convective flow terms 11
conversion 39
cooling jacket 97, 283, 431
countercurrent 214
– contactor 44 f.
– cooling 287
– extraction cascade with backmixing 137
countercurrent flow 515
– heat exchanger 214
countercurrent multistage equilibrium extraction 450
countercurrent multistage extraction 449
countercurrent packed column 471
countercurrent stagewise extraction cascade 453
curve fit 83, 263
cyclic fed batch 129
cylindrical coordinates 178

d

deactivating catalyst 268
dead zones 123
defining equations 16
degree of segregation 394
density 127
– influences 409
desorption of solute 480 f.

Subject Index | 611

destillation process examples 490
difference differential equation 480
difference equation 173
difference formulae for partial differential equations 219
differential column 130
differential contacting 45, 173
differential control constant 421
differential extraction column with axial dispersion 468
differential flow 173
differential mass transfer 43, 199
differential scanning calorimetry 119
diffusion 13, 521
diffusion and heat conduction 175
diffusion and reaction 179, 529, 533
– in a spherical bead 533
– split boundary solution 525
diffusion coefficient 528
diffusion film 462
diffusion fluxes 177, 534, 551
diffusion from the interior of the solid 521
diffusion model 551
diffusion process examples 521
diffusion rate 593
diffusion time 524
diffusion-dispersion time constant 65
dilution rate 128
dimensionless concentration 481
dimensionless equations 31, 270
dimensionless form 480, 536
dimensionsless groups 63
dimensionless kinetics 235
dimensionless model 315
– equations 31, 527
dimensionless solute concentration 481
dimensionless terms 407
dimensionless variables 464, 527 f.
discharge nozzle 412
discharge of molten steel 412
discrete control systems 78
dispersed phase 460
– holdup 153
dispersion model 193, 196, 198
dispersion number 197
dissolved oxygen 569
– concentration 462
– profile 572
distillation period 509
distributed parameter 8
distribution 391
– of residence times 123

diurnal variation 560
double Monod kinetics 592
driving force 97
drying 521
– of a solid 521
dynamic component balances 127
dynamic difference equation model for chromatography 209
dynamic diffusion and enzymatic reaction 529
dynamic method 462
dynamic models 2, 4
dynamic simulation 1, 2
dynamic tubular reactor 190, 332
dynamics of a shell-and-tube heat exchanger 511
dynamics of the cooling jacket 97
dynamics of the metal jacket wall 100

e

e-curve 274
Eddy dispersion coefficient 469
effective area for heat transfer 97
effectiveness factor 528, 536
effectiveness of the catalyst 525
eigenvalues 114
electrode membrane 463
electrode response characteristic 462
endothermic reaction 103
energy balance 169, 420, 472
– equation 95, 183, 413
energy balancing 22
enriching section 163
enthalpy 23
entrainment fractions 147
environmental examples 560
environmental impact 581
enzyme 525
equalisation basin 560
equation stiffness 90
equilibrium 129, 445, 494
– line 48
– oxygen concentration 463
– relationship 48, 131, 472, 480
equilibrium stage 44
– behaviour 136
– extraction 134
– extractor 447
estimation of rate and equilibrium constants 83
examples of chemical engineering processes 225
exothermic reactor 356

exothermic reaction 110, 283, 437
– sequence 253
exothermic semi-batch reactor 430
exponential and limiting growth phases 42, 126
exponential response 52
extraction cascade 136
– with backmixing and control 456
– with slow chemical reaction 139
extraction column 150
extraction from a solid 20
extraction vessel 130
extractor 468

f

Fanning friction factor 417
feed composition 503
feed control 430
feed phase 130
feed plate 164
feed-batch fermenter 128
feed-forward control 77
feedback control 68, 430
– system 420
fermentation 124
Fick's law 47, 175, 177, 552
field capacity 589
filling and emptying tanks 426
film heat transfer coefficient 100, 513
filter bed 481
filter washing 479
finite difference 531
– approximations 91
– elements 468, 522
– segments 178
finite differencing 176, 336, 511, 529f., 534
– of heat exchanger 216
– of tubular reactor 174, 191
finite elements 204
first-order decay 42
first-order lag 52, 428
– equation 463
first-order response 51
first-order time lag 420
flow phenomena time constant 64
flow term energy balance 26
flow terms material balance 12
fluidised bed 547
Fourier's law 175, 518
fractional conversion 39, 182
fractional holdup 459f.
fractional phase holdup 470

fractional yield 40
free-radical polymerisation 310
frequency of the oscillations 422
friction factor 414
frictional force 417

g

gain coefficient 136
gas absorption 154
– heat effects 199
– steady-state design 200
– steady-state simulation 201
gas separation by membrane permeation 475
gas–liquid contacting systems 153
gas–liquid interface 46
gas–liquid mass transfer 34, 385
gas–liquid mixing in a stirred tank 385
gas–liquid transfer 462
gas-permeation 475
– module 475
gas-phase reaction with molar change 344
gas-phase tubular reactors 186
general aspects of modelling 3
general heat balance 25
groundwater pollution 584
growth rate 42, 126

h

heat balance 428, 516
heat capacity 24
heat conduction 518f.
heat emissivity 519
heat exchange 214
heat exchanger 96, 511
– boundary conditions 218
– differential model 216
– dynamics 215
– steady-state 214
heat gain curve 112
heat loss 112
– by radiation 518
heat of reaction 25, 39
heat transfer 96, 511
– applications 213
– area 430
– coefficient 509, 514, 516
– from metal wall 511
– rate 100, 513
– steady-state tubular flow 213
– energy balance 26
– time constant 67

heating period 508 f.
heating in a filling tank 28
heavy metals accumulation 565
heavy phase 143
Heerden steady-state stability criterion 297
Henry coefficient 465
Henry's law 463
heterogeneous catalytic reactions 37
higher order responses 58
holdup 492
– distribution 460
hot spot effect 327
human irrigation 587
hydrostatic equation 144
hydrostatic pressure force 417

i
ideal gas law 463
immiscible liquid phases 130
implicit algebraic loop 155
information flow diagram 22, 131, 169, 171, 185, 18 f., 228
inhibitory substrate 543
initial conditions 126
inoculum 126
input rate 125
insecticide distribution 581
integral control 428, 458
– constant 421
integrated extraction 280
integration method 90, 600
integration parameters 90
integration routine 89
integration step length 91
intensity of segregation 396
interacting solute equilibria 133
interacting tank reservoirs 416
interfacial area 131
interfacial concentrations 48
internal energy 22 f.
interphase mass transfer 46
interphase transport 14
interstage flow rates 137
introductory MADONNA example 227
irrigation 585
ISIM 82
isothermal reactor with axial dispersion 335
isothermal reactor with complex reaction 265
isothermal tank 407
isothermal temperature conditions 406

isothermal tubular reactor 320
iterative calculations 504
iterative loop for the bubble point calculation 172
iterative procedure for parameter estimation 86

j
jacket cooling 83, 297, 437
jacket heating 99
jacketed batch reactor 253

k
KLa 462

l
lag in the system 424
lag phase 126
lake ecology 565
Langmuir isotherm 587
Laplace transformation 464
latent heat of vapourisation 428
leaching 591
lead 565
least squares 81
level control 424
limit cycles 115
limiting substrate 126
linearisation 113
liquid film 463
liquid impurities 479
liquid stream blending 409
liquid–liquid extraction 129, 280
– column dynamics 202
liquid–phase tubular reactors 185 f.
liquid–vapour equilibrium 505
lumped parameter 7, 8, 100

m
MADONNA 82, 90, 597
manipulation of the cooling-water flow rate 430, 432
mass transfer coefficient 47, 131
mass transfer process examples 442
mass transfer theory 43
mass transfer time constant 65
mass transfer to a continuous tank reactor 34
material balance 4, 10, 95
– equations 124
material balancing procedures 6
mathematical modelling 2
MATLAB 82

614 | Subject Index

maximum likelihood 81
mean square fluctuations 396
measured oxygen concentration response curve 462
measurement and process response 51
measurement dynamics 56, 463
measurement lag 57
measurement time constant 56
membrane permeabilities 475
metal balance 565
metal jacket wall 100
metal rod 518
methanol 507
Michaelis–Menten 528, 530
– equation 526
microbial growth kinetics 41
migration 584
mixer 142
– dynamics 143
mixer–settler 129
– cascade 147
– extraction cascades 142
– stage 148
mixing history 123
mixing in chemical reactors 394
mixing model 374, 378, 381, 390
mixture of hydrocarbons 494
modelling a non-isothermal, chemical reactor 22
modelling approach 3
modelling fundamentals 1
modelling of chromatographic processes 207 f.
modelling of cooling effects 99
modelling procedure 3
modes of reactor 93
molar feeding rate 105
molar quantities 106
mole ratios 202
molecular diffusion coefficient 47
momentum balance 31, 416
momentum transport 175
Monod equation 43
Monod-type equation 43, 535
Monod-type rate expressions 127
multicomponent differential distillation 494
multicomponent extraction 444
multicomponent mixture 158
multicomponent separations 165
multicomponent steam distillation 168
– semi-batch 508
multicomponent system 140, 508

multiple feeds 137
multiple steady states 91
multiple tanks in series 58
multisolute batch extraction 132
multisolute equilibria 133
multistage countercurrent extraction cascade 136
multistage extractor with backmixing 453
multistage multicomponent extraction cascade 130
mutual inhibition 555

n

N profile 572
n-decane 508
n-octane 508
naturally occurring oscillations 91
negative feedback 116
Nelder–Mead search algorithm 79
Newton's gradient method 79
nitrate 547
nitrification 547, 574
nitrite 547
nitrogen 475 f.
non-equilibrium differential contacting 130
non-equilibrium staged extraction column 150
non-ideal distillation 504
non-ideal flow 123
non-ideal liquid behaviour 504
non-ideal stirred-tank reactor 374
non-linear parameter estimation 79
non-linear systems, parameter estimation 82
nozzle mass flow 413
Nth-order reaction 270
numerical aspects of dynamic behaviour 79
numerical integration 88

o

objective function 352, 368
on/off control 70
optimal cooling 79
optimal feeding policy 350
optimisation 79, 600
order of reaction 37
organism balance 125
oscillating tank reactor 290
oscillations 421, 429
outlet 128

output rate   125
overall heat transfer coefficient   97
overall mass transfer coefficient   131, 446
overhead distillation   156
oxidation reaction in an aerated tank   250
oxygen   475 f., 533
– deficit   573
– diffusion   594
– enrichment   475
– limitation   533
– sag curve   569
– transfer coefficient   462
– uptake rate   535
oxygen electrode   462
– dynamic   463

p

parallel reaction in a semi-continuous reactor   29
parameter estimation   81–82, 600
parametic plot   600
partial differential equation   480
partial molar enthalpy   24
Peclet number   193, 480
penetration distance   533
perfect mixing   102, 123
– stirred tank   93
permeabilites, membrane   477
permeability coefficient   478
permeation rate   475, 478
perturbation variables   113
phase equilibria   45, 171
phase flow rate   144
phase volume   133, 144
phase-plane diagram   92, 546
PID control   246, see also proportional-integral-derivative
plant start up/shut down   107
plants ecology   585
plate efficiency   166
plate hydraulics   162
plug-flow   123, 455
– reactor   189
– tubular reactor   181
pollutant transport   584
pond ecology   581
porosity   481
– distributions   480
porous biocatalyst   529
porous solid supports   525
principles of mathematical modelling   2
probability density   82
– function   81

process control examples   420
process development in the fine chemical industry   119
process dynamics and response   51
product inhibition   526, 530
production rate   15, 125
– material balance   15
profit function   79
programming tips   598
proportional control   428, 458
– constant   421
– equation   425
proportional controller   116
proportional gain   116, 428
proportional plus integral control   456
proportional temperature controller   72
proportional-integral feedback controller   421
proportional-integral-derivative (PID) control   70, 246
pulse function   337
pulse input   339, 599
pure time delay   61

q

quasi-steady state   129, 155

r

radiation from metal rod   518
raffinate   141
rain   587
rate   64
– expressions   126
– of accumulation   11, 125
Rayleigh equation prediction   495
reaction calorimetry   118
reaction enthalpy   25
reaction, exothermic   247
reaction, fractional conversion   39
reaction heat term energy balance   27
reaction kinetics   103
reaction rate   35, 37, 398
– constant   38
reaction selectivity   40
reaction yield   40
reactor   124
– cascade   109, 287
– configurations   93
– control   107, 115
– energy balance   254, 301
– stability   110
– temperature   255
reboiler   490, 492, 498, 504

recovery of valuable filtrates 479
rectangular slab 525
recycle-loop reactor 268
reflux 505
– drum 490 ff., 499
– ratio 158, 491 f., 499, 502 f., 505 f.
relative volatility 157, 165, 491–492, 495, 498, 502
relaxation time 63
release of chemicals 581
residence time 128, 469
– distribution 381
response controller 71
response measuring element 55
response of a stirred tank reactor 55
response of the measuring instrument 57
reversible reaction 299, 305
Reynolds number 414
river ecology 572
root growth 586
RTD experiments 123
runaway scenario 118

s

sampled data 78
second-order kinetics 431
second-order measuring element 60
second-order response lag 463
sediments 565
segregation 394, 398
– batch reactor 397
– continuous reactor 397
– semi-continuous reactor 398
selectivity 39, 276
semi-batch 353
– continuous stirred tanks 93
– operation 94
– reactor 104, 106, 355, 401
– system 508
semi-continuous reactor 347, 350, 430
– adiabatic 29
semi-dimensionless form 531
sensitivity analysis 85
sequential reactions 276
setpoint 68 ff.
settler 142, 577
– dynamics 144
shell-and-tube heat exchanger 216, 511
short-cut flow 123
signal and process dynamics 53, 55, 57, 59, 61, 92
simple overhead distillation 156

simplified energy balance 27
simulation 129
– software 226 f.
– tools 226
single batch extraction 130
single solute batch extraction 442
slip velocity 153
sludge recycle 577
software links 226
soil ecology 584, 591
solid drying 522
solute 130
– transfer rate 131
solvent or organic phase 130
specific growth phase 42
specific growth rate 128
specific interfacial area 47
spherical balancing aggregate 592
spherical bead 534
spherical coordinates 178
spherical shell 534
split-boundary problems 88
split-boundary type 174, 527
spouted bed reactor 390
staged extraction columns 149
stagewise absorption 153
stagewise contactor 44
stagewise discontinuous extraction 451
stagewise distillation 156
stagewise mass transfer 43, 129
stagewise model with backmixing 137
stagewise processes 93
startup 500
steady-state 127, 418
– absorption column design 471
– balance 4, 503
– conditions 126
– criterion 115
– diffusion 179
– energy balance 326, 518
– gas absorption 199
– operation 95
– simulation 174
– split boundary problem 530
– tubular flow 213
– tubular reactor dispersion model 196, 315
– two-pass heat exchanger 515
steam distillation 508
steam tables 100
steam-heated stirred tank 427
steam-heating in jackets 99
step-change disturbance 51 f.

steps in model building 4
sterile feed 126
still pot 490f., 494
stirred-tank reactors 93, 95, 97, 99, 101, 103, 105, 107, 109, 111, 113, 115, 117, 119, 121, 123, 125, 127
stoichiometric coefficient 25, 36
stoichiometry 103
Streeter and Phelps curves 569
stripping section 163
substrate balance 125
substrate concentration 124
substrate uptake kinetics 43
superficial phase velocity 151, 470
superficial velocities 460
system stability 91

t

tank 124
– drainage 17
– flow examples 406
– mixing 409
– reactor 317
tanks-in-series 58, 123, 276, 424
– approximation of a tubular reactor 333
Taylor's expansion theorem 113
temperature conditions 406
temperature control 69
– for semi-batch reactor 430
temperature difference 97
temperature distribution 101
temperature measurement 420, 427, 430
– device 420
temperature profile 508, 518
theoretical plate behaviour 491, 498, 505
theoretical plates 490, 504
thermal capacity 100
thermal stability of a CSTR 283, 437
thermally safe process 365
Thiele modulus 528
thin slab 521
time constant 35, 63, 65, 67
– application 67
– for heater 421
– for transfer 466
– measurement 421, 466
time delay 61f., 145
– function 62
time delayed values 144
time integral criteria 76
toluene 501
total energy 23
total mass balance 125

total material balance 5, 16, 95
total reflux 507
tracer experiments 274
tracer information in reactor design 123
tracer response 333, 384
transfer function 62
– in series 435
– simulation 435
transfer rate relationships 131
transient holdup profiles in an agitated extractor 459
transport of heavy metals 565
transport streams 10
trial and error method controller tuning 74
tube 511
tubular reactor 7, 185ff., 317, 324, 329, 333
– batch reactor analogy 189
– boundary conditions 195
– chemical 180
– component balance dispersion model 181, 193
– energy balancing 184
tubular reactor model for the steady-state 315
tubular reactor with axial dispersion 193, 340
tuning the controller 458
turbulent fluctuations 395
two tank level control 424
two-film theory 46, 131
two-pass heat exchanger 515
two-solute batch extraction 445
– with interacting equilibria 444

u

ultimate gain method 422
– controller tuning 75
unmixed reactants to a tubular reactor 395
unsteady-state diffusion 175
unsteady-state heat conduction 178

v

vadose soil 584
van Laar equation 505f.
van't Hoff equation 39, 300, 306
vapour pressure 100
variable heat capacities 299
variable volume reactor 431
volumetric feed rate 105
volumetric flow rate 124

w
washing of filter cake   479
washout   128
waste air treatment   555
waste holding tank   19
wastewater   551
– discharge   573
water   507
– field capacity   585
– heater   420
– jacket   430
– table   584
water-methanol system   504
well-mixed   124
– region   147
Whitman two-film theory   46
work term   23

x
xylene   501

y
yield   39, 117
– coefficient   125

z
zero-order kinetics   536
zero-order reaction   531
Ziegler–Nichols method controller tuning   74
Ziegler–Nichols-criteria   429
zinc   565

Related Titles

P. S. Agachi, Z. K. Nagy, M. V. Cristea, A. Imre-Lucaci

Model Based Control
Case Studies in Process Engineering

2007
ISBN 978-3-527-31545-1

A. C. Dimian, C. S. Bildea

Chemical Process Design
Computer-Aided Case Studies

2006
ISBN 978-3-527-31403-4

T. G. Dobre, J. G. Sanchez Marcano

Chemical Engineering
Modelling, Simulation and Similitude

2006
ISBN 978-3-527-30607-2

Z. R. Lazic

Design of Experiments in Chemical Engineering
A Practical Guide

2004
ISBN 978-3-527-31142-2

I. J. Dunn, E. Heinzle, J. Ingham, J. E. Přenosil

Biological Reaction Engineering
Dynamic Modelling Fundamentals with Simulation Examples

2003
ISBN 978-3-527-30759-3

J. B. Snape, I. J. Dunn, J. Ingham, J. E. Přenosil

Dynamics of Environmental Bioprocesses
Modelling and Simulation

1995
ISBN 978-3-527-28705-5